Agricultural Sustainability
and Environmental Change
at Ancient Gordion

GORDION SPECIAL STUDIES

GORDION EXCAVATIONS FINAL REPORTS

MUSEUM MONOGRAPH 145

GORDION SPECIAL STUDIES VIII

Agricultural Sustainability and
Environmental Change at Ancient Gordion

John M. Marston

UNIVERSITY OF PENNSYLVANIA MUSEUM OF ARCHAEOLOGY AND ANTHROPOLOGY

PHILADELPHIA

Publication of this book has been aided by a grant from the von Bothmer Publication Fund of the Archaeological Institute of America.

LIBRARY OF CONGRESS CATALOGING-IN-PUBLICATION DATA

Names: Marston, John M., author.

Title: Agricultural sustainability and environmental change at ancient Gordion / John M. Marston.

Description: Philadelphia : University of Pennsylvania Museum of Archaeology and Anthropology, 2017. | Includes bibliographical references and index.

Identifiers: LCCN 2017006172| ISBN 9781934536919 (hardcover : acid-free paper) | ISBN 1934536911 (hardcover : acid-free paper)

Subjects: LCSH: Plant remains (Archaeology)--Turkey--Gordion (Extinct city) | Excavations (Archaeology)--Turkey--Gordion (Extinct city) | Agriculture, Ancient-Turkey--Gordion (Extinct city) | Sustainable agriculture--Turkey--Gordion (Extinct city) | Landscape changes--Turkey--Gordion (Extinct city) | Social change-Turkey--Gordion (Extinct city) | Gordion (Extinct city)--Antiquities. | Gordion (Extinct city)--Environmental conditions. | Environmental archaeology--Turkey--Gordion (Extinct city) | Social archaeology--Turkey--Gordion (Extinct city)

Classification: LCC DS156.G6 M37 2017 | DDC 630.939/26--dc23

LC record available at https://lccn.loc.gov/2017006172

Distributed for the University of Pennsylvania Museum of Archaeology and Anthropology
by the University of Pennsylvania Press.

Printed in the United States of America on acid-free paper.

Contents

Figures

(the color insert appears between pages 100 and 101)

Tables

Preface

This book has two aims: to develop an approach for the reconstruction of agricultural decision making using archaeological data, and to illustrate the value of that approach by synthesizing the results of 25 years of botanical research at Gordion to describe environmental and agricultural changes at that famous site. My central contention is that archaeological data permit not just the identification of agricultural practices but also the decision-making processes behind them, which, in turn, enables a deep understanding of how farmers and herders respond to environmental, social, and economic pressures. By understanding these relationships, we can better explore how unsustainable agricultural and land-use practices arise and how those practices affect productive landscapes and local ecologies over decades to millennia. I center this work at the site of Gordion due to its rich diachronic archaeological and paleoenvironmental records that allow a robust and multifaceted exploration of agricultural and environmental change at the site over a span of nearly 3,000 years.

The multidisciplinary argument presented here draws on various methods in environmental archaeology, though primarily paleoethnobotany and ecology. I draw on both ecological resilience thinking and behavioral theory to build a model for considering agricultural sustainability and decision making within its environmental and social contexts. As such, this volume speaks to multiple audiences: scholars of agriculture and human-environmental relationships, paleoethnobotanists, archaeologists, and ecologists. In addition, this volume and its attendant online data files provide the full, sample-by-sample results of botanical material excavated at Gordion over ten years. These data are now available for comparative research in Anatolia and the broader Near East.

Together with Naomi F. Miller's 2010 volume, this book presents the final results of flotation sample and hand-collected wood charcoal analysis for the 1988–2005 excavation campaign under Mary M. Voigt as Director of Excavations at Gordion. As excavations under C. Brian Rose have now been ongoing since 2013, this volume represents one part of the combined efforts to complete the publication of prior Gordion excavation campaigns and so should be read in conversation with other titles in the Gordion series appearing before and after, including publications to come from the new campaign of excavations.

This book also completes a personal journey for me. Although I joined the Gordion excavation only in 2002, the final year of full-scale excavation under Voigt's direction, I was offered the opportunity to undertake the analysis of botanical samples from 1993–2002 for my dissertation project. I ultimately completed analysis of this dataset as part of my doctoral work at the University of California, Los Angeles (Marston 2010). The richness of this dataset has sparked multiple article publications that utilize these data as a case study (Marston 2009, 2011, 2012a, 2012b, 2015; Marston and Miller 2014; Miller and Marston 2012) but the complete data have not been published in the intervening years. This book, together with its online appendices, makes available one of the richest archaeobotanical datasets in the Near East for future comparative and regional work by other scholars in the field.

I have been working in earnest on this project for more than a decade now, so there are many individuals who deserve thanks. This project owes much of its success to the quality work produced by Mary Voigt and her teams at Gordion over the years. Mary provided not only access to the samples analyzed here but also many hours of focused support in reconstructing the stratigraphic context of each botanical sample taken at Gordion over nearly a decade. I owe much to her careful reading and useful corrections to earlier publications on this dataset. Ken Sams and Brian Rose provided logistical and financial support as Directors of the Gordion Project. Many others at Gordion, including Richard Leibhart, Ben Marsh, Lisa Kealhofer, Peter Grave, Ayşe Gürsan-Salzmann, Andy Goldman, Brendan Burke, Julie Unruh, Jessie Johnson, Angie Elliott, Gareth Darbyshire, Shannan Stewart, Canan Çakırlar, Hüseyin Fırıncıoğlu, and Mecit Vural, provided support in the field and conversations that greatly enriched my thinking about the archaeology and plant ecology of Gordion.

Particular thanks are due to Naomi Miller. Not only did she train me in the identification of wood charcoal and Old World archaeobotanical samples, but she also provided me the initial opportunity to join the Gordion excavation and later served as an external examiner for my dissertation. We have worked together directly on several articles, but her prior analysis and publication of botanical samples from Gordion served as the foundation for my own work. Building from Naomi's work, especially her 2010 book, I am able to move beyond simple reconstructions of diet and environment and consider broader questions of decision making in human-environmental relationships at Gordion. In addition, her data expand the timeline of my samples into earlier periods, giving valuable depth to this study.

Analysis of these samples was done primarily at the Cotsen Institute of Archaeology at UCLA. I owe many there a debt of gratitude: Charles Stanish, Julia Sanchez, Elizabeth Klarich, Shauna Mecartea, Virginia Popper, Jeffrey Brantingham, Elizabeth Carter, David Scott, Ioanna Kakoulli, and the greatest thanks to John Papadopoulos, my dissertation advisor. John's expert editorial skills, prompt attention to my work when needed, and both financial and personal support were invaluable during my dissertation and career beyond. Some 23 UCLA undergraduates volunteered in the lab to help me sort these samples, making a significant contribution to this project.

Funding for my research comes primarily from the National Science Foundation, which provided a Dissertation Improvement Grant (BCS Grant #0832125)

for this project and a Graduate Research Fellowship that supported my early years of study. I also thank the Gordion Archaeological Project for research travel funds to Turkey and Philadelphia, the Cotsen Institute of Archaeology for additional research travel funding, the UCLA Graduate Division for conference travel funds and both a Portable Supplement Fellowship and a Dissertation Year Fellowship, and the Departments of Anthropology and Classics at UCLA for supporting me through several teaching apprentice positions. I thank the University of Pennsylvania Museum of Archaeology and Anthropology (hereafter Penn Museum) for several loans of the charcoal and flotation samples needed for this dissertation, and the Museum's Office of the Registrar for facilitating those loans. The United States Department of Agriculture supplied many seeds for my comparative collection through their Agriculture Research Service National Genetic Resources Program (ARS-GRIN) National Plant Germplasm System (NPGS), which greatly increased the quality of my collection, especially for economic plants. The Polatlı Meteoroloji İstasyonu of the Turkish Meteoroloji Bakanlığı generously supplied the climatic data used to model environmental variability at Gordion. The Gordion Project operates under the supervision and with the support of the Turkish Kültür ve Turizm Bakanlığı and the Museum of Anatolian Civilizations in Ankara, and with the kind support of the people of Yassıhöyük.

This book was completed at Boston University, where I have received support from the College of Arts and Sciences and a Peter Paul Career Development Professorship, which provided a semester of course release towards this publication. I owe thanks to my colleagues in the Department of Archaeology and friends and collaborators across campus, including especially my writing partners Catherine West, Carolyn Hodges-Simeon, Jennifer Talbot, Christopher Schmitt, Jonathan Bethard, and Anita Milman. Ethan Baxter and Denise Honn made possible the thesis project of Adam DiBattista (2014), which enriched my understanding of pastoral mobility at Gordion. I thank Penn Museum Publications, including its director Jim Mathieu and Gordion publications director Brian Rose, for their encouragement and support. Two anonymous reviewers supplied valuable and specific suggestions for revision of this text.

1

People, Environments, and Agriculture at Ancient Gordion

Human life is intricately connected to surrounding environments—physical, biological, social, and imagined—in myriad ways. This book addresses one such environmental interaction, between people and natural landscapes, through a place-based case study of the ancient city of Gordion over a span of nearly 3,000 years. In this volume, I seek to identify how social factors—including political and economic forces—and environmental factors—ranging from climatic to ecological and landscape change—influenced reciprocal interactions between the inhabitants of Gordion and their surrounding environment over this duration. My focus is on people and their processes of decision making with regard to activities that affected the landscape, the most significant of which was agriculture. Agriculture, here including both farming of plants and herding of animals, is a nexus between human economy and environmental response and provides a powerful lens for viewing reciprocal dynamics within the complex system of human-environmental interaction.

Why use archaeology, and why study Gordion, to explore human interactions with local environments? In this introductory chapter I lay out the benefits of a diachronic approach for identifying human impacts on environments and ways in which environmental change influences societies, and argue that archaeology provides a critical perspective on dynamics of these processes in the past. I also suggest Gordion as a uniquely valuable case study, as a result of both its social trajectory and history of excavation and analysis, which are summarized here. I conclude with an overview of the remaining chapters that lays out the datasets and arguments made throughout the book that address the core questions presented here.

People, Environments, and Agriculture

Questions

The topic of human-environmental interaction is so broad that it can be explored from multiple angles and theoretical perspectives. I restrict this study to two sets of related questions. First, what agricultural strategies were employed at Gordion, and when and why was one chosen over another? Second, what were the environmental implications of those strategies and other land-use practices, and what were the reciprocal effects of environmental change on human decision making?

Methods for identifying agricultural practices in the archaeological record include the analysis of plant (Jones 1987; Jones and Halstead 1995; Marston 2012a; Scarry 1993) and animal (Arbuckle 2012; Balasse et al. 2002) remains, as well as identification of agricultural features, such as raised fields and irrigation canals (Erickson 2006; Kirch 2006; Morehart 2012; Wilkinson 2003). Moving from the identification of agricultural practices to the determination of agricultural strategies, however, is a more complex challenge. I define agricultural strategies as practices employed in a deliberate manner, potentially including multiple practices (e.g., crop rotation, irrigation, seasonal transhumance) that are used in combination and utilized to varying degrees in response to changing cultural needs or environmental constraints. How do agricultural practitioners choose these strategies, and when do they choose one over another? Which factors, either social or environmental, result in changing strategies, and when do strategies remain resilient in the face of external change? Which strategies are mostly successful over the long term and can be termed sustainable, and which offer immediate benefits but increase long-term risks?

Agriculture leads to environmental change on diverse scales, both spatial and temporal, as documented worldwide in both the present day (Foley et al. 2005) and in the past (Redman 1999; Smith and Zeder 2013). In addition, other land-use practices, such as the acquisition of wood for construction and fuel, can lead to significant change in forest structure and result in geomorphological transformation (Bishop et al. 2015; Fall et al. 2015; Gremillion 2015; Marston 2009; Miller 1985; Willcox 1974), as can deliberate deforestation for agricultural land clearance (Clark and Royall 1995; Roos et al. 2010; Stinchcomb et al. 2011). Which agricultural and land-use practices are most significant in the transformation of vegetation communities and landscapes? How do varied environments respond differently to anthropogenic pressures? Over what timescales and spatial scales are these effects realized?

Environmental change also has implications for human communities, ranging from decrease in grassland value for animal grazing (Fırıncıoğlu et al. 2009; Marston 2015) to erosion of soils suitable for agriculture (Beach and Luzzadder-Beach 2008; Casana 2008; Marsh and Kealhofer 2014) and soil salinization (Redman 1999; Wilkinson 2003). Taken together, dramatic transformations in human landscapes have been deemed the causes of social change termed "collapse" (Diamond 2005; Tainter 1988, 2006a; Wills et al. 2014) or even "ecocide" (Hunt and Lipo 2010; Middleton 2012). What types and degree of environmental change lead to substantive responses in human societies? How do societies, and distinct segments of societies, respond differently to environmental change? What is the timescale of these interactions and responses?

Approaches

To address the questions above, I employ a theoretical framework that draws on two bodies of ecological theory: behavioral ecology and resilience thinking. Behavioral ecology deals broadly with human decision making in the context of immediate social and ecological environments, considering the role of economic and environmental factors in behavior from an evolutionary perspective. In contrast, resilience thinking addresses how and why change occurs in complex adaptive systems, including those with both ecological and social components. In exploring agricultural systems, there is much to be gained from a hypothesis-testing approach that focuses on variables that can be studied from both a resilience and behavioral ecology perspective. Such an approach marries the specific tools of behavioral ecology modeling with the general, analogic explanatory framework of resilience thinking. I use behavioral ecology to develop specific, testable metrics for agricultural decision making, while the framework of resilience thinking illuminates complex interrelationships between social and environmental change, including how and when one leads to the other.

Several datasets inform patterns of cultural, agricultural, and environmental change at Gordion. I draw on historical sources and decades of excavation at Gordion to identify changes in political economy, while the plant and animal remains excavated allow me to reconstruct agropastoral systems from the Late Bronze Age (ca. 1500–1200 BCE) to the Medieval Period (13th–14th centuries CE). To study environmental change, I utilize regional paleoclimate records, archaeological plant assemblages that indicate changes in plant ecology, and geomorphological studies that record sedimentary histories of erosion and alluviation. The challenge lies in integrating these datasets, which have different levels of temporal and spatial resolution, discrete quantification methods, and distinct levels of compatibility with the theoretical models employed. In order to reconcile these data and address the questions outlined above, I derive specific statistics from parameters employed in behavioral ecology to quantify decision making and use resilience frameworks to interpret causal relationships between social and environmental factors.

Choice of a Case Study

Central Anatolia is a semi-arid region with annual rainfall that in most years supports rainfed farming of cereal crops (today, mainly bread wheat and barley), although droughts are a regular occurrence and render agriculture a risky activity. Rainfall varies with elevation, creating a mosaicked landscape of

Table 1.1. The chronology of Gordion. YHSS (Yassıhöyük Stratigraphic Sequence) phasing was introduced by Voigt based on 1988–89 deep sounding excavations (Voigt 1994). Dates follow Voigt (2011:1074), with revised dates for the Medieval occupation, which remains poorly understood (Scott Redford, personal communication, July 2011), and the Roman period (Andrew Goldman, personal communication, August 2013); Early Bronze Age levels were not reached during Voigt's excavations.

YHSS Phase	Period Name	Approximate Dates
1	Medieval	13th–14th cent. CE
2	Roman	50 CE–early 5th cent. CE
3	Hellenistic	330–100 BCE
4	Late Phrygian (Achaemenid)	540–330 BCE
5	Middle Phrygian	800–540 BCE
6	Early Phrygian	900–800 BCE
7	Early Iron Age	1100–900 BCE
8/9	Late Bronze Age	1400–1200 BCE
10	Middle Bronze Age	1600–1400 BCE

steppe grasslands and wooded hilltops, as discussed in detail in Chapters 3 and 4. There is limited evidence for Paleolithic settlement of Central Anatolia, with sustained human occupation of the region not present until the Holocene (Sagona and Zimansky 2009). Agriculture has been the focus of economy in Central Anatolia since the onset of large-scale human colonization of the region and also the primary mechanism by which people have transformed its landscape, rendering Central Anatolia an ideal case study for studying human-environmental interaction through the perspective of agriculture.

The immediate Gordion region shows only limited evidence for Paleolithic and Chalcolithic (ca. 4000–3000 BCE) settlement, while by the Early Bronze Age (ca. 3000–2000 BCE), the city of Gordion and outlying sites were established (Kealhofer 2005b; Marsh and Kealhofer 2014; Voigt 2013). More or less continuous habitation of both Gordion and the greater region persisted until the late Roman period (early 5th century CE) (Kealhofer 2005a; Marston and Miller 2014). There is little evidence of post-Roman occupation in the region until Medieval resettlement of Gordion in the 13th–14th centuries CE, possibly in concert with a broader regional expansion of settlements in Central Anatolia during the 13th century (Baird 2001; Marsh and Kealhofer 2014; Marston 2012a:381). As a result of extensive excavation and regional survey since 1988, environmental archaeological and paleoenvironmental data are available from a nearly 3,000-year period at Gordion, producing a rich dataset with which to explore agricultural and environmental change.

Gordion also became entangled in the political and economic transformations that swept Anatolia between the Bronze Age and Medieval period (Table 1.1). Gordion was capital of a regional polity during the Phrygian period, a peripheral center of both the Hittite and Persian empires, and a small rural settlement under Hellenistic, Roman, and Medieval Islamic control. The city grew and shrank, both in physical size and economic and political importance, over its settlement history. As such, Gordion provides an opportunity to explore the environmental impact of empire and how agricultural strategies respond to periods of economic integration and relative autonomy. These variations in both environmental and social setting during its history, together with the extent of excavation and paleoenvironmental sampling at Gordion, make it an excellent case study to consider the role of agriculture in mediating complex systems of human-environmental interaction.

The History of Archaeological Investigation at Gordion

Finding Gordion

Gordion is referenced by many historical Greek and Roman authors and had been known to exist along the ancient Sangarios (modern Sakarya) River in Central Anatolia by classicists for centuries prior to its archaeological identification (Burke 2001; Roller 1984). Herodotus (14.2–3) describes the famous King Midas of Gordion, whose legendary wealth allowed him to set up a dedicatory offering at the sanctuary of Apollo at Delphi; Aristotle (*Politics* 1257b) and Ovid (11.85, 11.146) recount fanciful stories of his "golden touch" and donkey's ears. Alexander the Great's visit to the city where he cut the legendary Gordian Knot is detailed by multiple authors of the Roman period (Plutarch *Alexander* 18.1–2, Arrian 2.3, Curtius 3.2.11–18, Justin 11.7.3–16). Livy (38.12–27) and Polybius (21.33–39) wrote about the conquest of the Celtic Galatians at Gordion by the Roman consul Manlius Vulso, an important historical account that contributed to the discovery of the site and the interpretation of Hellenistic occupation in the region. Despite these accounts, however, no archaeological evidence for the existence of Gordion was known before the excavations of the Körte brothers in 1900.

Ancient remains uncovered by German railroad engineers building a rail line west from Ankara in November 1893 through the Sakarya Valley were first argued to belong to ancient Gordion by Alfred Körte (1897:4). Körte used the description of Manlius Vulso's march through Galatia in Livy's text (38.12–27) to argue convincingly that the location of these remains matched very well classical descriptions of Gordion's location along the ancient Sangarios River (Körte 1897). Together with his brother, Gustav, Alfred Körte spent three months in 1900 directing excavations at the occupation mound of Yassıhöyük (Turkish for "flat-topped mound") and in several earthen tumuli that dot the valley (Körte and Körte 1901; Körte and Körte 1904). The Körte brothers dug into both the eastern and western halves of the Yassıhöyük or *Stadthügel* ("City Mound," as they termed it, now called the Citadel Mound) and excavated five tumuli,

numbered Tumuli K-I through K-V (Fig. 1.1; Körte and Körte 1904).

The Körtes' excavation on the Citadel Mound focused on one roughly 20-meter-square trench in the southwestern section of the mound. Within this area, they uncovered a "temple" with painted relief terracotta plaques similar to those later excavated in greater number by Rodney Young during his exploration of Middle Phrygian layers of the eastern half of the mound (Glendinning 2005; Körte and Körte 1904:153–70). Pottery from this area included datable Greek imports and Hellenistic and Roman wares (Körte and Körte 1904:177–211). The tumuli, however, offered richer finds, including wooden furniture; iron and bronze serving stands, vessels, armor, and fibulae; and fine pottery vessels including beer-strainers—a notable element of elite Phrygian tablewares (Körte and Körte 1904; Sams 1977, 1994b). Tumuli K-III and K-IV are now dated to the early Middle Phrygian period, during the 8th century BCE, when Phrygia and Gordion reached their peak of power and wealth (Rose and Darbyshire 2011; Voigt 2002), while Tumuli K-I, K-II, and K-V are dated later in that period, as evidenced by the presence of datable Greek artifacts of the 7th and 6th centuries BCE (Sams 2005:10). Although the Körtes' excavations lasted only a single season, they provided a sufficient archaeological background for continuing work at the site 50 years later.

Rodney Young and Phrygian Gordion

Rodney S. Young, Professor of Classical Archaeology and Curator of the Mediterranean Section at the Penn Museum began excavations at Gordion in 1950 (Sams 2005:10–12). His 17 years of excavation at the site eventually opened up much of the eastern half of the Citadel Mound to Phrygian levels, exposing a system of public buildings burned in a massive fire (Young 1958a). The reason for his focus on this level was clear: 1) he believed it to be the most historically salient period of occupation at the site, including the reign of King Midas described by classical sources; 2) it was the best preserved level of the Citadel Mound, due to the fire that had destroyed buildings with their contents left in situ; and 3) it was the period to which he dated the largest and richest burial tumuli, including Tumulus MM, which Young believed to contain the re-

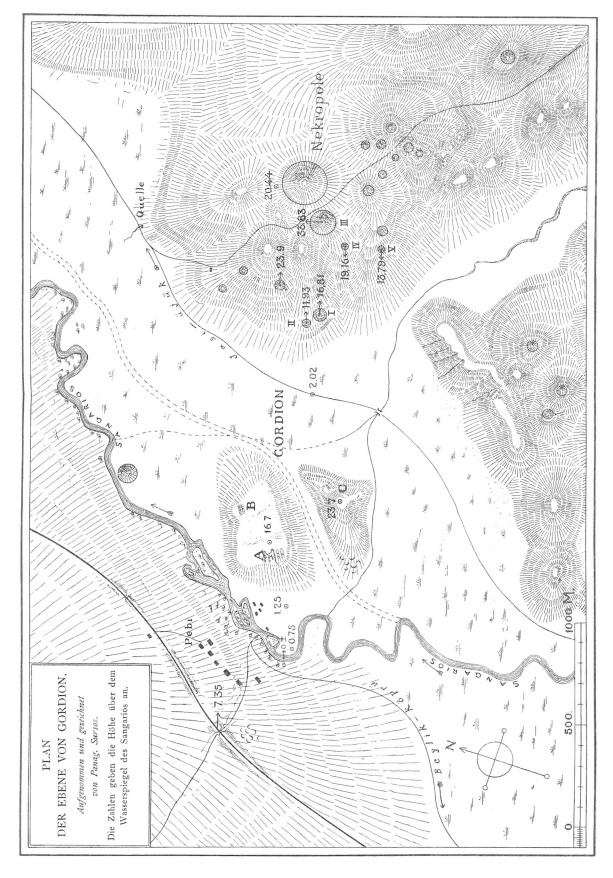

Fig. 1.1. Plan of Gordion showing excavation areas of the Körte brothers in 1900 (from Körte and Körte 1904: pl. 1).

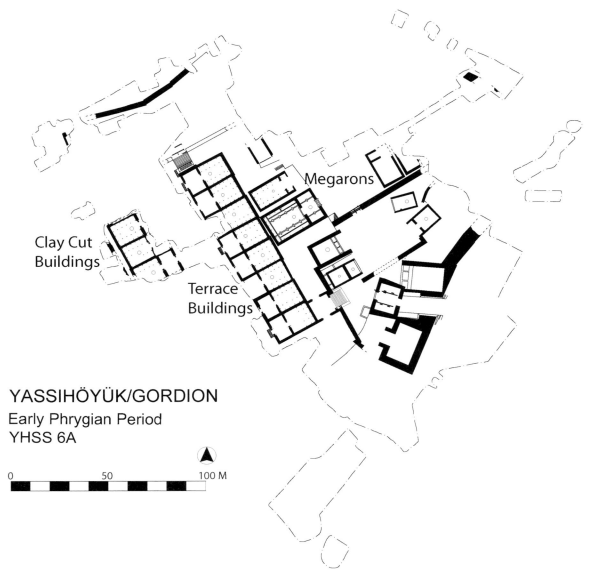

Fig. 1.2. Plan of Early Phrygian Gordion as excavated in 2005. Dotted lines represent the extent of excavation on the eastern half of the Citadel Mound. (Image courtesy of Mary M. Voigt.)

mains of King Midas. Once he had reached this level and identified it as Phrygian, Young removed the overlaying Medieval, Roman, Hellenistic, and Persian levels with minimal attention to their stratigraphy and plan (Sams 2005; Young 1958a). As a result, ongoing efforts to understand Achaemenid Persian and Hellenistic settlements at Gordion have needed to reconstruct the association of artifacts and architecture through archival research, relying extensively on excavation notebooks (e.g., Stewart 2010; Wells 2012).

Young identified two major construction levels during his excavations on the Citadel Mound: the "Phrygian" occupation (now termed Early Phrygian) and the overlying "Persian" level (later "Archaic" and now Middle Phrygian), which were separated by several meters of clean clay fill. The Phrygian citadel burned in a sitewide fire that Young associated with the historically attested invasion of Asia Minor by the Cimmerians in the early 7th century BCE (Voigt 2009; Young 1951, 1953). Young surmised that following destruction of their citadel by the Cimmerians, the Phrygian rulers' power was broken and the site lay abandoned for some 150 years until it was reoccupied by Achaemenid Persian rulers in the 6th century, who covered the burned layers with clay and started anew (Voigt 2009:225; Young 1956:264). The Citadel of the Persians (aka, the "New Citadel"), as Young presumed, had been robbed for stone by later occupants of the site and lacked the in situ floor deposits characteristic of the earlier "Old Citadel," making it less attractive for archaeological exploration for practical as well as historical reasons (Sams 2005).

During the 1955–73 seasons, excavation focused on exploring as large an area as possible of this Early Phrygian Destruction Level. The buildings uncovered included two facing rows of megaron-style (i.e., with a short antechamber and central hearth in the main room) buildings, termed the Terrace Buildings and Clay Cut Buildings, that contained evidence for food storage and preparation and textile manufacture (Fig. 1.2; Burke 2005; DeVries 1990; Sams 2005; Voigt 2005). Individual megaron-style buildings, termed Megarons, lay to the east of these structures and stood alone. The contents of these buildings indicated that they might have served as the political and possibly religious center of life at Phrygian Gordion, especially Megaron 3, the largest of these structures (Sams 1994a; Young 1957, 1960, 1962b).

Young also excavated a number of Phrygian tumuli, which he lettered A–Z (Fig. 1.3, see color insert). Not every tumulus was lettered and not every lettered tumulus was excavated. The three largest and richest tumuli (MM, W, and P) are the focus of one final publication (Young 1981), while 15 smaller inhumation tumuli have also been fully published (Kohler 1995). The cremation burials have yet to be published in detail, although they are described briefly in many of Young's annual reports in *Expedition* and the *American Journal of Archaeology* (see Young [1981:xxxv–xxxvi] for full references). The finds within the large tumuli were similar to those discovered by the Körte brothers in their Tumulus III, although greater in number, size, and variety. Tumulus MM is unquestionably a royal burial and the only fully intact burial chamber encountered by Young. Its excavation in 1957 brought great publicity to Gordion and is the focus of a large display at the Museum of Anatolian Civilizations (Anadolu Medeniyetleri Müzesi) in Ankara (Young 1958a,

1958b, 1981). The interior structure is a unique wooden building, the oldest preserved wooden structure in the world, with an elaborate roofing structure and a casing layer of juniper trunks that protected it from the crushing weight of the earthen mound above (Liebhart and Johnson 2005; Young 1960, 1981:85–100). Within this burial chamber, Young found the remains of the deceased laid out in an open coffin (Simpson 1990), wrapped in burial shrouds (Ellis 1981), with elaborate grave goods piled to the side, including furniture (Simpson 2007; Simpson and Spirydowicz 1999) and banqueting vessels containing the residues of food and drink from his funerary feast (McGovern 2000; McGovern et al. 1999).

Young assumed Tumulus MM to be the grave of King Midas, as it was the largest and most impressive burial, the skeleton was identified as a 61- to 65-year-old male (Young 1981:101), and its dating to the end of the 8th century BCE based on relative chronologies of metal fibulae and vessels and ceramic wares was coincident with attestations of the name "Mita of Mushki" in Assyrian sources during the 5th through 13th years of the reign of Sargon (717–709 BCE; Young 1981:269–72). Recent redating of the tomb using radiocarbon-adjusted dendrochronology puts the felling date of the tumulus logs at 740 +4/-7 BCE, with the highest probability of a date of 743–741 BCE (DeVries et al. 2003; Kuniholm et al. 2011; Rose and Darbyshire 2011). This redating, which is consistent with other radiocarbon and relative dates from the Citadel Mound that push back the date of the destruction of the Early Phrygian Destruction Level to ca. 800 BCE, places the construction of Tumulus MM within the subsequent Middle Phrygian period, contemporary with the New Citadel, and before the death of King Midas, who was still active against Sargon in 709 BCE (DeVries et al. 2003; Voigt 2005, 2009). One possibility is that this was the burial mound of King Gordias, the father of Midas, built during the peak of wealth and power at the site that is now thought to correspond to the Middle Phrygian period (Voigt 2007; Voigt and Young 1999).

Young's sudden death in 1974 in an automobile accident in Philadelphia brought to an end large-scale excavation at Gordion for 15 years. Preliminary reports had not yet been published on the 1969, 1971, and 1973 excavation seasons, nor had final publications been completed for excavations at the Citadel

Mound or of the tumuli. Keith DeVries succeeded Young at the University of Pennsylvania and took over responsibility for Gordion until 1988, when investigation of the site began anew under G. Kenneth Sams and Mary M. Voigt. DeVries published the results of Young's final few seasons and reviewed the research that took place over the intervening 15 years (DeVries 1990), a period that also saw the posthumous publication of Young's final report on the great tumuli (Young 1981).

Establishing Chronology

By 1987, several critical questions remained unanswered from Young's excavation campaigns. Chief among these was the need for a detailed chronology of the entire history of occupation at Gordion, several periods of which remained poorly understood. Additionally, little was known about the daily life of people at the site or about the distribution of population and land use in the surrounding Sakarya Valley. A new excavation program began in 1988 to begin to address these gaps in contemporary understanding of Gordion, led by Mary M. Voigt as Director of Excavations and G. Kenneth Sams as Project Director responsible for site conservation and the publication of Young's excavations. A deep sounding in two locations on the mound during 1988 and 1989 produced a stratigraphic sequence for the occupation of Gordion and yielded numerous animal and plant remains with which to address questions about subsistence economies and land use in the region.

Voigt identified ten stratigraphic levels of occupation through the deep sounding and used these to create the Yassıhöyük Stratigraphic Sequence (abbreviated YHSS) of ten phases, with YHSS 1 being the latest (Medieval period) and YHSS 10 the earliest (Middle Bronze Age) (Voigt 1994). Further refinements to the YHSS followed based on further excavation between 1993 and 2002, resulting in the combination of some phases (i.e., YHSS 8 and 9, the Late Bronze Age) and splitting of others (e.g., the Early and Late Hellenistic as 3B and 3A); this revised chronology with the most recent estimated absolute dates is provided as Table 1.1. The Early Bronze Age, which was sampled by Machteld Mellink during excavations below an Early Phrygian courtyard in 1961 (Gunter 1991; Sams 2005:13; Young 1962a:168),

was not reached during Voigt's excavations and so not numbered as part of this sequence.

Voigt's chronology offers several advantages over the earlier phasing produced by Young. The dates associated with the phases are the result of a combination of radiocarbon dating of sealed deposits of short-lived annual plants, avoiding the "old wood" effect (DeVries et al. 2003; Manning and Kromer 2011), combined with relative dating of stratigraphic levels based on imported pottery and distinctive artifact types (DeVries 2005; Voigt 2009) and historical documentation of verifiable events at the site, including the Persian conquest around 540 BCE (Darbyshire 2007), Alexander's visit in 334/3 BCE, and the destruction of Galatian Gordion by Manlius Vulso in 189 BCE (Dandoy et al. 2002; Sams 2005). More important, however, is the attention to stratigraphy in dating these sequences. This allows individual deposits to be dated precisely relative to other deposits and to create detailed subphasing for chronological periods of interest at Gordion (e.g., the Roman period [Goldman 2000, 2005]).

Also notable in Voigt's chronology is more accurate association of chronological periods with cultural affiliations. For example, her identification of subsequent phases of occupation during the Hellenistic period associated with Greeks and Celtic Galatians have enabled a more nuanced assessment of how changes in cultural and political systems at Gordion led to changes in subsistence and land-use practices at the site. I use Voigt's revised chronology (Voigt 2009) throughout this book and generally refer to phases in the text by name: e.g., Middle Phrygian.

While Young conducted operations at Gordion in a manner typical of 1950s classical archaeology, with a focus on architecture, diagnostic ceramics, art objects, and burials, Voigt introduced a broader suite of anthropological methods to archaeological investigation at Gordion. Her excavations adopted a holistic approach with the sieving of all deposits, full recovery of all macroscopic remains (including wood charcoal, animal bones, and non-diagnostic sherds), and taking flotation samples from a variety of contexts (Voigt 2005). In addition to site-focused work, the Gordion Regional Survey, under the direction of Bill Sumner and later Lisa Kealhofer, aimed to locate outlying sites and identify trends in settlement across the entire Gordion region in order to investigate the relationship between land use and political change across the region (Kealhofer 2005b; Marsh and Kealhofer 2014).

Detailed investigation of animal and plant remains from the 1988–89 Gordion excavations form the foundation for further environmental research at the site. Publication of zooarchaeological (Zeder and Arter 1994) and paleoethnobotanical remains (Miller 1999b, 2010) from the deep sounding has allowed integrated assessment of land use over time (Miller et al. 2009). Combined with unpublished results from later seasons of excavation at the site, this is one of the largest and most comprehensive environmental archaeology datasets from the Near East, a valuable resource for regional comparison and interpretation.

Urban Topography and Economy

The promising results of the 1988–89 seasons allowed Voigt to again secure substantial funding from the National Endowment for the Humanities to expand excavations at Gordion and shift to a broader focus on changes in the size, form, and organization of the urban settlement over time (Voigt 2005, 2011, 2013; Voigt et al. 1997). Rather than continuing investigation of the urban core of the eastern half of the Citadel Mound, Voigt shifted focus to three areas of the site that had been previously underexplored: the western half of the Citadel Mound; the Lower Town, which lay in the floodplain between the Citadel Mound and the remains of a defensive fortification known as the Küçük Höyük; and the Outer Town, which lay outside this defensive wall across the modern course of the Sakarya River (Fig. 1.4). Excavation on the Citadel Mound continued primarily in the Northwest Zone, but also in the Southwest and Southeast Zones of the mound, from 1993–2006 (Fig. 1.5).

Detailed accounts of stratigraphy from these seasons have yet to be published in full, although a number of interim accounts and publications of limited areas of the site exist. These include annual excavation reports published in *Kazı Sonuçları Toplantısı*, the proceedings volume of the annual symposium on archaeological excavation in Turkey (Sams and Burke 2008; Sams et al. 2007; Sams and Goldman 2006; Sams and Voigt 1995, 1996, 1997, 1998, 1999, 2003, 2004) and several synthetic works by Voigt (2002, 2005, 2009, 2011, 2013).

Middle Phrygian remains were excavated in the Northwest and Southwest Zones of the Citadel Mound, as well as in the Lower Town and Outer Town (Voigt 2002; Voigt and Young 1999). The broad distribution of these remains and an intensive surface survey of the areas west and north of the mound by Keith Dickey and Andrew Goldman demonstrated that the Middle Phrygian occupation was likely the peak of urban population at Gordion, covering an area of more than one square kilometer, although the subsequent Late Phrygian (Achaemenid) period maintained a similar urban extent (Voigt 2002:194). The clay layer separating the Early and Middle Phrygian citadels in the eastern Citadel Mound extended onto the western half of the mound, separated by a paved road at a lower level that had been discovered by Young during his final years of excavation (DeVries 1990:378). These twin mounds, separated by a road and interior circuit wall, appear to have been differentiated in use of space, with the eastern mound a public space and the western mound primarily residential (Voigt 2002; Voigt and Young 1999). Within the Lower Town, large ashlar buildings similar to those on the eastern half of the Citadel Mound and smaller domestic structures were present; domestic structures appear to cover much of the area of the Outer Town (Voigt 2002; Voigt and Young 1999). The Lower Town, approximately 51 hectares in area, was walled with mudbrick fortifications: both the Küçük Höyük and Kuş Tepe were fortified towers, preserved to great height due to the construction of Persian siege mounds against them when Gordion was taken around 540 BCE (Fig. 1.4; Darbyshire 2007; Sams 2005; Voigt 2011).

Much of the excavated material from the 1993–2006 seasons postdates the Middle Phrygian period, coming instead from the Late Phrygian period of Achaemenid control of Gordion and the subsequent Hellenistic (Greek and Galatian) and Roman occupation of the site. Relatively little Medieval material was encountered, and only in the Northwest Zone of the Citadel Mound (Sams and Voigt 2004), although the Medieval occupation of Gordion did extend across the Citadel Mound (Voigt 2002).

Late Phrygian (Achaemenid) settlement extent at Gordion indicates a large urban population, with the Lower and Outer Towns remaining primarily domestic in function (Voigt 2002). The political center

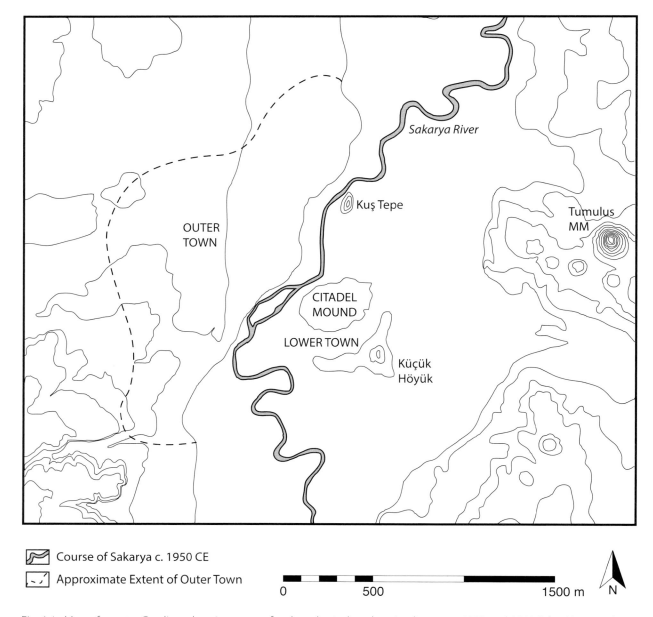

Fig. 1.4. Map of greater Gordion, showing areas of archaeological exploration between 1993 and 2002 (after Voigt and Young 1999:194). This map shows the course of the river as of 1950, before canalization.

appears to have shifted to the Mosaic Building in the southeastern area of the Citadel Mound, originally excavated by Young in 1952 and further explored in 1995 and 2006 (Glendinning 2005; Sams and Burke 2008; Sams and Voigt 1997; Voigt 2002; Young 1953). Achaemenid domestic architecture changes from that of the preceding Middle Phrygian period, with semi-subterranean structures as well as structures built at grade; the type of architecture varies by area across the site (Voigt 2002:194; Voigt and Young 1999:223–35).

Hellenistic settlement was located only on the Citadel Mound; it appears that the population of the city shrank substantially from the preceding period and no longer required settlement of the less defensible Lower and Outer Town areas (Voigt 2002). The later Hellenistic Galatian presence at Gordion, described in Roman historical sources, is attested archaeologically by several unique finds indicating a European origin for the inhabitants of the city. Several human bodies in the abandoned Lower Town, apparently killed by strangulation, and disarticulated bodies placed in association with animal bones suggest Celtic ritual sacrifice (Dandoy et al. 2002; Selinsky 2005, 2015; Voigt 2012). An abstract stone sculpture of a human head, similar to Celtic iconography from Europe, and a silver coin typical of the type used to pay mercenaries during the 3rd century BCE, when Celtic mercenaries were hired and brought to Anatolia, were associated with what appears to be a public structure in the Northwest Zone of the Citadel Mound, likely the center of the Galatian town of Gordion (Dandoy et al. 2002; Sams and Voigt 1999, 2004; Voigt 2012). The violent destruction of this building and associated craft production structures has been dated to the early 2nd century BCE, consistent with the 189 BCE date of Manlius Vulso's campaign against the Galatians recounted by Livy (38.12–27), but subsequent resettlement of the site, evident archaeologically, is not recorded in Livy's account (Sams and Voigt 1999; Voigt 2002).

Roman Gordion was also confined to the Citadel Mound, with burials in the unoccupied Lower Town below (Goldman 2000, 2005, 2007, 2010; Sams et al. 2007; Sams and Goldman 2006; Sams and Voigt 1996; Selinsky 2005; Voigt 2002). Excavations in the Northwest Zone between 1993 and 2002 exposed four phases of Roman occupation in that area (Goldman 2005) and subsequent excavations by Goldman in 2004 and 2005 demonstrated that the Roman settlement extended over much of the western half of the Citadel Mound (Goldman 2007; Sams et al. 2007; Sams and Goldman 2006). Remains of weapons and scale armor confirmed earlier theories that the Roman settlement at Gordion was primarily a military encampment, perhaps to protect the stretch of Roman road linking Pessinus with Ankara that was discovered adjacent to Tumulus MM (Bennett and Goldman 2009; Goldman 2007, 2010). Roman burials include adult inhumations in pits and brick-lined tombs, child inhumations in pits, and pit cremations (Sams and Voigt 1996; Selinsky 2005).

Medieval Gordion seems to have been larger than Roman Gordion, with settlement across the entire Citadel Mound during at least parts of the Medieval Islamic period (Sams and Goldman 2006; Voigt 2002). There may have been a small fortress on the highest point of the western half of the Citadel Mound at this time (Voigt 2002:195). Little is known about Gordion from historical records of the early Turkish period but archaeological remains indicate residential settlement, including houses, courtyards, and ovens, and the presence of a local metalworking industry, as evidenced by a large furnace containing iron slag (Sams and Voigt 2004).

Regional Archaeological and Environmental Survey

The Gordion Regional Survey, directed by Lisa Kealhofer, was an intensive surface survey program designed to investigate changing patterns of land use related to political changes at Gordion; nine transects were conducted throughout the Gordion region between 1996 and 2002 (Kealhofer 2005b). Field walkers collected all ceramic sherds, including plainware body sherds that were dated by matching fabric with a reference collection of sherds from excavated contexts at the site of Gordion itself. These results allowed reconstruction of settlement and land-use intensity in various areas of the Gordion region over time (Kealhofer 2005b), especially once placed in a greater regional context via elemental characterization of ceramic fabrics (Grave et al. 2009; Grave et al. 2012).

Most notably, the survey recognized that settlement in the Gordion region was focused on areas of the landscape close to permanent or seasonal watercourses; roughly ¾ of all sherds were found within 250 m of a water source (Kealhofer 2005b:144). Upland plateau areas were nearly devoid of evidence for occupation, in keeping with ethnographic research that suggests these areas were suitable only for extensive pastoralism until recent years (Gürsan-Salzmann 2005; Kealhofer 2005b). Low-lying floodplain soils also showed little evidence for settlement, likely the effect of substantial recent alluvial deposition in these areas (Kealhofer 2005b:144; Marsh 1999).

Two major trends in regional settlement are evident from these findings. During both the Middle to Late Bronze Age and the Roman period, settlement appears to have been dispersed throughout the landscape but relatively intensive. In contrast, during the Early and Middle Phrygian periods settlement was focused on only two creek valleys east and northeast of Gordion but intensively so in those areas. The Early Bronze Age, Early Iron Age, Late Phrygian, Hellenistic, and Medieval periods appear to have involved relatively low-density settlement across the Gordion region. These changing patterns of regional occupation have implications for understanding how political change at Gordion affected regional agricultural strategies (Kealhofer 2005b:148), a theme further developed in Chapters 5 and 6 of this volume.

Geological and ecological survey accompanied archaeological survey of the Gordion region. Ben Marsh undertook a series of geological cores in the Sakarya floodplain, investigated geological sections in the river cut of the Sakarya, and sampled soils across the survey region; these data allowed him to reconstruct the geomorphological history of the Sakarya River and erosion and alluviation in the Gordion region (Marsh 1999, 2005, 2012; Marsh and Kealhofer 2014; Voigt 2002). Marsh recognized that the Sakarya now flows through the area of the ancient city, while it used to flow to the east of the site (Fig. 1.4); this course change appears to have occurred after settlement ceased at the site, perhaps as recently as the 19th century (Marsh 1999). Evidence for increased alluviation begins around 700 BCE, suggesting increased human impacts on the surrounding landscape typical of the Beyşehir Occupation Phase (Marsh 1999; see further discussion in Ch. 3). The changing course and flow rate of the river is determined by erosion in upland areas that increased alluviation rates of the river; this erosion appears to have increased substantially after 600 CE but did not lead to increased river deposition, a feature of anthropogenic soil erosion systems (Marsh 2005:171).

Naomi Miller began a series of structured plant surveys in 1997 as part of an attempt to minimize ongoing erosion on Tumulus MM caused by overgrazing (Miller 1999a, 2000, 2010, 2012; Miller and Bluemel 1999). A barbed-wire fence was erected around the tumulus in 1996, permitting undisturbed regrowth of native steppe vegetation. Miller documented this process through annual surveys of vegetation on six faces of the tumulus and on other neighboring (unfenced) tumuli (Miller 2010: app. C). Aside from the archaeological and biological conservation implications of this work (which has nearly eliminated erosion on the tumulus; Erder et al. 2013; Miller and Bluemel 1999), this research provides a quantitative baseline for steppe regeneration useful for interpretation of archaeobotanical remains from Gordion (Miller 2012).

Reassessing Gordion

Although archaeological investigation of Gordion has taken place over more than 100 years, resulting in the publication of more than one hundred articles and over a dozen monographs, the integration of these studies remains incomplete. One challenge that remains is due to incomplete publication of the Citadel Mound excavations under Rodney Young, a task now being undertaken piecemeal in a series of dissertation and book projects (e.g., Dusinberre 2005; Gunter 1991; Roller 1987, 2009; Romano 1995; Sams 1994b; Simpson 2010; Stewart 2010; Wells 2012, and others in progress). Additionally, many classes of archaeological remains have not yet been fully studied or published from the 1993–2006 excavation seasons. Differences in recovery and documentation standards and excavation methodologies between the Young and Voigt excavations have rendered those datasets incompatible or marginally compatible in a variety of ways, which introduces challenges in reconciling the results. For example, the only botanical remains recovered and identified from the Young excavations are in situ deposits from the Early Phrygian Destruction Level and represent crop storage (Nesbitt 1989). Flotation samples from the 1993–2006 excavations span a range of depositional and use contexts but do not include any in situ caches of crop seeds. The differences in preservation circumstances would strongly bias any analysis that overlooks these sampling differences and makes it difficult to integrate results of the full sequence of archaeological deposits excavated at Gordion over the past 60 years.

One current effort to promote interpretation and publication of materials from Gordion is the Digital Gordion initiative, a project of the Penn Museum under the direction of C. Brian Rose (Penn Museum 2015). Digital Gordion aims eventually to digitize the entire Gordion archive, making original excavation notebooks and other records available online and cross-referencing those works in a spatial database to promote research and publication at Gordion. Such an approach offers great potential for modern spatial reanalysis of the site, akin to efforts at other sites excavated and published long before GIS was available, such as Olynthos (Cahill 2002). Making unpublished records from the Gordion excavations more available, both in digital format and as formal comprehensive monographs, is one important step towards the ultimate goal of creating an integrated reassessment of occupation at Gordion.

Overview of the Volume

Chapter 2 elaborates the theoretical framework of this study by considering how archaeologists have approached decision making and risk management in the study of past agricultural systems. The two bodies of theory, resilience thinking and behavioral ecology, have considerable explanatory potential in the study of agricultural systems. Despite a lack of prior engagement between these theoretical perspectives, I provide evidence that they are epistemologically compatible and uniquely suited to the consideration of risk management in the past.

The biogeography, ecology, and paleoenvironmental setting of Gordion are presented in Chapter 3. Drawing on published sources and firsthand ecological survey, I describe the plant ecology and phytogeography of the Gordion region today as an analog for interpreting past vegetation communities and human impacts on the extent, diversity, and spatial distribution of plants. I review the paleoclimatic data relevant for Central Anatolia and make an effort to reconstruct the paleoenvironmental setting of Gordion throughout its occupation sequence. I conclude by considering the likely limited influence of climate change on agricultural practices at Gordion.

Chapter 4 presents archaeological evidence for wood use at Gordion and changes in woodland structure in the region. Drawing on the analysis of wood charcoal fragments excavated at the site, I identify several trends in the use of woodland communities, likely caused by ongoing anthropogenic modification of both the spatial extent and species composition of regional woodlands. I reconstruct the diversity of wood use onsite through spatial and stratigraphic interpretation of charcoal samples, distinguishing 50 distinct activity areas and exploring the context of wood use in each. Finally, I consider general patterns of wood use sitewide, applying foraging theory to consider how inhabitants of Gordion made decisions about which woodland patches to exploit and which wood types to gather for both fuel and construction.

In Chapter 5, I turn back to agriculture and present a series of methods, derived primarily from behavioral ecology, for reconstructing agricultural decision making in the archaeological record. I apply these methods to the analysis of 220 flotation samples, the results of which are combined with an additional 252 previously published by Miller (2010). Drawing on this robust diachronic dataset, and incorporating preliminary data on changes in animal economy, I identify agricultural practices employed at Gordion, consider how they changed through time, and assess their environmental impact. Animal grazing, as reconstructed from the seeds present in dung burned as fuel, appears have to have had the greatest impact on local vegetation structure and led to significant landscape change during later periods of occupation at Gordion.

Chapter 6 provides a synthesis of these results and returns to the broader questions of the study, taking a resilience perspective to explain agricultural decision making and environmental change at Gordion, and arguing for consideration of agricultural sustainability in long-term perspective. Here I consider cultural and environmental systems to be coupled and explore reciprocal relationships between these systems to offer general insights about human decision making applicable to other archaeological settings. I conclude by extending these findings to the contemporary world and suggest potential avenues for further integration of archaeology into environmental policy today.

2

Modeling Agricultural Decision Making and Risk Management

In order to understand human behavior in the past, it is necessary to utilize both particularistic historical analyses—to reconstruct what actually happened within its historical context—and social theory—to contextualize a particular case study in comparative perspective. In this book, I reconstruct the history of agriculture, land use, and environmental change at Gordion in order to understand the underlying social processes that led to the specific land-use practices documented in the archaeological record. I draw on theory from the social and behavioral sciences and from ecology to provide an integrated perspective on environmentally situated agricultural decision making in the past.

In this chapter, I introduce two broad bodies of theory—human behavioral ecology and resilience thinking—that offer distinct, but complementary, approaches to understanding agricultural choices. I argue that these two theoretical perspectives are logically and conceptually compatible, and that the strengths of each compensate for the limitations of the other. In particular, I focus on risk management as a theoretically consilient explanatory principle behind agricultural decisions in many regions—especially those characterized by a higher degree of environmental variability—from which it is possible to derive specific archaeological expectations that are testable with botanical and faunal remains, as detailed in the later chapters of this volume.

Modeling Agricultural Decision Making in Archaeology

A primary concern of archaeologists for nearly a century has been understanding patterns of human behavior in the past (e.g., Childe 1934; Steward 1955).

What we now term "processual" approaches to archaeological inquiry had a central aim of identifying and systematizing these patterns, or processes, of human behavior (e.g., Binford 1967; Flannery 1969; Watson et al. 1971). The problem that still faces archaeologists interested in behavioral patterns, however, is that we need to understand not just behaviors but also the decisions that motivate them. Even in contemporary society, understanding decision making is a challenge and, thus, a central aim of many branches of scientific inquiry, from information science (e.g., Case 2012; Khazraee and Gasson 2015; Pirolli and Card 1999; Sandstrom 1994) to human ecology (e.g., Henrich and McElreath 2002; Smith and Winterhalder 1992; Winterhalder and Smith 2000). The study of decision making is even more challenging for archaeologists, who cannot conduct experiments or interviews with their research subjects who lived and died, usually in historical anonymity, centuries to millennia ago.

In this context, exploring agricultural decision making at least constrains this problematic topic of inquiry to a single realm of study: agricultural practices and the decisions that led to their invention, adoption, propagation, and abandonment. These themes suggest specific datasets that we might collect, such as material evidence for agricultural crops, tools, and storage and trade systems. Even still, understanding decision making in the past requires robust theoretical frameworks that allow us to infer human thought processes from what remains of their tangible outcomes. Coupled with innovative applications of well-collected data, it then becomes possible to identify different behaviors and infer the decisions behind them using a combination of archaeological, paleoenvironmental, and ethnographic information.

Many inferential frameworks have potential for explicating human behavior and decision making in the past, but in this chapter I briefly outline only three, focusing on those that readily accept archaeological datasets and have been demonstrated to offer new systematic insights into past agricultural practices: cultural niche construction, human behavioral ecology, and resilience thinking. All are based on evolutionary, ecological principles and, thus, explain behavior in terms of both historical development and contemporary environmental adaptation.

Niche construction acknowledges that organisms not only adapt to their environment but also actively construct it through patterns of behavior (Laland and O'Brien 2011; Lewontin 1983; Odling-Smee et al. 2003; Smith 2014). In changing their environment, organisms shape their own ecological niches, thus changing the evolutionary forces acting upon them. Although well studied in animal behavior (e.g., the beaver), cultural niche construction by humans has only recently been investigated in detail. It is obvious that in domesticating plants and animals, humans have transformed our food sources with profound ecological and evolutionary consequences (Rindos 1984; Smith 2007, 2011, 2014, 2015; Zeder 2015; Zeder et al. 2006). Considerable archaeological data also identify the manipulation of wild food resources, both plant and animal, by diverse human groups (e.g., Anderson 2005; Deur and Turner 2005; Doolittle et al. 2004; Fowler and Rhode 2011; Groesbeck et al. 2014; Lepofsky and Caldwell 2013; Lepofsky and Lertzman 2008; Smith 2009b, 2013). While cultural niche construction offers important new insights into domestication processes and the origins of agriculture, and is theoretically well suited to explore the environmental transformations that accompany agricultural practices, it has not yet been applied to cases of developed agricultural systems. In the future, cultural niche construction may prove particularly useful in explaining patterns of agricultural innovation (Brite and Marston 2013; Bruno 2014; Gremillion 1996; Van der Veen 2010) and the role of cultivated wild foods in agricultural societies (Gremillion 2004; M.L. Smith 2006; Sullivan et al. 2015; Sullivan and Forste 2014; VanDerwarker et al. 2013). The current literature, however, provides few points of ready engagement using niche construction to model decision making in fully agricultural societies, such as the economies of the post-Neolithic Near East. Thus, in this volume, I instead engage with two

bodies of theory more developed in the study of agricultural decision making: human behavioral ecology and resilience thinking.

Human behavioral ecology deals broadly with human decision making in the context of immediate social and ecological environments (Borgerhoff Mulder and Schacht 2012; Smith and Winterhalder 1992; Winterhalder and Smith 2000). Fundamentally, much of behavioral ecology research focuses on tradeoffs between different behavioral strategies and their evolutionary implications. Derived ultimately from ethology, the study of animal behavior, human behavioral ecology has been used to address a range of topics, including cooperation (Bliege Bird et al. 2002; Fehr et al. 2002; Gintis et al. 2003; Gurven 2006; Madsen et al. 2007; Richerson et al. 2003; Smith and Bliege Bird 2005), distribution of populations (Kraft and Baum 2001; Kraft et al. 2002; Madden et al. 2002), engaging in risky or expensive behaviors (i.e., "showing off" [Bliege Bird et al. 2001; Codding and Jones 2007; Smith and Bliege Bird 2000; Smith et al. 2003]), and foraging, primarily for food but also other necessary resources. It is foraging theory that offers the best engagement with questions of subsistence related to agricultural systems. In recent years, foraging theory has been applied to the origins of agriculture (e.g., Barlow 2002; Gremillion et al. 2014; Kennett and Winterhalder 2006; Piperno 2011; Winterhalder and Goland 1997) and decision-making processes related to agricultural (Gremillion 2002, 2006; Marston 2011; VanDerwarker et al. 2013) and land-use practices (Gremillion et al. 2008; Marston 2009). Due to the variety of models available in behavioral ecology and their ready adaptation to the types of data recovered archaeologically, human behavioral ecology and, specifically, foraging theory offer considerable potential for understanding agricultural decision making in the past, as discussed in further detail below.

Resilience thinking offers a novel approach to understanding decision making. Rather than focusing on explicitly evolutionary frameworks of adaptation, resilience thinking addresses how and why change occurs in complex adaptive systems, both ecological and social (Gunderson et al. 2009; Gunderson and Holling 2002; Walker and Salt 2006). In resilience thinking, both social and ecological systems are modeled through a common set of concepts and terminology, centering on the analogic model of the adaptive cycle

(Holling 2001; Holling and Gunderson 2002). Although a relatively new body of theory in archaeological reasoning and explanation (e.g., Redman 2005; Redman and Kinzig 2003), resilience thinking has already contributed to understanding agricultural decision making in the past (e.g., Marston 2015; Nelson et al. 2012; Peeples et al. 2006; Rosen and Rivera-Collazo 2012). Through a resilience perspective, questions of scale and interaction between ecological and social systems are highlighted, which clarifies causal linkages between variables and illuminates explanations for specific behaviors and environmental changes observed in the archaeological record. Resilience thinking and human behavioral ecology offer compatible and complementary models for reconstructing past agricultural systems and the human decision-making processes behind them.

Resilience Thinking and Behavioral Ecology

Although both derived from ecological theory, behavioral ecology and resilience thinking have shared little conceptual history and are rarely applied to similar questions or datasets. While some archaeologists have embraced one or the other approach (more commonly human behavioral ecology), there is no body of archaeological literature that combines or integrates the two. I suggest that both perspectives offer valuable and distinct solutions to archaeological problems, but also that the two theoretical frameworks are conceptually compatible and that an integrated application of the two provides a richer understanding of past human decision making and its consequences than either alone. Behavioral ecology has been critiqued for oversimplifying the reality of human agency (Joseph 2000; Smith 2009a) and lacking sufficient explanatory power to describe significant human-environmental interactions, such as the transition to agriculture (Smith 2015; Zeder and Smith 2009), while resilience thinking is so broad and adaptable that it can become purely descriptive and lose analytical value, especially when applied to social systems (Brown 2014; MacKinnon and Derickson 2013). With judicious use and in appropriate combination, the limitations of each can be mitigated, yielding a powerful interpretive framework for studying human decision making in the past.

Here I review the core conceptual basis of resilience thinking and human behavioral ecology, then turn in the following section to points of intersection and engagement, as well as tension and distinction, between the two. I conclude by suggesting a framework for integrating these theoretical approaches in archaeological analysis, one that I adopt in the study of land use and agriculture at Gordion that follows in Chapters 4 and 5.

Core Principles of Resilience Thinking

Resilience thinking, also termed resilience theory or the theory of adaptive change, emphasizes that both continuity and change are processes inherent to ecosystems (Holling 1973). It embraces the complexity of ecological and social systems, as well as interactions between the two, and views change as a continuous and natural component of such systems. Resilience thinking attempts to understand how and why change occurs in complex adaptive systems, whether ecological, social, or linked social-ecological systems (Folke 2006; Gunderson and Holling 2002; Miller et al. 2010; Walker and Salt 2006). Resilience thinking can be applied to a wide variety of adaptive systems because it provides a set of flexible models based on ecological principles that can be applied as analogies to relate specific social and ecological systems to one another and to abstract models (Carpenter et al. 2001; Gunderson and Holling 2002; Holling 2001). Chief among these are the concepts of the adaptive cycle, a general model for cycles of growth and decline within a system, and the panarchy, a series of adaptive cycles nested in time and space (Holling et al. 2002b). It is these that represent the applications of resilience thinking to archaeological problems to date (e.g., Allcock 2013; Peeples et al. 2006; Redman and Kinzig 2003; Rosen and Rivera-Collazo 2012). In contrast, I argue that a more fruitful application of resilience concepts to understanding human decision making from archaeological remains is a focus on how resilience thinking illuminates the effects of scale on adaptive systems, especially scalar and cross-scalar interactions resulting in mismatches, thresholds, and legacy effects (Marston 2015). To understand how resilience conceptualizes scale, however, we first need to consider the adaptive cycle and panarchy.

The adaptive cycle concept builds upon the traditional ecological model for ecosystem succession

(Holling et al. 2002a). In ecological succession, plant and animal communities are found in two phases: exploitation, where pioneer species expand to fill available ecological niches, and conservation, where populations of climax species dominate and stability is the defining characteristic (Bazzaz 1996; Clements 1916; Whittaker 1953). Release events disturb climax communities and provide new openings for exploitation phases and pioneer species. Resilience provides an explicit mechanism for systemic change by offering a fourth phase, reorganization, during which existing systems change in small or large ways and new systems can be created (Fig. 2.1; Holling and Gunderson 2002). From a resilience perspective, during the exploitation phase, new or disturbed habitats are colonized, while the conservation phase emphasizes the slow accumulation of energy and biomass within a habitat or social system. The release phase is often described, especially in human systems, as one of "collapse," wherein energy and biomass are released and become less organized, often suddenly (as in a forest fire). The phase of reorganization is one of innovation and restructuring, where new species may appear, new habitats may arise, or new social structures may form. The function of the adaptive cycle analogy is to describe in general terms the ways in which ecosystems and societies change in

predictable ways, with growth, decline, and reorganization as typical stages. While these phases appear equal in size on a diagram (Fig. 2.1), a system may pass through each over millennia or seconds; the exploitation to conservation phases are typically the slowest, while release to reorganization may be extremely fast (Holling and Gunderson 2002).

Change in any adaptive system can follow this cycle and lead to reproduction of the same system, or break off at any stage (typically during release and reorganization) and create a new system, distinct in space, time, and/or organizational scale. The panarchy concept emphasizes the connections between successive adaptive cycles, rather than their potentially more obvious differences, and so serves as a useful metaphor for understanding long-term patterns of cultural or environmental change within a region, where historical legacies influence current conditions. The panarchy, in particular, offers a valuable model for archaeologists who trace continuity in cultural elements or material signatures across social boundaries within a region or between descendant communities that otherwise possess distinct cultural traits (Redman 2005).

Time, space, and organizational scale provide the dimensions of change for successional processes as described above, and interactions across these scales

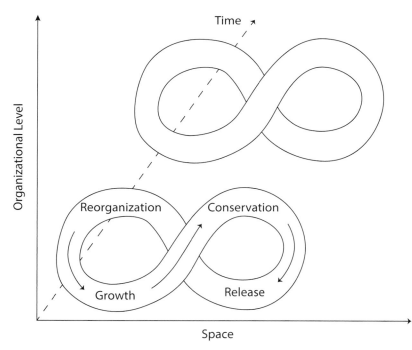

Fig. 2.1. Generalized model of a panarchy consisting of two adaptive cycles (after Holling 2001:394, 397). Note that the three axes of time, space, and organizational scale do not imply that a progression necessarily moves in one or the other direction along these axes, but rather that adaptive cycles operate over these three dimensions and exist at distinct points in time, space, and organizational scale within a single descendent panarchy.

are critical for understanding feedback processes that can lead to transformative change within systems. Such cross-scalar effects arising from these interactions are among the most revealing of how social and ecological systems interact (Cumming et al. 2006; Liu et al. 2007a; Liu et al. 2007b; Marston 2015). Three classes of scalar effects can be distinguished from a resilience perspective: mismatches, thresholds, and legacy effects.

Mismatches arise from interactions between variables that operate at different polarities within one of these scales, e.g., fast versus slow processes, such as the relationship between predation rate on a species and the birth rate of that species, or at incompatible rates across two scales, e.g., organizational hierarchies misaligned with the spatial distribution of a managed resource. What distinguishes mismatches from variables operating at compatible speeds or scales is that mismatches occur when variables "are aligned in such a way that one or more functions of the social-ecological system are disrupted, inefficiencies occur, and/or important components of the system are lost" (Cumming et al. 2006:3). Mismatches, especially those that persist for long periods of time or involve severely misaligned variables, may lead to systemic changes of varying degrees. Results of such mismatches include species extinctions (Liu et al. 2007b), deforestation (Peeples et al. 2006), and erosion (Marston 2015).

Thresholds are tipping points in a system where a small change that crosses a threshold value leads to substantial subsequent change, whether within the system (perhaps to an alternative stable state, or from one phase of an adaptive cycle to the next) or to a new adaptive cycle within a panarchy (Groffman et al. 2006; Scheffer 2009). They are sometimes reversible, and sometimes not (Scheffer and Carpenter 2003; Walker and Meyers 2004). Contemporary societies may even be facing a massive, global scale threshold event (Barnosky et al. 2012) or set of events (Rock-ström et al. 2009; Steffen et al. 2015). Thresholds are often more evident in retrospect than at the time they are crossed and may be unpredictable or invisible in advance, absent large-scale simulation models (Scheffer and Carpenter 2003). However, once a threshold event has been identified, it may be possible to use that knowledge to foresee future thresholds in analogous circumstances (Folke et al. 2004; Groffman et al. 2006): for example, eutrophication of lakes and gulfs

has been well studied and now specific threshold values of phosphorus levels leading to turbid water states can be predicted with great specificity in certain lake systems (Zhang et al. 2003), together with increasing variance that may presage a state change (Carpenter and Brock 2006). More complex are efforts to model multiple thresholds and the effects of interaction between them, including one threshold event triggering multiple other thresholds in a cascading fashion (Kinzig et al. 2006).

Legacy effects are the result of past processes, within either ecological or social systems, on the current state of a place. Many of these are the result of human-environmental interaction, including soil salinization, deforestation, and overfishing; such legacy effects are slow to change and potentially irreversible, imposing constraints on future ecosystems and societies (Jackson et al. 2001; Liu et al. 2007a; see also Butzer 1982:258). An extreme legacy effect is a "poverty trap," in which both diversity and resilience have been eliminated from a system and the impoverished state departs from the adaptive cycle entirely, forcing complete social or ecological change: desertification and soil loss in periglacial regions are ready examples (Holling 2001:400–401; Holling et al. 2002b:95–96). Legacy effects form part of a landscape of options available to a society, thus greatly affecting decision making by subsequent inhabitants of a region and rendering an understanding of the continuity of change within a region of critical importance for understanding human decision making.

Resilience thinking offers a framework for understanding how variables affect one another, especially across scales and between human and ecological systems, and provides a rationale for explaining the role of historical continuity within a regional or cultural tradition of practice. Resilience relates social and ecological systems in an explicit but consistent framework, allowing cross-scalar comparison and consideration of causal relationships and feedbacks between mutually interacting variables. As scale and scalar processes (including mismatches, thresholds, and legacy effects) have great explanatory power for understanding change in linked social-ecological systems, scale is the critical element of resilience thinking for understanding human decision making, and offers a valuable point of intersection with human behavioral ecology.

Human Behavioral Ecology and Foraging Theory

Behavioral ecology arose from ethology, the study of animal behavior, by incorporating evolutionary theory to explain the relationship between individuals and their environment, aiming at an understanding of the ultimate causes of behavior (*sensu* Mayr 1961). As documented by Alcock (2006:7), the transition within animal behavioral studies from proximate causes (ethology) to ultimate causes (behavioral ecology) took place primarily in the 1960s and 1970s. This included the earliest applications of economic and mathematical models to address behavioral questions, including development of optimal foraging models (Charnov 1976; Emlen 1966; MacArthur and Pianka 1966; Maynard Smith and Price 1973; Parker and Stuart 1976). Although several foundational works in animal behavioral ecology also included some discussion of human behavior (e.g., Alexander 1974:367–74; Dawkins 1976; Denham 1971:93–94; Wilson 1975), anthropologists rather than biologists wrote the first publications explicitly focused on the evolutionary study of human behavior. The earliest of these appeared in the mid-1970s (Beaton 1973; Hartung 1976; Wilmsen 1973), followed by the first journals dedicated to the field (*Ethology and Sociobiology*, now *Evolution and Human Behavior*, and the *Journal of Social and Biological Structures*, now the *Journal of Social and Evolutionary Systems*) and the first edited volumes (Chagnon and Irons 1979; Winterhalder and Smith 1981). Bibliographic survey demonstrates a more than six-fold increase in human behavioral ecology publications from the 1970s to the 1980s (Winterhalder and Smith 2000:52), corresponding to the expanding diversity of evolutionary ecology as applied to questions of human behavior. While behavioral ecology is used within biological anthropology to address a broad range of topics including altruism and cooperation (Gintis 2000; Gintis et al. 2003; Gurven 2004, 2006; Trivers 1971), territoriality (Fretwell 1972; Kraft and Baum 2001), and signaling (Bliege Bird and Smith 2005; Bliege Bird et al. 2001; Boone 1998; Gintis et al. 2001; Grafen 1990), the majority of archaeological applications of behavioral ecology apply concepts related to foraging behavior (Bird and O'Connell 2006).

Foraging theory addresses how animals find and select prey under a number of environmental and social conditions. Foraging models compare alternate foraging strategies, giving insight into why specific strategies might be followed by certain individuals or groups, but not others. Models have been tested in animals through experimental manipulation of food availability or environmental conditions (e.g., Milinski and Heller 1978; Parker and Stuart 1976), as well as through repeated observation of foraging behaviors in a species (e.g., Belovsky 1978; Krebs et al. 1977; Zach 1979). Foraging models have been particularly valuable in human behavioral ecology studies due to the importance of foraging systems in subsistence patterns, including the transition to agriculture (Gremillion 2014; Gremillion et al. 2014; Gremillion and Piperno 2009; Kennett and Winterhalder 2006), and the significance of foraging in structuring patterns of daily practice (Winterhalder and Smith 2000). Foraging models have been applied successfully to limited classes of non-food resources as well (e.g., Elston and Brantingham 2002; Marston 2009; Sandstrom 2001), although this area of inquiry has been pursued only in the last two decades.

All formal foraging models in behavioral ecology are based on a key assumption, termed the optimization assumption, wherein it is assumed that organisms are designed to be maximally efficient due to their evolutionary history. Thus, natural selection forms the background for the underlying economic logic regarding decision making, in much the same way that evolutionary history is the basis of theoretical frameworks for explaining social learning (Boyd and Richerson 1985, 2005; Richerson and Boyd 2005). For this reason, foraging theory is often referred to as optimal foraging theory, as models assume that organisms prefer efficient and optimal returns.

Several variables must be identified in order to construct a useful foraging model based on the optimization assumption. These include identification of (1) the currency of the model which is to be optimized, (2) the range of possible alternative strategies available to the organism in question, and (3) consideration of evolutionary and environmental constraints (Maynard Smith 1978; Winterhalder and Kennett 2006). Alternative strategies and constraints are fundamentally interrelated, since the constraints of evolutionary history and local environment define the effectiveness of alternative foraging strategies. The currency of any Darwinian evolutionary model should be reproductive fitness: i.e., which strategy results in the most offspring. Fitness is rarely assessed in behavioral ecology studies due to its abstract nature (but see Smith et al. [2003] for a notable exception) and instead correlates of fitness are tested. The currency used most frequently is the long-term rate of energy intake (Pyke et al. 1977; Smith 1991; Stephens and Krebs 1986), but this may not be the most appropriate foraging currency for every instance (Zeder 2014, 2015); overall survival (Milinski and Heller 1978) or intake of a specific nutrient (e.g., sodium [Belovsky 1978] or protein [Keegan 1986]) may be a more appropriate variable to model, and it may be necessary to model several variables simultaneously, e.g., through linear programming (Belovsky 1978; Gremillion 2014).

Constraints are often assumed in the construction of a model, including assumptions about the environmental knowledge of the forager, how prey or patches are encountered and handled, and whether the forager is capable of modifying some constraints (Stephens and Krebs 1986:10–11, 38). Models that fail to accurately incorporate constraints relevant to foraging behavior, such as the degree of information available to the forager, may require further manipulation and refinement of the model as it is tested (Gremillion 2002; Pulliam 1975). The identification of previously unknown constraints that significantly modify foraging patterns is often a primary conclusion of case study tests of foraging theory (e.g., Belovsky 1978; Bird et al. 2013; Gremillion 2002; Marston 2009). Among the foraging models available in behavioral ecology, however, three are most frequently and effectively used.

Three Foraging Models

The three classes of models that have been the most widely applied in behavioral ecology studies of foraging patterns are the diet breadth, patch choice, and risk models, all of which have been applied to both animal and human foraging behavior. Diet breadth models are the most straightforward, considering only a single currency (caloric intake) with minimal constraints and addressing the basic question of which prey types a forager should choose to exploit in a given environment; for this reason, these models are alternately termed prey choice models (Emlen

1966; Krebs and Davies 1993; MacArthur and Pianka 1966; Stephens and Krebs 1986). The basic diet breadth model assumes that a predator encounters prey sequentially, and on each encounter can choose to pursue (handle) the prey or ignore it in search of something better. The likelihood of pursuit is a calculation based on the caloric value of the prey, the rate at which it is encountered, and how long it takes to pursue and process it (handling time). It is then possible to predict which potential prey items should be taken, and which ignored, upon encounter. An outcome of the mathematical form of this model (as described by Krebs and Davies [1993]) is that a prey type joins the optimal diet based on the frequency (or, rather, scarcity) of higher ranked prey, not based on its own frequency. Prey should either be taken whenever encountered, if they are in the optimal diet range, or never, if they are not. This is often referred to as the "zero-one" rule (Stephens and Krebs 1986:20–21), yet random variance in the diet threshold can lead to prey that are outside the optimal diet being taken occasionally, and prey that are inside the optimal diet occasionally being ignored, leading to a better match with observed dietary patterns where prey may be pursued a proportion of the time (Hawkes et al. 1982; Krebs et al. 1977; Krebs and McCleery 1984; Werner and Hall 1974). Although in humans, diet breadth models have been applied primarily to understand subsistence patterns among foraging populations (e.g., Bird and Bliege Bird 2002; Bird et al. 2013; Broughton 1994, 1997; Hill et al. 1987; Smith 1991), they also have been applied to explain domestication (Gremillion 2004, 2014; Piperno 2011; Piperno and Pearsall 1998; Winterhalder 1986) and crop choice among agricultural populations (Barlow 2002, 2006; Diehl and Waters 2006; Gremillion 1996; Keegan 1986). Limitations of the explanatory value of these applications to agricultural systems have led some scholars to alternative forms of behavioral ecology modeling, such as risk minimization (e.g., Gremillion 2004; Marston 2011).

The patch choice model applies to foragers operating in a coarse-grained environment in which patches of resources are randomly distributed in the environment and require travel time to move from one to the next. The other major assumption of the model is that patches offer a diminishing rate of return to foragers (they become "depressed") such that foragers must eventually move on to another patch to optimize the long-term rate of energy gain (Charnov 1976:129; Parker and Stuart 1976; Stephens and Krebs 1986:28). Early tests of these models with animals (Cowie 1977; Orians and Pearson 1979; Parker 1978) led to further consideration of prey selection, processing, and transport decisions important for human foragers (e.g., Cannon 2003; Jones and Madsen 1989; Keegan 1986; Lupo 2007; Metcalfe and Barlow 1992; Rhode 1990). Patch types can include different habitats (Bird et al. 2005; Kaplan and Hill 1992; O'Connell and Hawkes 1981; Smith 1991), animal versus plant resources (Metcalfe and Barlow 1992; Zeanah 2004), and agricultural and foraged foods (Gremillion 2006; Keegan 1986). The currencies and constraints applied in patch choice models tend to follow closely those used in the diet breadth model, with frequency-adjusted caloric return as the primary currency of interest.

Risk models differ significantly from both diet breadth and patch choice models in that caloric returns are not presumed to be maximized; instead, the currency optimized is probability of survival. In the context of behavioral ecology, risk describes probabilistic variation in foraging returns, as it does in economics (Cashdan 1990a:2–3; Knight 1921; Smith and Boyd 1990; Stephens 1990; Stephens and Krebs 1986:128; Winterhalder 1990). This differs from definitions of risk used in geography, hazards research,

and agronomic and many ethnographic studies of food production, where risk equals "risk of loss" (e.g., Fleisher 1990; Göbel 2008; Goland 1993; Halstead and Jones 1989; Shutes 1997). A related concept is that of "uncertainty," which is distinguished from risk in the economic literature to refer to chance occurrences that cannot be accurately ascribed a predicted probability, or situations of incomplete information more generally (Cashdan 1990a; Clark 1990; Knight 1921; Smith and Boyd 1990; Stephens 1990; Stephens and Charnov 1982). The distinction between risk and uncertainty is central to economic research but often conflated in anthropological research, especially empirical case studies, presumably due to the inability of humans to predict real-world risk with complete accuracy. Risk and uncertainty do remain distinctive concepts in the behavioral ecology literature, with omniscience an assumption of optimal foragers and an extensive literature on how individuals deal with incomplete information in the real world (Stephens 1990; Stephens and Krebs 1986).

Risk can be beneficial for foragers under adverse circumstances, when mean returns fail to reach a threshold required for survival. Under those conditions, foragers seek risk, as documented in animal models (Caraco 1981, 1982, 1983; Caraco et al. 1980). The Z-score model (Stephens 1981; Stephens and Charnov 1982) maximizes the probability of surviving over a period of time based on which foraging

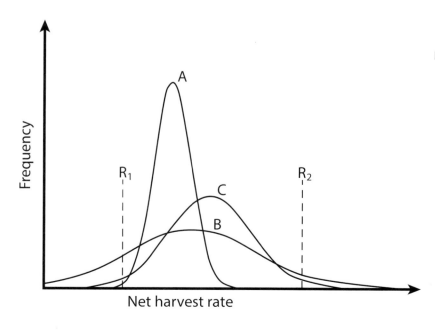

Fig. 2.2. The Z-score model for three resource options (A, B, and C) and two possible values for the required minimum energy for survival (R_1 and R_2). Note that when the means of the options are greater than R_1, low variance (option A) is preferred. In the second case, where means are all below R_2, high variance (option B) is preferred (after Winterhalder and Goland 1997: fig. 7.3).

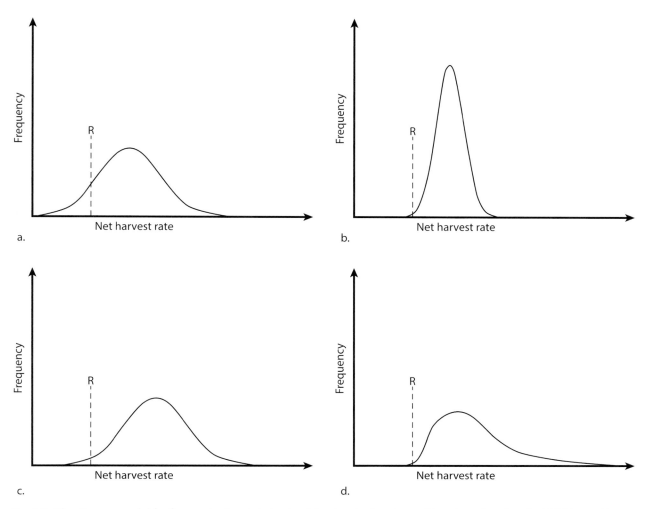

Fig. 2.3. The *Z*-score model for four agricultural strategies: (a) a standard system with a starvation threshold (R) lower than the mean of production (base case), (b) a standard curve with reduced variance resulting from diversification, (c) a right-shifted curve with a higher mean resulting from overproduction, and (d) a right-shifted and left-truncated curve resulting from irrigation (from Marston 2011: fig. 2).

strategy is chosen when the means and variances of foraging options differ. The important implication of the *Z*-score model is that it predicts seemingly non-optimal choices in diet, as shown graphically in Fig. 2.2, where the highest mean choice (C) is never chosen, although a diet breadth model would assume this to be the preferred strategy. Thus, the *Z*-score model has implications for diet breadth and can predict increases in diet breadth for risk-minimizing foragers beyond what an optimal diet model would predict (Gremillion 1996; Marshall and Hildebrand 2002; Winterhalder 1986). Methods for risk reduction include food sharing (Bliege Bird et al. 2002; Bliege Bird et al. 2012; Gurven 2004, 2006; Winterhalder 1986),

storage (Chesson and Goodale 2014; Gallant 1989; Winterhalder and Goland 1997), and spatial diversification of food production (Goland 1993; McCloskey 1976). In prior work (Marston 2011), I developed a comprehensive framework for distinguishing forms of spatial, temporal, and crop diversification, as well as intensification strategies, as alternative solutions to minimize risk using the *Z*-score model (Fig. 2.3). This framework was also used to generate specific testable archaeological correlates for these strategies. Risk-sensitive models allow a greater range of options for relating environmental factors to human subsistence behaviors than is permitted through simple prey and patch choice models, so I return to these expectations in Chapters 5 and 6 in considering agricultural decision making by inhabitants of Gordion.

Reconciling Behavioral Ecology and Resilience Thinking

Limitations and Critiques of Behavioral Ecology Modeling

A variety of critiques of behavioral ecology have been raised over the last 40 years, ranging from those that focus on the evolutionary logic for the premise of optimization (Lewontin 1979; Zeder 2014) to those that address the limited explanatory power of behavioral ecology models as applied to archaeological remains (Madsen 1993; Smith 2009a). Recent critiques of the diet breadth model have been particularly well articulated (Smith 2015; Zeder 2015). Because models are inherently abstractions of reality, with that attendant lack of realism and precision, it remains necessary to test and appraise models to determine their ultimate value, even though well-designed models may explain and even predict core elements of complex phenomena, such as human behavior (Winterhalder 2002). We can distinguish three distinct lines of critique specific to optimizing foraging models, relating to: 1) the evolutionary logic behind optimizing models, 2) the selection of an appropriate currency to measure, and 3) the goodness of fit between model expectations and real-world datasets. I address each in turn, beginning with the evolutionary validity of the basic logic for optimization.

Optimization is not directly derived from Darwinian evolutionary theory, but rather economic theory (Gray 1987:72; MacArthur and Pianka 1966:603; Smith 2015:14), and yet optimizing principles are found in many areas of evolutionary reasoning (Beatty 1980; Lewontin 1979; Maynard Smith 1978). In dealing with humans, the idea that microeconomic decision making might result in distinct evolutionary forces leading to certain adaptive behaviors should not be controversial. Thus, despite the mixed heredity of optimization within evolutionary theory, it remains valid to the extent that it represents ideal abstractions of real-world systems and can be used as the heuristic premise for model building and testing,

with the caveat that continued assessment of the logic and fit of those models is necessary (cf. Beatty 1980; Lewontin 1979; Maynard Smith 1978; Smith 1983; Winterhalder 2002). Continued efforts in model testing, including logical analysis and evaluation of a model's core assumptions, are still appropriate (Pyke 1984:525–27; Winterhalder 2002:210–11), as has been done recently in exploring the Broad Spectrum Revolution of the Near East (Zeder 2012) and the origins of agriculture worldwide (Smith 2011, 2015; Zeder 2015). That some of these studies suggest severe limitations of, or even the invalidity of, the optimizing assumption precludes immediate acceptance of the assumption and attendant models (i.e., the diet breadth model) as general principles structuring human behavior (Smith 2015; Zeder 2012, 2015). Recent arguments that optimal foraging is one body of theory that has broad evolutionary support and explanatory power in reconstructing the origins of agriculture (Gremillion et al. 2014) do not account for this critique. I suggest instead that optimization should find its greatest use as a hypothetical ideal expectation that has the potential to explain many (but not necessarily all) adaptive behaviors, rather than as covering law or theory (Beatty 1980), which generally lack explanatory power in archaeological inquiry (Smith 2015; Zeder and Smith 2009). A cautious, case-by-case use of optimizing models is more appropriate for studies of human behavior, past or present. Recent particularistic and case-based work among the Martu of Australia offers a model for the frugal application of foraging theory to a contemporary human population (e.g., Bird et al. 2009; Bird et al. 2013; Bliege Bird and Bird 2008; Scelza et al. 2014).

A second challenge raised against optimality models relates to the currency chosen. The default approach in foraging models is to focus on caloric intake as a proxy for reproductive success, the true currency of evolutionary success. It has been noted that caloric optimization does not explain real world data about human foraging patterns very well (e.g., Zeder 2012) and instead alternative currencies that might be maximized should be considered. Risk minimization is one such alternative; by minimizing risk, overall survival (and reproduction) is maximized (Caraco 1982; Caraco et al. 1980; Marston 2011; Milinski and Heller 1978; Zeder 2012:250–51). A variety of nutrients beyond simple calories may also serve as the proximate currencies considered by foragers to achieve the ultimate goal of increased survival and reproduction (Belovsky 1978; Hockett and Haws 2003; Keegan 1986). In some cases, apparently non-optimal behavior leads to the identification of additional currencies, such as resources that can be traded or sold, thus creating a parallel second currency that is sometimes the driving force behind subsistence decisions (e.g., Smith 1991:221–22).

The final critique of optimizing foraging models is that real-world data on animal, and especially human, behavior rarely fit the expectations of these models precisely. Frequently, non-optimal foraging behavior is identified in ethnographic (e.g., Hawkes et al. 1982:391; Levi et al. 2011:172–73; Smith 1991:217, 221–22) and archaeological (e.g., Zeanah 2004:13–14; Zeder 2015) datasets. Some of this lack of fit has been attributed to real world margins of error, including stochastic variation in prey size and availability, as well as imprecise forager discrimination of prey (Hawkes et al. 1982:391; Krebs et al. 1977; Krebs and McCleery 1984:98–99; Stephens and Krebs 1986:20–21; Werner and Hall 1974). Atypical conditions during specific hunting bouts can change the logic of foragers and bring prey that usually fall outside the optimal diet range into consideration (Smith 1991:217, 221). Still, examples persist of distinct preferences in prey selection that are clearly not optimal, such as the infrequency of camel hunting among the Martu of Australia despite its caloric returns vastly exceeding other subsistence options (Bird et al. 2013). Such tests of foraging models ought to lead to re-evaluation of the proposed hypothesis (Stephens and Krebs 1986:207–8; Winterhalder 2002). If the expectations of an optimizing model are not met, then perhaps all critical dynamics of the behavior of interest have not been included in the model, or the model hypothesizes optimization of the wrong variable; but also, potentially, the basic behavior predicted in the model fails to explain the data—i.e., optimizing foraging returns is not the decision-making process behind the foraging behavior (Pyke 1984). In the case of Martu camel hunting, social pressures regarding meat sharing appear to play a role in reducing the desirability of camel hunts (Bird et al. 2013:162–63). Whereas, with the widening diet breadth in the Early Holocene, resource abundance, rather than scarcity as proposed by the diet breadth model, appears to have

led to initial cultivation and eventual domestication of animals and plants (Zeder 2012, 2015). As some studies show a good fit between model predictions and observed datasets, but others do not, these results challenge the use of optimizing foraging models to explain general patterns of behavior.

Given these three limitations of optimizing foraging models and the generally inferential, rather than deductive, structure of archaeological theory building and problem solving (Fogelin 2007; Salmon 1976; Salmon and Salmon 1979; Smith 1977; Sullivan 1978), I argue that direct and literal application of behavioral ecology models is best applied to explain discrete case studies and specific archaeological problems, rather than as general explanatory models for broad scale behaviors, such as the origins of agriculture. In such broader cases, however, the logical structure and variables associated with foraging models may be of strong explanatory value when appropriately structured within inferential and inductive theoretical frameworks. The tension in the literature between proponents of behavioral ecology modeling (i.e., Gremillion 2014; Gremillion et al. 2014; Gremillion and Piperno 2009; Piperno 2011) and critics (i.e., Smith 2009a, 2015; Zeder 2014, 2015; Zeder and Smith 2009) can be resolved through careful case-by-case, even "particularistic" (Smith 2015:5), applications of behavioral ecology thinking to help solve problems and distinguish between alternative decision-making structures in the past.

Behavioral ecology models for risk management escape several of these critiques. The relevance of optimizing the currency of caloric intake and the argument that this is a flawed assumption that fails to explain archaeological evidence are not relevant to risk management, which utilizes a broader currency: survival. In addition, risk models necessarily involve time series of data, as it is temporal (as well as potentially spatial) variability in return that drives decision making. Unlike diet breadth models, which focus on sequences of instantaneous decisions (pursue prey A or move on), risk models incorporate future planning and memory of the past. From a risk perspective, a decision can only be made in the context of past experience, aligning these models closely with theories of social learning (Boyd and Richerson 1985, 2005) and the emphasis on temporal scale and sequence presented in resilience thinking, as well as theoretical work on the role of temporal processes of innovation and memory in human decision making (Alcock 2002; Halbwachs 1992; Schiffer 2011; Van Dyke and Alcock 2003).

On the Limited Application of Resilience Thinking in Archaeology

Although resilience thinking as an explanatory principle relating human behavior to environmental change has exploded in popularity across the social and environmental sciences over the past two decades, it has found little purchase in archaeology. Although the term "resilience" is increasingly found in archaeological publications, in common archaeological parlance, resilience is a term used in dialogue with the literature on collapse, demonstrating human agency in responding to and constructing environmental change (e.g., Dunning et al. 2012; McAnany and Yoffee 2010; Middleton 2012; Tainter 1988, 2006a). In contrast, archaeological applications of resilience thinking as an explanatory framework mainly focus on applications of the adaptive cycle to describe diachronic change in societies and environments (Allcock 2013; Peeples et al. 2006; Redman and Kinzig 2003; Redman et al. 2009; Rosen and Rivera-Collazo 2012). In this mode, resilience is conceptualized as a property of cultural systems that explains the persistence of societies and material traditions over long spans of time, even through upheavals from within or without (e.g., changes in rainfall and streamflow patterns [Nelson et al. 2012; Redman et al. 2009]). These uses are compatible, although only approaches that explicitly embrace the tenets of resilience thinking theorize the concept of resilience. Given current societal interest in how people have affected past environments and adapted to environmental change, why is resilience thinking, which offers a model for relating these concepts across time and space, so limited in archaeological application, even as the term "resilience" grows in popularity?

I suggest this may be due in part to the novelty of the theoretical paradigm of resilience thinking to archaeologists. Archaeological publications in resilience only began to appear in the early 2000s (e.g., Redman and Kinzig 2003). Although several social sciences are well represented in the development of resilience thinking, including sociology, geography, and economics, anthropology (the discipline in which the majority of American archaeologists are trained) is not generally a participant in the resilience literature. For example, in *Panarchy,* the core synthetic text of resilience thinking, there is one, brief mention of archaeological data (Gunderson and Holling 2002:92) and just two of cultural anthropology (Gunderson and Holling 2002:246, 320); none of the 20 authors are anthropologists or archaeologists. *Ecology and Society*, the official journal of the Resilience Alliance and the leading publication devoted to resilience thinking, has a 99-person editorial board (Ecology and Society 2015). Only two are archaeologists (Charles Redman and Sander van der Leeuw, both of Arizona State University) and four others are cultural or biological anthropologists (at the University of Arizona, Colorado State University, Rutgers University, and the University of Georgia); in addition, two scholars with science backgrounds regularly publish on archaeological topics in collaboration with archaeologists (Ann Kinzig and John Anderies, both of Arizona State University). I conclude that outside of Arizona State University, there may be little opportunity within the United States to study resilience thinking when gaining an education in anthropology, especially with a focus in archaeology; this is even more true for archaeology in other countries where interdisciplinary education is less common, and where no archaeologists who publish regularly on resilience topics teach.

Beyond this demographic issue, perhaps more importantly, resilience thinking has failed to speak to the interests of many archaeologists, in part because it invokes a new set of terminology and references an unfamiliar literature. While broader societal narratives about the importance of certain topics may influence the adoption of specific theoretical or methodological innovations within a community of practitioners (Khazraee and Gasson 2015), it is difficult to argue that resilience thinking is out of sync with contemporary societal (and archaeological) discourse on human relationships with environments. Instead, I propose that the adoption of resilience thinking in the archaeological literature has been restricted by an initial emphasis on the adaptive cycle, which appears intuitively obvious to archaeologists who study long-term patterns of societal growth and decline, and such cycles have been discussed for many years (cf. Butzer 1982). Many of the early publications of resilience thinking in archaeology were targeted at broader academic communities already concerned with resilience (e.g., Redman and Kinzig 2003), rather than at archaeologists.

A better case for the archaeological adoption of resilience thinking was presented initially by Redman (2005) and has since been expanded by other authors writing in anthropological journals (e.g., Chase and Scarborough 2014; Hegmon et al. 2008; Marston 2015; Nelson et al. 2012; van der Leeuw 2009). In particular, these publications direct our attention to variables that hold greater saliency with archaeologists: scale, identification of causality in complex systems, and differential vulnerability and resilience among segments of society. Temporal, spatial, and organizational scale form the dimensions of adaptive change (Fig. 2.1) and scalar processes, especially mismatches, have great explanatory power in archaeology (Marston 2015). Similarly, resilience thinking can help archaeologists assign causality within periods of rapid, or even apparently simultaneous, change in linked social-ecological systems, such as periods of climate variation, soil erosion, and social collapse. The causal relationships between social and environmental change can be elucidated from a resilience perspective, contributing to archaeological explanation (Fisher 2005; Redman 2005:74–76). Similarly, attention to societal vulnerability and differing degrees of resilience within and among societies helps to explain patterns of differential social change and illuminates the lives of individuals in the past (Hegmon et al. 2008; Iannone et al. 2014; Nelson et al. 2010; Schoon et al. 2011).

A more robust understanding of the theoretical tenets and applications of resilience thinking, such as demonstrated in the publications cited above, will help to illustrate the explanatory value of resilience thinking in archaeological problem solving. In my view, the adaptive cycle has been unconvincing to many archaeologists, and concrete case studies that focus narrowly on specific elements of resilience, such as scale, causality, and vulnerability, are needed to clarify how archaeologists can make use of this new body of theory. Resilience thinking can generate specific expectations about human-environmental interactions that we might see in the archaeological record, and compatible bodies of theory, such as behavioral ecology, can be used to model those interactions, leading to hypothetical implications testable with archaeological datasets. Resilience thinking can also be used

in a purely inductive framework, helping to explain results and generate new testable implications based on those interpretations.

Integrating Resilience Thinking and Behavioral Ecology

There is much to be gained from a hypothesis-testing approach to explaining the past by focusing on variables that can be studied from both a resilience and behavioral ecology perspective. Such an approach marries the specific, and even mechanistic, tools of behavioral ecology modeling with the general, analogic explanatory framework of resilience thinking. The ready testability of behavioral ecology expectations and the depth with which such models have been applied to archaeological datasets to date, counters the generality of resilience thinking—which can be applied at multiple scales—and its early state of theoretical development within archaeology. In particular, I suggest that three key variables offer ready points of intersection for these two theoretical perspectives: scale, causality, and risk.

Scale, especially across space and time, provides the axes of variation for adaptive cycles as nested within a panarchy (Holling et al. 2002b) and also the key variables that center decision making in the diet breadth, patch choice, and risk models for foraging behavior (Stephens and Krebs 1986). Organizational scale describes a range of social practices that can increase or diminish adaptive capacity within human and environmental systems, such as landscape management practices related to fuel acquisition or animal husbandry. Considering the role of organizational frameworks from local to distant decision making gives insight into how social groups adopt and maintain landscape management practices (e.g., Lepofsky and Kahn 2011; Marston 2012a; Redman et al. 2009). Mismatches within a single scale or across scales are identified as sources of systemic perturbation and change within a resilience perspective, but also explain challenges to and deviations from formal foraging models. For example, cases of incomplete information or rapidly changing environmental parameters, where a forager does not have access to complete knowledge about the current foraging environment, are predicted to deviate from standard optimal foraging models—although perhaps in predictable ways. These provide a framework for understanding the relative benefit of social learning versus trial and error (Henrich 2004; Richerson et al. 2001; Stephens 1990).

Explaining causal relationships between variables, especially seemingly coincident social and environmental change, can be a challenge for archaeologists. While finely resolved chronological sequences and logical chains of argumentation allow archaeologists to explain, for example, the causal relationships between the desertification of the Sahara and mobile pastoralism in that region (Linseele et al. 2014), other apparently synchronic sets of change, such as the relationship between early Holocene climatic change and the origins of agriculture (Gremillion et al. 2014; Richerson et al. 2001; Zeder 2008, 2015), remain contentious and the subject of debate. Refined theoretical frameworks allow better understanding of changes that appear coincident given the limits of precision for available dating methods. Resilience thinking provides a mechanism for parsing synchronic environmental and social change through understanding underlying scalar processes. Rather than just describing change, resilience concepts help to explain processes of change, providing a logical framework for understanding how complex adaptive systems respond to changes in specific variables (Holling 2001; Liu et al. 2007a; Marston 2015). Behavioral ecology uses a model-building approach to simplify complex interactions of variables and provide predictions about patterns of cause and effect among those variables. Due to the ready testability of these predictions, it is possible to assess how well conceptual models explain the system of interest, and to revise those models accordingly when expectations do not fit the data available.

The Role of Risk

While the theoretical frameworks of behavioral ecology and resilience thinking provide the logical tools for interpreting agricultural decision making in the past, I propose that human responses to both environmental and social risk are key to understanding decision making in most archaeological contexts, including that of Gordion. Reducing or mitigating risk is a primary concern of farmers and herders in areas where agriculture is impacted by climatic and/ or environmental fluctuation, which is most of the world today (Cashdan 1990b; Casimir 2008; Fleisher 1990; Pope 2003) as it was in the past (Gallant 1991; Garnsey 1988; Halstead and O'Shea 1989a). Risk is of particular importance for farmers in semi-arid environments, where variable rainfall and potential water shortages present a significant constraint on agriculture of staple crops, such as wheat and maize. As such, risk offers a ready perspective on why certain agricultural decisions may be made within their contemporary social and environmental contexts.

Human behavioral ecology and resilience thinking offer distinct perspectives on the role of risk in agricultural decision making. In part this is due to different definitions, where risk in economics and behavioral ecology equals probabilistic variance in returns, while in other fields risk means chance of loss (Cancian 1980; Cashdan 1990b; Halstead and O'Shea 1989b; Stephens 1990). In resilience thinking, definitions of risk are often tied explicitly to the concept of vulnerability, or susceptibility to harm from a hazard, which is a property of social, ecological, and linked socio-ecological systems (Adger 2000, 2006; Berkes 2007; Füssel 2007; Stern et al. 2013; Turner II et al. 2003). Risk is seen as an environmental variable that affects human populations in proportion to their vulnerability; in this context, risk is most often defined as in hazards research as "chance of loss" (Berkes 2007; Füssel 2007:160; Turner II 2010; van der Leeuw 2008). Of note is that the same strategies that reduce environmental risk may reduce a population's vulnerability: e.g., mobility (Anderies and Hegmon 2011). In fact, vulnerability assessment often focuses explicitly on mechanisms to reduce risk (often, both environmental and economic) and enhance resilience (Adger 2006).

Risk is a key conceptual element in foraging models used within behavioral ecology, as it explains how foragers respond to variable returns in resources needed for survival. Risk has been modeled explicitly to explain human behavior from archaeological datasets, whether in terms of agriculture (Marston 2011), foraging (VanDerwarker et al. 2013), site location and fortification (Zori and Brant 2012), or landscape change (White et al. 2014). Behavioral ecology models explain when risk should be avoided and suggest how risk can be managed. Resilience thinking also explicitly considers risk, both as stochastic variation and as chance of loss or substantial change. System connectedness, a property described by the adaptive cycle, is a measure of how rapidly or strongly changes in one variable affect another within a system; systems with high connectedness are characterized by primarily internal variability and are resistant to external perturbations (Holling and Gunderson 2002). Overconnectedness, however, can lead to system rigidity and increase the risk of that system "collapsing" into the next phase of the adaptive cycle (often the "release" phase) or falling into a "rigidity trap" from which it cannot escape (Hegmon et al. 2008; Holling et al. 2002b). This rigidity and the associated concept of vulnerability have been used to explain risk of societal collapse and transformation among past societies (e.g., Chase and Scarborough 2014; Dunning et al. 2012; Nelson et al. 2012; Rosen and Rivera-Collazo 2012; Schoon et al. 2011). Conceptualizing risk within a resilience perspective helps to frame and contextualize behavioral ecology risk models within a broader social setting. Additionally, it aids understanding of how social changes affect the resilience and riskiness of agricultural systems independent of environmental variation affecting returns from subsistence activity.

In the work that follows, I adopt the strict definition of risk used in behavioral ecology and economics, probabilistic variance, because it provides a clear set of expectations for behavioral responses to environmental change and, thus, insight into agricultural decision making. Although I focus on risk management as a response to environmental variation, I also consider both environmental and social processes that increase the chance of crop failure and loss. Behavioral ecology logic is used to argue that absent exceptional circumstances, these two definitions of risk (probabilistic variance and chance of loss) have the same practical implications and that risk management strategies, thus, often aim to reduce chance of loss by minimizing variance in subsistence returns. Using this logic, in Chapter 5 I draw on historical, ethnographic, and archaeological research to identify specific strategies for agricultural risk management and develop specific archaeological correlates for these risk management practices that can be tested using archaeological plant and animal remains. Ultimately, these measures provide insights into decision making in light of en-

vironmental, economic, and political change during the occupation history of Gordion, and allow us to consider linked processes of social and environmental change across multiple scales in Central Anatolia.

Modeling Agriculture at Gordion through Resilience Thinking and Behavioral Ecology

In this chapter, I have argued that resilience thinking and behavioral ecology present two powerful theoretical frameworks for reconstructing human decision making in the past, especially with regard to agricultural practices, and that the two perspectives are both theoretically and practically compatible. Points of intersection between these theoretical approaches include a focus on scalar processes and risk in structuring human decision making, leading to effective forms of explanation for causal relationships among multiple variables within social and ecological systems. In the chapters that follow, I draw on these bodies of theory to reconstruct and explain human-environmental interactions at Gordion, namely those that result from wood acquisition, farming, and animal herding within the region of the city.

I begin by reconstructing changes in the local landscape (Ch. 3), regional woodlands (Ch. 4), and agricultural systems (Ch. 5), using a number of paleoenvironmental and paleoethnobotanical datasets from Gordion and the surrounding area. I then utilize behavioral ecology models to identify the behaviors and decision-making processes represented by those data, with regard to wood acquisition (Ch. 4) and agropastoral practices (Ch. 5). In Chapter 6, I adopt a resilience perspective on the environmental, social, and economic history of Gordion to hypothesize why those particular behaviors were chosen, rather than alternatives, and why they led to specific cultural and environmental changes. I conclude by reflecting on the methodological and theoretical implications of this study and future directions for archaeological consideration of agricultural sustainability and cultural resilience, both in the Near East and beyond.

3

Biogeography and Paleoclimate of the Gordion Region

The Gordion region is varied in topography, soil type and density, and vegetation cover, a landscape that has evolved over many centuries of human interaction with and modification of the soil, vegetation, and hydrology of the area. People continue to change the physical and phytogeography of the region through both intentional (mining, dredging rivers, habitat protection and restoration) and incidental (grazing, irrigation) land use activities. This chapter presents information about the present biogeography of the Gordion region, which is integrated with regional paleoclimatic proxy data to reconstruct landscapes of the past. As landscape change is the result of interaction between climate and human activity, in the final section of this chapter I reconstruct local landscapes during the occupation period of Gordion and draw out the factors most responsible for those changes. This analysis provides context for the interpretation of archaeological plant remains from the site as presented in the final three chapters of this volume.

Present Biogeography

Biogeographic survey of the Gordion region exists in various forms. Turkish government-sponsored surveys include detailed topographic maps for military use (Türk Silahlı Kuvvetleri 1997) and geological maps (Erentöz 1975). Remote sensing data includes imagery from the Shuttle Radar Topography Mission, Landsat Thematic Mapper, and aerial photographs from the Turkish military taken during the 1950s and the 1980s. Within the immediate vicinity of Gordion, Ben Marsh has conducted a thorough geomorphological study (Marsh 1999, 2005, 2012; Marsh and Kealhofer 2014) and Naomi Miller has surveyed regional vegetation patterns since 1988 (Miller 1999a, 2010;

Miller and Bluemel 1999). I undertook forest surveys on all major mountains in the region between 2006 and 2008. Climate data presented below comes from the closest government weather station, in the neighboring city of Polatlı, and includes monthly rainfall and temperature data from 1964 to 2006. Together, these various data sources provide complementary information on the recent geographic, climatic, and botanical history of the region.

Physical Geography

Gordion sits in the floodplain of the Sakarya River, just upstream from its confluence with the Porsuk River (Fig. 3.1), although several of the burial tumuli of Gordion sit on prominent ridges in the surrounding landscape. The Sakarya floodplain, at about 680 m above sea level (masl), is the lowest terrain in the area. The highest peaks in the immediate area include Çile Dağı (1440 masl) to the northeast and Dua Tepe (1100 masl) to the southeast, as well as the more distant peaks near Sivrihisar (1800 masl) and Mıhalıççık (1530 masl) to the west and north. Large upland plains extend west and southwest from Gordion above the Porsuk floodplain.

Bedrock in the area of Gordion is composed of a thick marl level and a more recent basalt intrusion, which is concentrated in the mountains to the east of Gordion. The marl is a sedimentary stone created by lake deposits in the early Pleistocene (1.0–2.5 million years ago), while the basalt is an igneous stone created by subsequent volcanic activity that uplifted and folded marls as it formed (Marsh 2005:161–62; Pamir and Erentöz 1975). Marsh (2005) describes the differences in soils formed from these parents, especially in relation to water retention, as a substantial factor in agricultural potential and land use patterns in the region. Marl soils

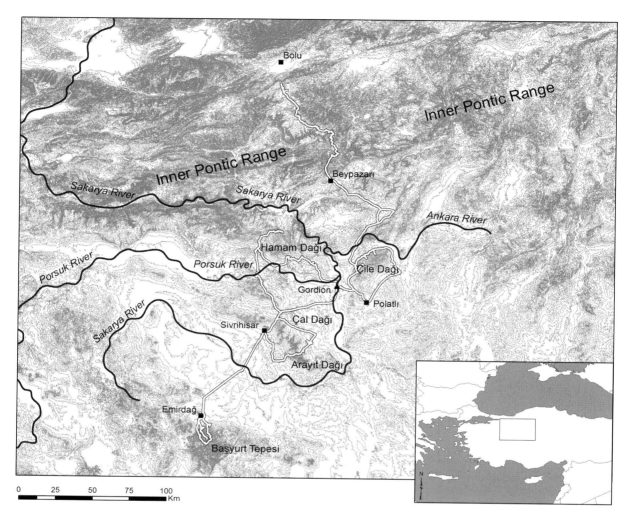

Fig. 3.1. Topographic map of the Gordion region showing major rivers (black line) and mountain areas. Elevation data derived from SRTM, courtesy of NASA; contour lines are 50 m. Squares denote modern cities. Outlined white line indicates informal transect path in area woodlands.

are pale, alkaline, and low in nutrients and moisture, while red basalt soils hold more nutrients and water and are often more productive for rainfed agriculture (Marsh 2005:162–64). Gypsum outcrops are prominent in the marl bedrock topography and include the gypsum ridge of Kızlarkayası north of Gordion, which rises 100 m above the floodplain below, and lower gypsum ridges to the south of the site. The main soil types, including marl-derived alluvium, basalt-derived alluvium, upland residual soils from marl and basalt bedrock, and gypsum soils, affect native plant growth as well as agriculture. Many plant species are restricted to specific soil types, as discussed further below.

Water sources in the area include two major rivers, the north-flowing Sakarya, which empties into the Black Sea near Istanbul, and the east-flowing Porsuk, which joins the Sakarya 4 km north of Gordion. Multiple perennial streams feeding the rivers were a feature of the landscape before post-Bronze Age alluviation covered much of the bedrock in the area, moving streams underground (Marsh 1999). Springs appear at the base of basalt massifs where water is forced to the surface between porous basalt and impermeable marl layers (Marsh 1999:164); many of these have been recently improved by the Turkish government with pipes leading to concrete watering troughs for herds.

Climate

The Polatlı station of the Turkish Meteorological Service provided the Gordion project with complete local weather data (rainfall, temperature, cloud cover, wind, etc.) from 1964 through the summer of 2007, in addition to monthly rainfall totals from 1929 to 2009. This station, approximately 17 km east-southeast of Gordion and nearly 200 m higher in elevation (885 masl), has a climate similar to that of Yassıhöyük but receives more rainfall, a function of its higher altitude. Despite the differences in absolute rainfall, however, patterns of interannual precipitation variation from Polatlı are broadly applicable to Yassıhöyük as well.

Figures 3.2 and 3.3 are based on mean monthly data collected at the Polatlı Meteorological Station between 1964 and 2005. The temperature data is averaged across 42 years of measurements, and the precipitation data across 41 years. Temperature data shows that temperatures peak in July and August with average daily highs around 30°C (86°F) and nightly lows near 15°C (59°F). January is the coldest month, with average daily temperatures around freezing (Fig. 3.2). The summer is also the driest time of the year, with mean monthly precipitation totals below 15 mm in July, August, and September (Fig. 3.3). The wet season stretches from the winter months through early June, with April and May delivering over 45 mm of rain per month on average. Snow falls most frequently in January and February, but can appear as early as November or late as April (unpublished data from the Polatlı Meteorological Station).

Monthly rainfall data recorded in Polatlı since 1929 show interannual variability in precipitation over 79 years (Fig. 3.4). Precipitation years are calculated most meaningfully from dry season to dry season, so the data are divided into July–June rainfall years numbered by the calendar year in which they conclude. Over these 78 years, precipitation amounts ranged from 212.5 mm (rainfall year 1945) to 521.3 mm (rainfall year 1963), with a mean of 349.4 mm and a standard deviation of 64.0 mm. The data have little skew and closely follow a normal distribution.

Despite relatively substantial interannual variation in precipitation, these numbers indicate that

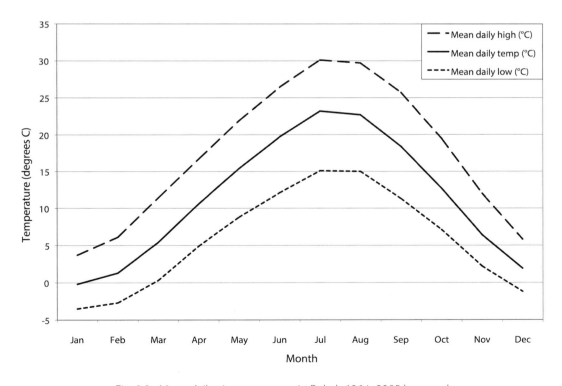

Fig. 3.2. Mean daily air temperature in Polatlı, 1964–2005 by month.

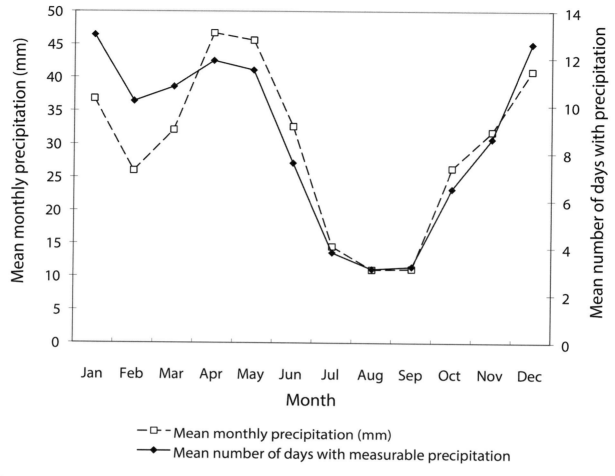

Fig. 3.3. Mean monthly precipitation in Polatlı, 1964–2005. Data include both mean total monthly precipitation and mean days with measurable precipitation per month.

rainfed (non-irrigated) agriculture can be a successful strategy in the Polatlı area, as cereal cultivation in the Near East generally requires 250 mm/year of precipitation (Miller 2010:11). Gordion, however, lies 200 m lower than Polatlı, and correspondingly receives less rainfall. Miller (2010:11) notes that early summer storms commonly drop rain on the higher mountains around Gordion but miss the Sakarya floodplain itself. Unfortunately, since rainfall data has not been recorded at Yassıhöyük, the quantitative relationship between rainfall levels at Polatlı and Gordion cannot be calculated, but a regional study that correlates farming and elevation in Central Anatolia suggests that 200 m in elevation ought to result in a significant decline in annual precipitation, with significant implications for farmers (Marston and Branting 2016).

Phytogeography

Local botanical survey as part of the Gordion archaeological project was initiated by Miller in 1988, and formal transects of steppe vegetation on Tumulus MM began in 1997 after the tumulus was fenced (Miller 1999a). The data presented here for steppe vegetation include unpublished information from Miller's surveys, which I assisted with in 2002, 2006, 2007, and 2008. Forest survey data presented here include some information from Miller's earlier work, but are primarily based on my own observations made during forest visits in 2006, 2007, and 2008. The closest forested mountains, Çile Dağı and Hamam Dağı, were visited by Miller in previous years, and more formally surveyed by me between 2006 and 2008, while

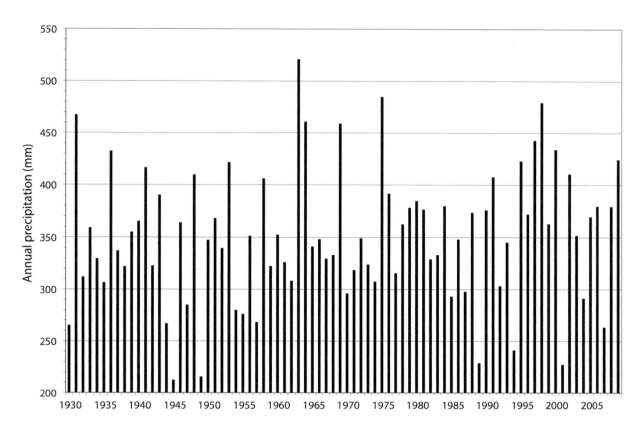

Fig. 3.4. Annual precipitation in Polatlı, 1929–2009. Data are presented by rainfall year from July–June, numbered by the calendar year in which they end. Polatlı is 17 km east-southeast of Gordion and nearly 200 m higher in elevation.

Arayıt Dağı and Başyurt Tepesi were first investigated in 2008.

Both formal and informal transect survey methodologies were employed, as well as unsystematic investigation of some areas. Most areas were surveyed through informal transects by compiling a running plant list of species identified along the course of a continuous route across or around a specified geographic feature. Survey methodology differed between forested and open areas. In forest areas, the transect was made by car, with periodic stops to investigate the local trees and ground cover. In steppe areas, transects were made on foot following the natural topography. All 2008 transects were tracked using a Garmin eTrex Vista HCx handheld GPS unit and are plotted in Figures 3.1 and 3.5 (see color insert); transects in previous years covered approximately the same areas. In forest areas, trees were recorded but ground cover was not systematically identified, while open areas were surveyed comprehensively. Areas sur-

veyed through informal transects include Hamam Dağı, Çile Dağı, Başyurt Tepesi, Arayıt Dağı, the inner Pontic range south of Bolu, the Sakarya riverbank, and the gypsum steppe.

Formal transects were used to investigate Tumulus MM, which is fenced, and Tumulus P, which is not protected from grazing, to provide a comparative perspective on steppe grassland structure in an ungrazed and overgrazed state. Survey on Tumulus MM included six transects up the SW, NW, NE, E, SE, and S faces of the tumulus. Each transect originated at a set point along the outer fence and terminated at the peak of the tumulus. Every 10 m along the course of each transect, a circular hoop inscribing an area of one square meter was placed on the ground. Within the hoop, the rough percentage of ground cover and estimated degree of slope were recorded and all plants inside were identified and recorded. Plants that dominate the area are noted as such. Additionally, all plant species in the immediate area around the hoop that were not observed within it were recorded. This process was repeated every 10 m to the top of the tumulus (Miller 1999a). A similar methodology was employed for Tumulus P, except that a single transect was made over the tumulus, with hoops placed every 5 m, going up the N side of the tumulus and down on the S side, which was backfilled after its excavation in 1955 (Young 1957).

Several areas were not systematically surveyed. This includes observations on the lower slopes of Çile Dağı and Hamam Dağı, and agricultural areas throughout the region. Agricultural areas were not transected because agricultural fields are monotype patches without strong geographical trends in placement due to recent irrigation systems; wild plant floras are limited by liberal application of herbicides.

The results of each transect are presented below, beginning with forested regions. Some of these data were collected by Miller before my involvement in the project, but most represent our combined effort. Any observations made by Miller alone are noted as such, while all uncited material represents our collaborative work. Throughout this volume all botanical names and plant taxonomy follow those adopted in the *Flora of Turkey* (Davis 1965–2000); note that several genera and families have changed names and relationships since the inception of the *Flora* in the 1960s (e.g., the Chenopodiaceae are now part of the Amaranthaceae) but I use terminology consistent with the *Flora* regardless.

Çile Dağı

The peak of Çile Dağı lies 22 km NE of Gordion, with foothills extending to within 10 km of the site (Fig. 3.1). The underlying bedrock is primarily basalt, and the mountain has a reddish soil that sets it apart from other hills in the area (Erentöz 1975). In previous years, Miller visited the area around Avşar, which is characterized by a low "oak carpet" scrub forest, with dense patches of downy oak (*Quercus pubescens*) under one meter in height (Miller 1999b:16; Fig. 3.6). The low height of these trees appears to be the result of grazing by goats; in recent years, as grazing has declined, this "carpet" has grown in height. Scrub forest around Avşar begins around 1000 masl in elevation, and is dominated by *Q. pubescens* with some prickly juniper (*Juniperus oxycedrus*) present as well. This area also contains several isolated and field edge trees: elm

(*Ulmus minor*), wild plum (*Prunus divaricata* ssp. *divaricata*), wild pear (*Pyrus elaeagnifolia*), hawthorn (*Crataegus* sp.), and wild almond (*Amygdalus orientalis*). Willow (*Salix alba*) and poplar (*Populus nigra*) grow along valleys and by streams. Within the village of Avşar itself, numerous cultivated trees are grown, including walnut (*Juglans regia*), mulberry (*Morus nigra*), sour cherry (*Prunus cerasus*), willow (*S. alba*), and other species of *Prunus* (apricot, peach, sweet cherry).

In 2008, we surveyed a transect up a stream valley on the NW face of Çile Dağı, 4 km from Avşar (Fig. 3.1). This area includes an earth dam, creating a small lake, and several small fields under cultivation within the valley, but is more remote and less visited than the area around Avşar. Along the transect, trees appeared just below 900 masl, and we reached 950 masl at the top of the transect. The species diversity of woody plants was much greater than at Avşar. In the waterlogged areas around the lake, cattails (*Typha* sp.) and reeds (*Phragmites australis*) dominated, with willow (*Salix alba*) and tamarisk (*Tamarix* sp.) just behind. On the valley floor, we identified willow (*S. alba*), poplar (*Populus nigra*), apricot (*Prunus armeniaca*), and bramble (*Rubus* sp.) and *Ononis spinosa* shrubs. On the valley slopes, we found wild almond (*Amygdalus orientalis*), sumac (*Rhus coriaria*), Russian olive (*Elaeagnus angustifolia*), ash (*Fraxinus excelsior*), hawthorn (*Crataegus* sp.), wild plum (*Prunus divaricata* ssp. *divaricata*), almond (*Prunus dulcis*), prickly juniper (*Juniperus oxycedrus*), *Colutea* sp., downy oak (*Quercus pubescens*), and pine (*Pinus nigra*). This is the closest that pine grows to Gordion today, roughly 20 km from the site. Pine, oak, and wild almond were the most numerous trees on these slopes. Our transect was discontinued due to the stream overtaking the road, but from this location the peak of Çile Dağı was visibly forested with dense oak canopy forest, presumably including *Quercus cerris*, the only oak species in this region that grows tall and forms a canopy forest. Our identification of this forest as primarily oak is supported by local villagers, who consistently referred to the forest as the "meşe ormanı" (oak forest); the forest is also noted as such on the 1:100,000 scale map (Türk Silahlı Kuvvetleri 1997).

Hamam Dağı

The forested mountain range around the town of Mıhalıççık is known as the Sündiken Dağları, and the

Fig. 3.6. "Oak carpet" forest of *Quercus pubescens,* near Avşar.

easternmost peak of Hamam Dağı lies 33 km NW of Gordion (Fig. 3.1). The substrate is primarily metamorphic schists (Pamir and Erentöz 1975:60). The dominant vegetation cover along the mountain ridge in this area is climax canopy pine forest, but much of that appears to be replanted as monoculture in previously clear-cut areas, an impression confirmed by the 1:100,000 scale map, where part of this forest is marked as "çam ağaclandırma sahası"—"pine plantation area" (Türk Silahlı Kuvvetleri 1997). The footslopes of Hamam Dağı extend within 25 km of Gordion and are wooded as well. Surveys in this area in 2007 and 2008 identified three distinct forest types.

Scrub forest cover first becomes evident around 1000 masl on the southern slopes of the mountain, but perhaps as low as 900 masl in well-watered valleys to the east. This scrub is similar to that described from Çile Dağı, dominated by downy oak (*Quercus pubescens*), prickly juniper (*Juniperus oxycedrus*), as well as the other juniper species found in the region (*J. excelsa*; often termed "Greek juniper"), although only as short (mostly <4 m) trees and shrubs. Minor components of this forest type include juniper (*J. excelsa*), Russian olive (*Elaeagnus angustifolia*), hawthorn (*Crataegus* sp.), wild pear (*Pyrus elaeagnifolia*), barberry (*Berberis* cf. *crataegina*), and Christ's thorn (*Paliurus spina-christi*). Above this forest type, from roughly 1100–1300 masl, is an oak forest dominated by downy oak (*Quercus pubescens*) but with some larger individuals of *Q. cerris*. This may be a secondary growth that arises when the climax pine forest is clear-cut. The final forest type, observed mainly above 1300 masl, is the climax pine forest, dominated by closed canopy *Pinus nigra*, with a sparse understory of downy oak (*Quercus pubescens*), prickly juniper (*Juniperus oxycedrus*), hawthorne (*Crataegus*), barberry (*Berberis*), and wild pear (*Pyrus elaeagnifolia*). Many pines reach heights of 20 m or more, and logging of pine trees remains a major industry around Mıhalıççık.

Arayıt Dağı

Two mountain peaks rise south and east of Sivrihisar: Çal Dağı and Arayıt Dağı (Fig. 3.1). The two peaks have different geologies, with the substrate of Çal Dağı primarily acid intrusives and that of Arayıt Dağı marble and limestone (Erentöz 1975). This may account for the differences in plant cover between the two peaks. The lower slopes of Arayıt Dağı were more densely wooded than those of Çal Dağı. We completed a transect in 2008 over Arayıt Dağı from NE to SW by car, climbing from about 1050 to 1350 masl.

The eastern foothills of Arayıt Dağı are wooded, and many trees are located along field edges just below. The variety of cultivated trees suggests that these hills were likely planted as orchards in the past, although it is possible that some domestic tree species spread up from the fields below. In this area we identified domestic walnut (*Juglans regia*), grape (*Vitis vinifera*), almond (*Prunus dulcis*), apple (*Malus domestica*), and sour cherry (*Prunus cerasus*) among wild species including hawthorn (*Crataegus* sp.), Russian olive (*Elaeagnus angustifolia*), *Colutea* sp., wild pear (*Pyrus elaeagnifolia*), prickly juniper (*Juniperus oxycedrus*), and oak (*Quercus cerris*). Oak is the dominant wild tree on these hillsides and forms dense forests at higher elevations.

Approaching the ridge of Arayıt Dağı, around 1350 masl, oak forest abruptly transitions to an open juniper woodland, with a band of scrub *Juniperus oxycedrus* giving way to an open forest of large *J. excelsa* on rocky soil. Few trees appear closer to the peak, around 1500 masl. These *J. excelsa* are substantially larger than those on Haman Dağı, reaching 9 or 10 m in height, beginning to approach the size of the juniper trunks found in the casing of the Tumulus MM tomb structure (Kuniholm et al. 2011; Liebhart and Johnson 2005; Young 1981; Fig. 3.7). Additionally, these junipers still have lower branches extending to the ground, unlike those on Hamam Dağı, suggesting that they are not trimmed for fuel or grazed. Other trees in this area include willow (*Salix alba*) and poplar (*Populus nigra*) growing at the base of small ravines.

Başyurt Tepesi

The peak of Başyurt Tepesi rises above the town of Emirdağ to a height of 2280 masl, with the peak lying 120 km SW of Gordion (Fig. 3.1). This is the highest mountain within a 150-km radius of Gordion, with the exception of the inner Pontic range to the north. The primary substrate of the massif includes levels of marble, limestone, quartzite, and schist (Pamir and Erentöz 1975:59). We drove through three major valleys on the north side of the mountain, reaching a maximum elevation of 1900 masl.

Approaching the first valley we encountered juniper forest just above 1000 masl before the village of Tezköy. This forest was entirely juniper (*Juniperus excelsa*), which also appeared as field trees in the cultivated area just below. Other field edge trees included hawthorn (*Crataegus* sp.), oak (*Quercus cerris*), apricot (*Prunus armeniaca*), sour cherry (*Prunus cerasus*), wild plum (*Prunus divaricata* ssp. *divaricata*), domestic pear (*Pyrus communis*), wild pear (*Pyrus elaeagnifolia*), and Russian olive (*Elaeagnus angustifolia*). Ascending the valley to 1400 masl we found increasingly dense *Juniperus excelsa* forest, with many trees 8–10 m tall. In this forest we identified very few other species, but did find a singularly large prickly juniper (*J. oxycedrus*) tree approximately 5 m in height. The forest changed along the valley floor and near springs, where oak (*Quercus cerris*), prickly juniper (*J. oxycedrus*), ash (*Fraxinus excelsior*), wild rose (*Rosa* sp.), and willow (*Salix alba*) were prominent. This transect continued to 1500 masl when the road became impassible.

The second transect followed the valley ascending from the village of Dereköy. An open juniper (*Juniperus excelsa*) forest extended from Dereköy (1040 masl) to Balcam and ended around 1300 masl, with a few individuals of oak (*Quercus cerris*) intermixed and extending up to 1450 masl. Above this level no trees were seen until the ridge at 1550 masl that provided a view into a valley opening to the east. This valley had dense oak (*Q. cerris*) forest at this elevation, extending higher up the mountainside. Following the ridge up to the west, we reached 1900 masl without seeing any more trees. A few stunted hawthorn (*Crataegus* sp.) specimens were observed, none of which exceeded 0.5 m in height. The tree line for Başyurt Tepesi appears to lie around 1500 masl on its northern slopes, and perhaps 1700 masl to the east. The alpine vegetation is dominated by an undetermined type of low Fabaceae bush, mixed with other plants, including a species of *Astracantha*, an unknown type of Lamiaceae, mullein (*Verbascum* sp.), nettle (*Urtica* sp.), and a perennial grass that is likely a species of *Bromus*.

Fig. 3.7. Tall, straight juniper (*Juniperus excelsa*) on Arayıt Dağı.

lies in the valley between them, along which the Ankara-Istanbul highway runs. The inner range, south of Bolu, reaches peaks of 2200 masl and its slopes extend south to the town of Beypazarı. Our transect followed the road from Beypazarı to Bolu, through the village of Kıbrıscık, reaching a maximum elevation of 1650 masl above Beypazarı and 1600 masl approaching Bolu (Fig. 3.1).

Forest cover began almost immediately outside of Beypazarı, which lies at the edge of the Central Anatolian Plateau at an elevation of 700 masl. A mixed oak-dominated forest appeared around 800 masl in the steep river valley that lies north of Beypazarı (Fig. 3.8, see color insert). The primary tree of this forest is downy oak (*Quercus pubescens*), with pine (*Pinus nigra*), a wild probable *Prunus* sp., prickly juniper (*Juniperus oxycedrus*), sumac (*Rhus coriaria*), wild pear (*Pyrus elaeagnifolia*), barberry (*Berberis* sp.), and *Colutea* sp. on the slopes. Trees in the valley bottom include tree of heaven (*Ailanthus altissima*, only close to Beypazarı, either deliberately planted or ruderal spread from the city), domestic pear (*Pyrus communis*), domestic apple (*Malus domestica*), walnut (*Juglans regia*), poplar (*Populus nigra*), and willow (*Salix alba*). Perennial wild barley (*Hordeum bulbosum*) grows in dense stands above 1200 masl, where the forest transitions to a much more open woodland and alpine meadow (Fig. 3.9).

Continuing down a valley to the west that terminates at the village of Çatallı, we encountered trees again around 1500 masl, including juniper (*Juniperus excelsa*), oak (*Quercus cerris*), wild pear (*Pyrus angustifolia*), hawthorn (*Crataegus* sp.), and wild rose (*Rosa* sp.). Dense oak forest begins at 1400 masl and continues to 1200 masl, just above Çatallı. This forest is also logged, more likely for fuel than construction wood, based on the size of stacked logs observed.

Inner Pontic Range

The northern Pontic Mountains, which form the northern boundary of the Central Anatolian Plateau, are divided into two ranges and the city of Bolu

Fig. 3.9. Alpine meadow north of Beypazarı.

Beyond the alpine meadow a closed-canopy pine forest begins at 1550 masl, descending to 1350 masl on the north-facing slope of the Kıbrıscık valley. This forest is almost entirely pine (*Pinus nigra*), with some wild rose (*Rosa* sp.) and hawthorn (*Crataegus* sp.) along the roadside. The majority of the Kıbrıscık valley contains an open woodland of mixed forest, with Christ's thorn (*Paliurus spina-christi*), downy oak (*Quercus pubescens*), pine (*Pinus nigra*), prickly juniper (*Juniperus oxycedrus*), juniper (*J. excelsa*), *Colutea* sp., wild pear (*Pyrus elaeagnifolia*), ash (*Fraxinus excelsior*), and wild rose (*Rosa* sp.). The road returns to pine forest at 1350 masl north of Karaköy. This forest contains larger trees than in the earlier section of pine forest, with sizes rivaling those of Hamam Dağı (heights ~30 m). This forest, interrupted by occasional meadows along stream banks or as the result of logging, continues until about 13 km south of Bolu, when it transitions into fir forest.

The transition from pine (*Pinus nigra*) to fir (*Abies nordmanniana*) is not strictly altitudinal. The first fir was observed around 1400 masl and dominates after 1450 masl, although pine was present at 1600 masl earlier in the transect. The fir forest is dense and the trees are very large, with most more than 30 m high and some over 40 m. Perhaps pine is overshadowed by

The open woodland/alpine meadow occurs on the mostly flat mountaintop from Karacören to Karagöl at elevations of 1250–1650 masl. The dominant ground cover consists of *Phlomis armeniaca*, *Astracantha* sp., *Vicia* sp., and low perennial grasses, including fescue (*Festuca* sp.), probable *Bromus* sp., and perennial wild barley (*Hordeum bulbosum*). Trees in this area take the habit of spreading bushes, mainly hawthorn (*Crataegus* sp.), but also wild rose (*Rosa* sp.), barberry (*Berberis* sp.), prickly juniper (*Juniperus oxycedrus*), wild pear (*Pyrus elaeagnifolia*), and pine (*Pinus nigra*) towards the slopes. This area is heavily grazed, as evidenced by abundant piles of sheep/goat dung and an interview with a local shepherd.

the larger fir, so unable to compete in undisturbed fir canopy forest. From the transect peak at 1600 masl, the fir forest extends down the slopes towards Bolu.

Sakarya Riverbank

The Sakarya riverbank is wooded along its entire course near Gordion and also is home to several other plant species that are only found along the waterway (Fig. 3.10, see color insert). Willow trees dominate the banks, both *Salix alba* and a weeping willow, which may be an import. Other trees along the banks include poplar (*Populus nigra*), tamarisk (*Tamarix* sp.), and plane (*Platanus orientalis*), as well as dense stands of reeds (*Phragmites australis*). A list of additional plants that grow along the banks and floodplain of the Sakarya are inventoried in Appendix A.

Protected Steppe

Several areas of protected steppe have been surveyed in the Gordion region. The most robust data exist from Tumulus MM, which has been protected since 1997 and where vegetation continues to expand in biodiversity and total ground cover (Fig. 3.11a, see color insert). Miller (2010:13) also surveyed an isolated patch of relict upland steppe through 2006, when it was grazed and farmed for the first time. Species present in these two areas of healthy steppe are inventoried in Appendix A; they are characterized by perennial grasses (*Festuca*, *Stipa*, *Bromus*, and *Poa* species) and high degrees of biodiversity. No trees are present among the lowland steppe vegetation.

Unprotected Steppe

Tumuli aside from MM are protected from agriculture by Turkish law, but they are still available for grazing. As some of the few areas of good-quality soil left uncultivated, they support a similar botanical profile as Tumulus MM, but with lower biodiversity and ground cover due to constant grazing pressure (Fig. 3.11, see color insert). Surveys of Tumulus P and Tumulus T7 (the second tumulus to its east, not lettered by Young's team) yield the species lists reported as overgrazed steppe in Appendix A. Characteristic species include wild rue (*Peganum harmala*), camelthorn (*Alhagi pseudalhagi*), wormwood (*Artemisia*

sp.), and short tufts of the same perennial grasses present in protected steppe, especially *Poa bulbosa*, likely due to its small size and ability to spread vegetatively, which allows it to expand even under close grazing that removes seed heads from most grasses.

Gypsum Steppe

Many gypsum outcrops and ridges surround the site of Gordion. The most notable are the south gypsum ridge and north gypsum slope leading up to the prominent outcrop of Kızlarkayası. Both were part of the ancient landscape of Gordion and have tumuli positioned on top of the ridges. The slopes of these ridges are not farmed today and likely have never been cultivated since the gypsum soil is nutrient-poor and dry, and the slopes are steep. They are some of the few areas grazed by flocks today as the dry uplands become increasingly used for irrigation agriculture (Gürsan-Salzmann 2005); most gypsum steppe areas are heavily overgrazed today (Fig. 3.11b, see color insert).

Species composition of the gypsum steppe differs substantially from that of the moister soil on the tumuli surveyed. Some species are only found on gypsum soils, and some prominent steppe species are missing from these areas as well. A full list of species identified on the south and north gypsum slopes is given in Appendix A; characteristic plants include wild thyme (*Thymus* sp.), *Astracantha*, *Onobrychis*, *Salsola*, and *Acantholimon* species. No trees or shrubs are found in these areas.

Agricultural Areas

Regional agricultural patterns have changed substantially in recent years with the advent of large-scale motorized irrigation in the 1990s (Miller 2011). Water is now pumped up several kilometers from the Sakarya and Porsuk rivers to ridges over 100 m above the river plain; earlier irrigation and mechanization efforts since the 1950s were primarily constrained to lowland areas (Gürsan-Salzmann 2005:177). This has also changed the mix of agricultural crops grown in the area. Formerly, rainfed cereal (primarily barley) agriculture and grazing were practiced in those upland areas with good soils, with cereals (primarily wheat) and irrigated crops grown at lower elevations closer to water sources. Gürsan-Salzmann (2005:177) notes sesame, cumin, lentil, bean, and chickpea as other upland agricultural crops.

Today, agriculture in the Sakarya and Porsuk river valleys is focused on wheat (Fig. 3.12a, see color insert), sugar beet, and onion production, with the latter increasingly popular in recent years. Sugar beets and onions require frequent irrigation (Fig. 3.12b, see color insert), and wheat yields are higher when irrigated as the grass is growing. Other cultivars observed since 2006 in lowland areas include barley, chickpea, and sunflower, and more rarely cabbage, tomato, squash, eggplant, and alfalfa. In upland areas, the agricultural products include wheat and barley, chickpea, sunflower, and occasional onion and sugar beet fields where irrigation water is available. On the footslopes of Başyurt Tepesi, closer to Afyon, poppies and a few grape vines are grown as well. Fields may also be left fallow and harvested as hay.

Trees often grow along field edges and are occasionally left to grow in the centers of fields. This includes a mixture of wild species and cultivated fruit trees. In the area surveyed, wild field trees include hawthorn (*Crataegus*), elm (*Ulmus glabra* and *Ulmus minor*), Russian olive (*Elaeagnus angustifolia*), and wild pear (*Pyrus elaeagnifolia*), with oak (*Quercus cerris*), juniper (*Juniperus excelsa*), and wild plum (*Prunus divaricata* ssp. *divaricata*) at higher elevations. Domestic species planted along field edges include apricot, pear, apple, almond, sour cherry, and walnut. These same species are planted in villages for their fruit; also seen in villages are mulberry (*Morus nigra* and *M. alba*) and box elder (*Acer negundo*). Several other plants grow as field weeds among crops and along field edges, especially those that are irrigated. These are inventoried in Appendix A.

Paleoclimates of Central Anatolia

Paleoclimatic proxy data provide ways of estimating past temperature and rainfall through a variety of indirect measures. The most valuable of these for studying recent climate change in the Near East include oxygen stable isotope studies of marine and lake sediments and calcareous cave formations, pollen cores from lakebeds and other laminated soils, and a variety of geomorphological data, including lake laminae, streambed behavior, and the presence and sorting of alluvial deposits. Each provides a different perspective on local and/or regional climate change with different levels of temporal and spatial specificity.

Of the various methods used to reconstruct paleoclimate, the most straightforward is $\delta^{18}O$ isotope studies. The stable isotopes ^{16}O and ^{18}O occur naturally in a fixed ratio on Earth of roughly 500:1 (Cronin 2010). The ratio of these isotopes in water, however, is affected by evaporation and dilution rates, because $H_2^{16}O$ evaporates more readily than its heavier cousin. In seawater, changes in $\delta^{18}O$ (ratio of ^{18}O to ^{16}O relative to a standard, here modern seawater) are caused primarily by the size of the polar caps and glaciers worldwide, so $\delta^{18}O$ serves as a proxy for worldwide temperature and humidity. The calcareous structures left by deep-sea-dwelling organisms, when deposited in layers on the seafloor, preserve a record of these worldwide climate fluctuations. In lake water, on the other hand, $\delta^{18}O$ levels are affected by local (rather than just global) trends of evaporation and dilution, so they preserve records of local and regional aridity and humidity (Roberts et al. 2010). The best sources of $\delta^{18}O$ in the Eastern Mediterranean include marine cores, lake laminae containing calcareous animal shells, and cave formations, known as speleothems, which have laminar growth lines formed by rainfall percolating into the cave, which can be dated precisely using the uranium-thorium (U-Th) radiometric technique (Wong and Breecker 2015); new developments in ion microprobe spectrometry even allow seasonal and interannual variations in rainfall to be observed and reconstructed (Orland et al. 2009).

Pollen data are more problematic to interpret because they potentially represent a broader suite of natural and anthropogenic processes of landscape change. Pollen is produced by all seed-producing plants, but differences in grain size, quantity produced, and the floral mechanism for dispersal between species can result in substantial variation between the pollen profile and the actual floristic profile of an area (Pearsall 2000:324–30). Different types of pollen are distinguishable to different taxonomic levels, but many can be identified to genus. Pollen becomes useful archaeologically when deposited in a lake or other wet area in which it can be preserved in laminar layers in an anoxic environment. A core taken through these fine-grained sediments is sampled systematically for pollen and for various carbon-containing structures (charcoal, shell, sediments) that can be dated to provide temporal reference points for the core (Faegri et al. 1989). Dates are often extrapolated for undated portions of the core by averaging the distance of a point from the nearest dated points in the core; this assumes constant deposition rates between dated levels and can be a source of error in the interpretation of pollen cores. Cores are usually divided into vertical zones corresponding to substantial changes in the frequency of one or more pollen types. Since each pollen type is usually expressed as a percent of total pollen, multiple pollen types often change in tandem. Significant changes are interpreted as the result of local vegetation change, which may be caused by climatic, geological, ecological, or anthropogenic factors. Certain large-scale changes, such as a shift from C_3-pathway to C_4-pathway plants, which outcompete C_3 plants in drier environments, or variations in the ratio of arboreal to non-arboreal pollen, may be a result of regional climatic change (Rosen 2007:22).

Geomorphological data are more varied in methodology than the approaches described previously, and may include lakebed coring, examination of geological profiles, reconstruction of stream- and riverbed movements, soil analysis, and other geochemical techniques, including elemental composition and magnetic susceptibility. As a whole, these methods record fluctuations in water flows and standing water levels through datable changes in geological strata. Anthropogenic effects, such as overgrazing, can lead to erosion or deflation of soils that are visible in the geological record (Goldberg and Macphail 2006). Climate change leads to variations in rainfall and above- and below-ground water transport systems, so identifying changes in local hydrology permits the reconstruction of rainfall levels over time. Lake levels are a particularly useful marker of both past humidity and temperature because they are the direct result of changes in the local balance between precipitation and evaporation.

Relating these disparate data sources is a complex process that is best accomplished on a regional scale, despite the fact that many sources of paleoclimatic data record both local and regional climatic trends and anthropogenic influences. In attempting to reconstruct the paleoclimatic history of Gordion, I integrate multiple data sources to produce regional patterns of climate change from which I suggest trends that would have affected Gordion's landscape. I review the relevant paleoclimatic data sources for Central Anatolia and, to a more limited extent, the broader eastern Mediterranean, organized by methodology and focusing on the past 4,000 years, equivalent to the Late Holocene period as defined climatically and including the entire occupation period of Gordion as explored archaeologically to date.

Isotope Data

The highest resolution oxygen isotope data for the arid Near East comes from the Soreq Cave (Israel) speleothem studies of Bar-Matthews and colleagues (Bar-Matthews and Ayalon 2004, 2011; Bar-Matthews et al. 1997; Bar-Matthews et al. 1999; Orland et al. 2012; Orland et al. 2009). The Soreq sequences, from multiple speleothems (calcareous stalactites and stalagmites), cover more than 60,000 years and are dated with great precision by U-Th dates obtained using thermal ionization mass spectrometry (TIMS) (Bar-Matthews et al. 1999). The relative precision of this sequence allows for the identification of centennial-scale fluctuations in rainfall throughout the Holocene and, at higher resolution, even changes in seasonal variations in rainfall (Orland et al. 2012; Orland et al. 2009). Ratios of $\delta^{13}C$, which also record changes in regional humidity, show a similar pattern to $\delta^{18}O$ fluctuations within these speleothems, lending additional confirmation to the validity of these paleoclimatic data (Bar-Matthews and Ayalon 2011). Bar-Matthews and colleagues (1997:161) used the recorded relationship between modern rainfall levels and the resulting $\delta^{18}O$ levels in Soreq Cave water to calibrate the Late Quaternary speleothem $\delta^{18}O$ levels to approximate contemporary rainfall levels. The resulting transformed data for the Late Holocene in the Levant are reproduced as Figure 3.13 (Bar-Matthews and Ayalon 2004).

One limitation of the Soreq data is the distance of the cave from Central Anatolia (Fig. 3.14). The speleothem records closest to Gordion come from Sofular Cave in northern Turkey and also provide important paleoclimatic data, but unfortunately record climatic trends from the moist, continental climate of the Black Sea coast, which is distinct from that of Central Anatolia and the arid Eastern Mediterranean as a whole (Fleitmann et al. 2009; Göktürk et al. 2011).

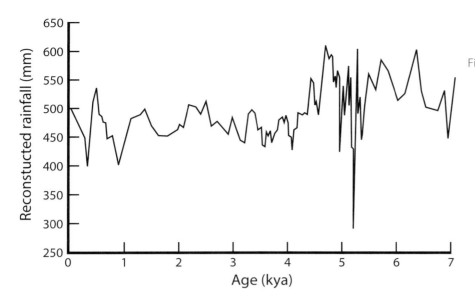

Fig. 3.13. Reconstructed rainfall data for the Late Holocene northern Levant from Soreq speleothem cores (after Bar-Matthews and Ayalon 2004).

The Jeita Cave speleothem record, from coastal Lebanon, encodes climatic trends similar to those evident in the Soreq core but the Jeita sequence from the Late Holocene (5200–1100 BP) has limited sampling resolution and presents data of uncertain interpretative value (Verheyden et al. 2008).

Seafloor cores taken in the eastern Mediterranean Sea provide complementary paleoclimatic data. Two cores from the southeastern Mediterranean offer high-resolution stable isotope values from siliceous testae, or shells, of marine foraminifera (namely the microorganism *Globigerinoides ruber*) for the past 3,600 years (Schilman et al. 2001). The authors argue that these marine data reflect the evaporation/precipitation ratio rather than temperature fluctuations over this period and, thus, represent differences in rainfall levels (Schilman et al. 2002:187). These results have much greater temporal specificity than previous marine core studies and can be superimposed over the Soreq Cave speleothem $\delta^{18}O$ values for the comparable time period to evaluate the similarity of sea-surface and terrestrial evapotranspiration balance over this period (Schilman et al. 2002). Figure 3.15 shows the relative levels of $\delta^{18}O$ between the two sequences displayed in overlapping scales. The peaks and valleys, corresponding to humid and arid periods, appear to be similar in timing, though not always in extent, between these sources. Several additional cores in the Mediterranean, Aegean, and Black Seas are reviewed by Nicoll and Küçükuysal (2013:127–28), although

their temporal resolution is lower, rendering them less useful than the Mediterranean cores for this study.

Contemporary data from terrestrial sources includes several other studies from the Levant, including a stable isotope analysis of Holocene land snails in the Negev Desert (Israel) by Goodfriend (1991, 1999). These $\delta^{18}O$ values are chronologically less precise than those from marine cores and the Soreq speleothem presented earlier and do not distinguish climatic variability over the Late Holocene (Goodfriend 1999:506). More useful are cores from Lake Kinneret (also known as the Sea of Galilee, Israel), where cores representing the century of sediment deposition ending in 1994 were used as a reference for longer cores extending back 3,300 years (Dubowski et al. 2003). Sediment carbon and nitrogen, as well as $\delta^{13}C$ and $\delta^{18}O$ levels from phytoplankton remains, were measured and used together to infer changes in input volume from the Jordan River and corresponding regional rainfall levels (Dubowski et al. 2003).

A new source for terrestrial isotopic data is carbonized plant remains from archaeological sites, including both seeds (Fiorentino et al. 2008; Roberts et al. 2011b) and wood charcoal (Masi et al. 2013b; Masi et al. 2013c). Changing values of $\delta^{13}C$ over time—when controlled for radiocarbon age—indicate periods of significant aridity, including the so-called 4.2 kya event (Fiorentino et al. 2008), a relatively rapid regional trend towards aridity ca. 2200 BCE that is implicated in the collapse of the Akkadian empire,

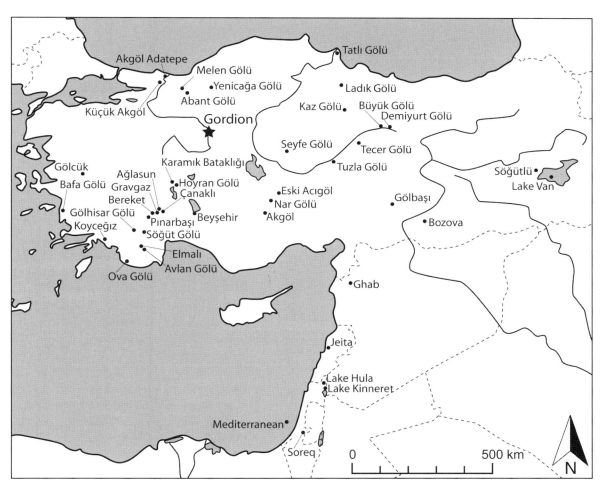

Fig. 3.14. Map of Near East with locations of all paleoenvironmental cores discussed in the text.

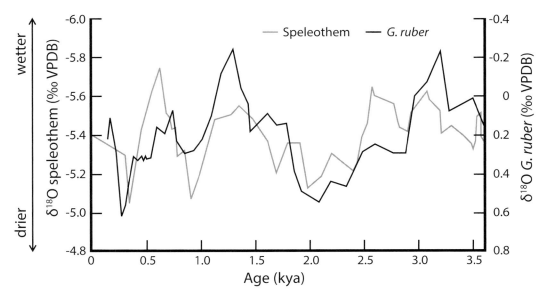

Fig. 3.15. Oxygen isotope levels from Soreq Cave speleothems and Mediterranean seafloor deposits (after Schilman et al. 2002: fig. 4). More negative values indicate wetter conditions.

as well as the Indus Valley civilization, among others (Cullen et al. 2000; Dalfes et al. 1997; Finné et al. 2011; Roberts et al. 2011b; Staubwasser et al. 2003; Weiss et al. 1993). Carbon isotopic data from wood charcoal from Arslantepe, in eastern Turkey, correspond well with data from Soreq Cave, Lake Van, and several Anatolian lakes (Roberts et al. 2011a), indicating generally wetter conditions prior to 2800 BCE and greater-than-present-day aridity ca. 2200 BCE (Masi et al. 2013b). Interestingly, it appears that different tree species record these climatic events on offset temporal scales, possibly having to do with correlations between seasonality of climate change and growth patterns of different species (Masi et al. 2013c). These same isotopic analyses on seeds can also serve as a marker of distinct agricultural regions (Fiorentino et al. 2012) and indicate differences in watering and/or fertilization practices between crops (Bogaard et al. 2007; Fiorentino et al. 2015; Fraser et al. 2011; Masi et al. 2013a).

Levantine paleoclimatic data with sub-millennial precision can be compared directly to produce a synthetic perspective of Late Holocene climatic change in that region. Oxygen isotope data from Soreq Cave speleothems, Mediterranean seafloor cores, and Lake Kinneret cores have comparable levels of temporal precision and are correlated in Table 3.1. The phases noted in the Kinneret sequence do not correspond precisely to the phases identified for the Soreq and Mediterranean sequences, but are roughly comparable during most time periods. The combined data point to a humid period in the Levant prior to 1000 BCE, an arid period following until sometime between 550 and 50 BCE, then data conflict on whether the phase until 350/650 CE was humid (Migowski et al. 2006; Schilman et al. 2002) or arid (Orland et al. 2009); the sources agree on increasing aridity until 1150 CE. The Kinneret data (Schilman et al. 2002:188) show aridity continuing to the present, while the Soreq and Mediterranean data identify a brief humid phase from 1050–1350 CE, which corresponds to the historically documented Medieval Climate Anomaly (also termed the Medieval Warm Period or Medieval Climatic Optimum). Together, however, these data show that humid and arid periods alternated throughout the Late Holocene Levant, as they did across the eastern Mediterranean, although the climate was generally more stable than in the earliest parts of the Holocene,

especially during the Pleistocene/Holocene transition (Roberts et al. 2011b; Rosen 2007:97–102).

Recently, paleoclimatic isotope data have become increasingly well represented in Anatolia, particularly from lake cores. An integrated, multidisciplinary study of Eski Acıgöl in southeastern Central Anatolia produced climatic data from the entire Holocene, although the Late Holocene is less well represented than earlier periods (Roberts et al. 2001). Adjusted $\delta^{18}O$ levels reach their highest levels during the Late Holocene, peaking during the first millennium BCE, which corresponds with the most saline sediments and the lowest lake levels of the sequence. This implies lower rainfall within the lake's watershed and may also represent an increase in temperature that led to increased evaporation (Roberts et al. 2001:733). Recent data from Gölhisar Gölü agree on a moist Middle Holocene that was followed by relatively more arid conditions after 3000 BCE (Eastwood et al. 2007). This core also indicates relatively moister conditions prevailing between ca. 750 BCE–650 CE than exist at present, a trend that differs from Eski Acıgöl and Lake Van further east (Roberts et al. 2011b). The brief arid period of ca. 550–50 BCE seen in the higher resolution data from the Levant is not evident at Gölhisar Gölü, but perhaps at Eski Acıgöl (Roberts et al. 2011b) and Lake Van (Wick et al. 2003).

Lake Van, in southeastern Turkey, has produced a new multiproxy paleoenvironmental record spanning the Holocene (Wick et al. 2003). Oxygen isotope and geochemical data (Mg/Ca ratios) indicate that maximum Holocene humidity was attained between 6400–2150 BCE, followed by a period of dryer-than-present climate between 2150–150 BCE; humidity since 150 BCE has been similar to the present day (Wick et al. 2003:667, 670).

The most recent isotopic datasets come from Nar Gölü, located on the southeastern periphery of Central Anatolia (Fig. 3.14). Oxygen isotope data dating from ca. 200 CE indicate seasonal changes in precipitation patterns over the past two millennia (Dean et al. 2013; Jones et al. 2006). The period 300–560 CE was characterized by dry summers (Jones et al. 2006:362) but wet winters (Dean et al. 2013:42), producing greater-than-present seasonal differentiation in humidity, while wetter summers and high-precipitation winters between 560–800 CE led to increased lake levels. The period of the Medieval Climate Anomaly

Table 3.1. Summary table of paleoenvironmental evidence from the Holocene in the Levant matched up with Gordion archaeological phases; parenthetical periods are not documented in the Gordion region (after Rosen 2007: fig. 5.7; Gordion phasing after Voigt 2009 and Kealhofer 2005b).

Date	Gordion Archaeological Phase	Soreq Core ∂^{18}O	Ghab Pollen	Hula Pollen	Dead Sea Levels	N. Israel Geomorphology
2000	YHSS 1	more arid	decline in evergreen oak · increasing olive	fluctuation of olive and oak		
1000	YHSS 1	more moist / more arid / more moist			very low	floodplain alluviation
0 BCE/CE	YHSS 2 / YHSS 3 / YHSS 4 / YHSS 5	more arid			very high low / high level	
1000	YHSS 6-7 / YHSS 8/9 / YHSS 10	more moist		olive decline	low level	
2000	Early Bronze Age	very dry / moist		olive decline		
3000	Early Bronze Age	dry / moist / dry	forest and maquis expansion	oak expansion	high level	floodplain alluviation
4000	Chalcolithic	moist				colluvium
5000	Chalcolithic	very dry			lower level	greater stream activity / colluvium
6000	(Pottery Neolithic)	very wet / very dry / very wet	shift from forest to maquis · more olive	more olive		buried soil in Hula and Harod Valleys
7000	(Pottery Neolithic)	fluctuation			rising level	
8000	(Pre-Pottery Neolithic)		forest readvance · more grassland	forest parkland		
9000	(Pre-Pottery Neolithic)	increasingly warmer/wetter				

(ca. 950–1350 CE) was generally wetter at Nar Gölü (Dean et al. 2013; Jones et al. 2006) and other sites in the eastern Mediterranean (Roberts et al. 2012), while the Little Ice Age (ca. 1400–1900 CE) was a more arid period at Nar Gölü. Sedimentological and geochemical data from a longer (~14,000 year) core of varved layers at Nar Gölü suggest a dry Late Bronze Age to Early Iron Age transition, followed by relatively moist and stable conditions through the Hellenistic period, and then a wet, but variable, climate period during Roman rule and increased climate variation from the later Byzantine period through the early modern era (Allcock 2013).

The primary challenge in integrating these data lies in variable temporal resolution between these cores, as well as the fact that some reflect environmental change on a more locally restricted spatial scale than others. Although multiproxy examination of lake catchments improve analytical resolution and allow increased distinction of human versus climatic impacts on vegetation and hydrology (e.g., Allcock 2013; Eastwood et al. 2007; Roberts et al. 2001), only regionally integrated studies of multiple sites can identify larger scale climatic trends. Recent work in Turkey (Nicoll and Küçükuysal 2013), the eastern Mediterranean (Roberts et al. 2011b; Rosen 2007), and the Mediterranean basin as a whole (Finné et al. 2011; Roberts et al. 2011a; Roberts et al. 2008; Roberts et al. 2012) provides a more comprehensive perspective on likely paleoclimatic change in areas where local isotopic data are lacking, such as Gordion.

Pollen and Diatom Data

In contrast to the limited number of isotope studies available from Anatolia, numerous pollen and diatom cores from various areas of Turkey have been published over the past three decades. Many of these cover the Late Holocene period and permit the analysis of regional climatic and vegetation changes throughout the occupation period of Gordion. Pollen and diatom data for the Late Holocene are available from the following cores in southwestern Anatolia: Ova Gölü (Bottema and Woldring 1984; van Zeist and Bottema 1991:83–84), Köyceğiz Gölü (Bottema and Woldring 1984; van Zeist and Bottema 1991:81–83; van Zeist et al. 1975), Söğüt Gölü (van Zeist and Bottema 1991:79–81; van Zeist et al. 1975), Pınarbaşı

(Bottema and Woldring 1984), Beyşehir Gölü (Bottema and Woldring 1984; van Zeist and Bottema 1991:75–79; van Zeist et al. 1975), Gölhisar Gölü (Bottema and Woldring 1984; Eastwood et al. 2007; Eastwood et al. 1998; Eastwood et al. 1999), Elmalı (Bottema and Woldring 1984), Avlan Gölü (Bottema and Woldring 1984), Gravgaz (Bakker et al. 2012a; Bakker et al. 2012b; Vermoere 2004; Vermoere et al. 2002a; Vermoere et al. 2000; Vermoere et al. 2002b; Vermoere et al. 2001, 2003), Ağlasun (Bakker et al. 2012b; Vermoere 2004), Çanaklı (Vermoere 2004; Vermoere et al. 2002a; Vermoere et al. 2003), Bafa Gölü (Knipping et al. 2008), Gölcük (Sullivan 1989), and Bereket (Kaniewski et al. 2007a; Kaniewski et al. 2007b). Cores from Central Anatolia include Akgöl Adabağköyü (Bottema and Woldring 1984; van Zeist and Bottema 1991:73–75), Karamık Bataklığı (van Zeist et al. 1975), Hoyran Gölü (van Zeist et al. 1975), Tuzla Gölü, Seyfe Gölü (Bottema et al. 1994), Eski Acıgöl (Roberts et al. 2001), and Nar Gölü (England et al. 2008; Woodbridge and Roberts 2011). Northern Anatolia includes cores from Melen Gölü (Bottema et al. 1994), Abant Gölü (Bottema et al. 1994; van Zeist and Bottema 1991:92–95), Akgöl Adatepe (Bottema et al. 1994), Yeniçağa Gölü (Bottema et al. 1994; van Zeist and Bottema 1991:89–92), Küçük Akgöl, Demiryurt Gölü, Kaz Gölü, Ladık Gölü, Büyük Gölü, and Tatlı Gölü (Bottema et al. 1994). More distant data are available from eastern Turkey, including Lake Van (Rossignol-Strick 1995:909–10; van Zeist and Bottema 1991:59–65; van Zeist and Woldring 1978; Wick et al. 2003), Bozova, Gölbaşı (van Zeist et al. 1968), and Söğütlü (Bottema 1995; van Zeist and Bottema 1991:60–64), and from the Levant: Ghab (NW Syria) (Rossignol-Strick 1995:906, 908; van Zeist and Bottema 1991:101–3), Lake Kinneret (Israel) (Baruch 1986; van Zeist and Bottema 1991), and Lake Hula (Israel) (Rossignol-Strick 1995:908–9; van Zeist and Bottema 1991:104–6). The location of these cores is plotted on Figure 3.14.

Climate Change

Although each of these cores includes data from the Late Holocene, their utility as a record of climate change in Anatolia during that period is often questioned (e.g., Bottema and Woldring 1990; Bottema et al. 1994:66; Eastwood et al. 2007). The primary weakness of pollen cores in paleoclimatic studies is that they reflect local vegetation histories, which may be more influenced by human activity than by climate change. The prime example of this is the Beyşehir Occupation Phase, named after its presence in the pollen cores from Beyşehir Gölü in southwestern Anatolia (Bottema and Woldring 1984, 1990; Eastwood et al. 1998; van Zeist et al. 1975). This distinct palynological profile, present in many cores across Anatolia, has been identified at different times in different places in the Eastern Mediterranean and is seen as the result of anthropogenic land clearance and agricultural activity.

Another limitation to the use of pollen in reconstructing the paleoclimate of Gordion is that no pollen core has been recorded within 120 km of the site; the closest core sites lie in the Black Sea range north and northwest of Gordion, where a different climate and vegetation type prevail (Bottema et al. 1994), or hundreds of kilometers to the east and south (England et al. 2008; Roberts et al. 2001; Woodbridge and Roberts 2011). Since the pollen of many plant species are produced and preserved only on a local scale, the records from sites hundreds of kilometers from Gordion likely do not accurately reflect the vegetation around the city during the contemporary period. The use of pollen and diatom studies here, therefore, is purely relative: climate change markers identified in profiles to the north, south, west, and east of Gordion may indicate climate change in Central Anatolia as well, and landscape changes due to human activity may be regional in scale as well, especially those associated with known cultural or demographic transitions documented historically or archaeologically.

Climate change in the Late Holocene is indicated by several individual pollen markers from multiple cores. Roberts and colleagues (Eastwood et al. 1998:78; Roberts 1990:61) have suggested that large quantities of olive pollen at some, but not all, mountain lakes in Southwestern Anatolia indicates olive production in that region, which is further supported by stone presses at neighboring archaeological sites that are likely to have been used in olive oil production (Vermoere et al. 2003). Data from the vicinity of Sagalassos confirm the intensity of olive production (Kaniewski et al. 2007a; Vermoere et al. 2003). For olive trees to grow at altitudes of up to 1400 masl, 400 m higher than its altitudinal limit today, the temperature would have had to be 3–4° C warmer in this period (ca. 50 BCE–300 CE;

Kaniewski et al. 2007b). Counter to this, is the theory of Bottema and Woldring (1984, 1990:262–64) that the decrease in levels of Asteraceae pollen, characteristic of open steppe environments, during the same period is the result of decreased airflow from the Central Anatolian Plateau that carries Asteraceae pollen to the lakes of southwestern Anatolia. They theorize that reduced summer temperatures in Central Anatolia or increased precipitation rates in the southwestern mountains would be required to change the pollen concentrations in this way. Unfortunately, it is difficult to reconcile these two contrasting theories, especially considering that the time period under discussion includes the Beyşehir Occupation Phase in much of western Anatolia, a period during which human modification of the landscape is at its maximum in these areas, tending to overwhelm climatic trends with anthropogenic signals. One recent, high-resolution, and well dated core from Gravgaz in southwestern Turkey suggests that climatic changes may have influenced agricultural patterns in highland environments: increasing aridity between 640–940 CE is associated with decreased local population, increased herding, and minimal cereal production, while the following period of the Medieval Climate Anomaly (ca. 940–1280 CE) was wetter and saw a return of cereal agriculture to the region (Bakker et al. 2012a). The relationship between climate-influenced landscape change, human behavior, and anthropogenic landscape change is complex and reciprocal, as well as variable over even short spans of time and space, which add additional complications to attempts at regional paleoenvironmental synthesis.

Cores taken from the Central Anatolian Plateau include Tuzla Gölü (Bottema et al. 1994), Seyfe Gölü (Bottema et al. 1994), Nar Gölü (England et al. 2008; Woodbridge and Roberts 2011), and Eski Acıgöl (Roberts et al. 2001). The cores from Hoyran and Karamık Batalığı (van Zeist et al. 1975) are from higher elevation steppe-woodland sites along the northern slopes of the Taurus mountain range. The cores from the plateau, especially Tuzla and Seyfe Gölü, are likely to best approximate the vegetation of the Gordion area, as the cores come from non-mountainous areas (ca. 1100 and 900 masl) (Bottema et al. 1994:21) and have similar open steppe vegetation communities today (Bottema et al. 1994). Unfortunately, neither core contains any datable material, so the span of time they represent is unknown. Bottema and Woldring (1994:47) reconstruct the base of the Tuzla Gölü core as around 1000 BCE. Both cores show a major contribution of Asteraceae pollen in all periods for Seyfe Gölü and in the earlier pollen zones of Tuzla Gölü. Pine comprises the majority of arboreal pollen, although pine does not grow in the immediate vicinity of either coring site today. Likely this pollen was produced in adjacent mountainous areas, where pine is common, as pine pollen is highly mobile (it can be transported hundreds of kilometers) and prolific. The high levels of pine pollen in the lowest and upper pollen zone of the Tuzla Gölü core may represent the natural state of forests in the neighboring mountains; the decrease in arboreal pollen in the intermediate period may be due to taphonomic processes in the lake sediment formation and preferential preservation of Asteraceae pollen rather than actual vegetation changes (Bottema et al. 1994:46). Akgöl Adabağ, which lies at around 1000 masl in the marshes near Lake Konya, is missing data from the later Holocene due to a sedimentary hiatus and a lack of preserved pollen (van Zeist and Bottema 1991:73–75).

Nar Gölü, a recently cored site in Cappadocia (Fig. 3.14) with abundant isotopic (Dean et al. 2013) and geochemical (Allcock 2013) data for changes in lake levels and regional humidity over time, also includes published pollen (England et al. 2008) and diatom data (Woodbridge and Roberts 2011). Unfortunately, the published pollen and diatom data originate from the earlier (and shorter) cores, which only date back to 300 CE. Pollen data indicate a Byzantine agricultural system in early parts of the core, ending with a period (ca. 670–950 CE) of secondary woodland growth, presumably corresponding to political and demographic instability due to Arab invasions of Anatolia, and followed by evidence for a locally dominant agropastoral system through the early modern era (England et al. 2008). Diatom data from Nar generally agree with the oxygen isotope data cited above (Dean et al. 2013; Jones et al. 2006) and indicate a drought period from ca. 400–540 CE, followed by a humid period resulting in a freshwater lake from 540–800 CE (Woodbridge and Roberts 2011:3386–87). Somewhat drier conditions returned from 800–950 CE, while the period of the Medieval Climate Anomaly (here, ca. 950–1400 CE) was moist. The period that follows involves a decoupling of the diatom and $\delta^{18}O$ levels, especially during the Little Ice Age (1700–

1900 CE), possibly the result of a lake level that was somewhat lower (as implied by $\delta^{18}O$ levels indicating an arid period) but not low enough to trigger changes in lake conductivity that would be recorded in the diatom record (Woodbridge and Roberts 2011:3388).

Pollen data from Eski Acıgöl, which lies very close to Nar Gölü, are closely calibrated with paleolimnological data from a second core in the lakebed. Throughout the Late Holocene percentages of oak pollen decrease steadily, while herb and grass pollen characteristic of a steppe environment remain relatively constant (Roberts et al. 2001:732). Increasing aridity can be seen in the decline of pistachio and other mesophilous tree pollen, correlating with evidence of lake-level regression. The authors note that human intervention is the dominant trend in the Late Holocene pollen:

> The fact that pine, which here represents long-distance pollen transport, becomes the dominant AP [arboreal pollen] type, reflects the opening up of the landscape during the late Holocene. Even so, it is likely that the human-induced trend towards a more open herb-steppe was aggravated by the tendency towards late-Holocene aridity indicated by the palaeolimnological record. The initial shift towards a more open, cultural landscape probably took place towards the end of the Early Bronze Age. (Roberts et al. 2001:733)

This coincides with the first major Bronze Age site in the area and the advent of complex regional polities in Central Anatolia. More concrete evidence of human intervention in existing vegetation communities, however, can be seen in the regional pollen profile change known as the Beyşehir Occupation Phase.

Human Impact:
The Beyşehir Occupation Phase

Although climate change continued to affect vegetation in Anatolia throughout the Late Holocene, human activity led to greater change that is readily identifiable in pollen cores from many areas of Anatolia and the Near East. The Beyşehir Occupation Phase (BOP) was first identified in cores from southwestern Anatolia, beginning with the first core from Beyşehir

Gölü published in 1975 (van Zeist et al. 1975). Since then, the BOP has been identified in many other cores, including Beyşehir II, Elmalı, Pınarbaşı (Bottema and Woldring 1984), Gölhisar (Bottema and Woldring 1984; Eastwood et al. 1998; Eastwood et al. 1999), Söğüt, Koyceğiz (van Zeist et al. 1975), Gravgaz (Bakker et al. 2012a; Vermoere et al. 2000; Vermoere et al. 2002b), Gölcük (Sullivan 1989), Ağlasun (Bakker et al. 2012b; Vermoere 2004), and Nar Gölü (England et al. 2008). In the Levant, cores from Lakes Hula and Kinneret (Baruch 1986; van Zeist and Bottema 1991), especially the latter, show similar pollen signatures at roughly the same time. Several cores from Greece also show a Beyşehir-type pollen change around 1250 BCE (Bottema and Woldring 1990).

The BOP is defined by a decrease in the arboreal to non-arboreal pollen ratio; an increase in the presence of grass, Chenopodiaceae, Asteraceae, and/or *Plantago lanceolata* pollen, all indicative of open and disturbed land cover; and an increase in fruit-bearing tree pollen (e.g., walnut, olive, chestnut) relative to dominant forest types (oak, pine, etc.) (Bottema and Woldring 1990). Not all of these variations occur in every core: for instance, the cores from northern Greece show an increase in arboreal pollen during this period, but the composition changes dramatically as pine and fir are replaced with deciduous oak, a typical pattern of forest succession suggesting primary forest clearance (Bottema and Woldring 1990:261; van Zeist et al. 1975). The BOP also appears at somewhat different times even in cores from the same region (Eastwood et al. 1998:75), indicating local differentiation in the temporal and spatial extent, as well as timing, of human intervention in local ecological dynamics. Nevertheless, all three of these pollen trends (decrease in arboreal pollen of primary forest species, increase in grasses and herbs, increase in fruit-bearing trees) point to extensive forest clearance by humans as the most likely cause for the BOP. People may have cut down trees for fuel, perhaps for iron smelting, or simply to clear land for agriculture and arboriculture (Bottema and Woldring 1990; van Zeist and Bottema 1991).

The end of the BOP, sometime around 700 CE in most cores, is correlated with widespread abandonment of highland settlements in southwestern Anatolia (Bakker et al. 2012b; Roberts 1990) and a more generalized decline in evidence for intensive agriculture. These patterns suggest a demographic shift, pos-

sibly to more mobile ways of life, perhaps due to the unstable political climate associated with declining Byzantine political control of Anatolia (Bakker et al. 2012b:262; England et al. 2008:1238). Deteriorating climatic conditions with increased aridity may also have played a role in this change in land-use patterns (Bakker et al. 2012a). The change in pollen back to a more "natural" forest composition after the BOP reflects a decreased human footprint on the landscape that permitted pine forest (or, in some cases, secondary oak forest, e.g., at Nar [England et al. 2008]) to regrow, albeit with a somewhat different steady-state forest composition due to permanently changed soil conditions (Roberts 1990).

As a result of these studies, the idea of the BOP has been extrapolated to describe human-initiated deforestation and erosion associated with intensified agricultural behavior from the Late Bronze Age through the Byzantine period across the Eastern Mediterranean (e.g., Baruch 1990; Marsh 1999). This term is problematic when describing human impact on landscapes outside of forested areas because it has different effects on local pollen dispersal and rarely involves arboriculture (e.g., at sites like Gordion [Marsh 1999]), but the widespread nature of the phenomenon reinforces its relevance to paleoenvironmental studies in the Eastern Mediterranean. The BOP remains the single most evident paleoenvironmental marker of anthropogenic disturbance across the region during the Late Holocene.

Geomorphological Data

Geomorphological studies of climate change in the Near East primarily focus on evidence for changes in water and soil transport systems (Dusar et al. 2012; Rosen 2007). Records of past erosion and alluviation, the formation and abandonment of river and stream terraces and floodplains, and changes in lake levels all reflect changes in water balance and hydrology of a region, as well as the effects of human impact on the landscape. The advantage of studying the remains of past water systems is that they can be detected in many landscapes, unlike pollen, which can only be recovered from waterlogged contexts, and that they are often associated with other proxy climate records, which allows various paleoclimatic indicators to be cross-referenced for better resolution.

Geomorphological studies of paleoenvironments are numerous in Anatolia. These include work with lake laminae and alluvial sequences (Allcock 2013; Kaniewski et al. 2007b; Knipping et al. 2008; Kuzucuoğlu et al. 2011; Roberts 1990; Roberts et al. 1999; Roberts et al. 2001; Wick et al. 2003) as well as records of alluviation and erosion in currently dry areas (Boyer et al. 2006; Kaniewski et al. 2007b; Marsh 1999, 2005; Wilkinson 1999). Lake levels in the Levant are also of interest as a potential indicator of rainfall levels on a broader regional scale (Bookman et al. 2004; Enzel et al. 2003; Migowski et al. 2006; Rosen 2007), as are alluvial cores (Kaniewski et al. 2010). Many of these studies include additional types of proxy paleoenvironmental data, such as pollen and oxygen isotope levels, which allow precise correlation of changes in water systems with other paleoclimatic phenomena.

Fortunately, extensive geomorphological fieldwork has taken place at Gordion, providing excellent data on local geomorphological change during the period of occupation at the site (Marsh 1999, 2005, 2012; Marsh and Kealhofer 2014). Marsh's (1999) focused study of the alluvial history of the Sakarya River describes five phases of the river's evolution from the Pleistocene to the present day. Two phases of river flow (Sakarya II and III) date to the occupation period of Gordion. The previous (Pleistocene to Early-Middle Holocene) Sakarya I river underlies settlement at the site and was covered with the Sakarya II paleosol layer and subsequent cultural material. The Sakarya II river apparently formed during the earliest occupation phase of the city and the paleosol is directly dated with 94% confidence to 2142–1926 BCE (Marsh 1999:166–67, 2005:167, recalibrated with OxCal 4.0). It was gravelly and relatively straight and steep compared to later riverbeds. The Sakarya III river, which began in the mid-first millennium BCE, is aggrading and brought as much as five meters of silty loam to parts of the Gordion settlement. The aggradation appears to have begun abruptly, and can be dated by a tree stump rooted in the paleosol that was buried in the Sakarya III fill. The radiocarbon date for the outer rings of this stump was 2460 ± 50 BP, unfortunately corresponding to the flattest part of the radiocarbon curve and yielding a 95% confidence range of 762–411 BCE (Marsh 1999:167, recalibrated with OxCal 4.0).

Data from the excavated portions of Gordion provide additional confirmation to the geomorphological data obtained from coring and examination of the river channel profiles. Marsh notes that "structures found beneath Sakarya III loam are always collapsed" and identifies areas where the river eroded the outer city walls and led to the formation of swampy areas in previously settled portions of the Inner City (Marsh 1999:169–70, 2012:41–43). The latest occupation in the Lower Town occurred in the Late Phrygian period and ceased after 330 BCE (Voigt 2002); this may be related to the continuing aggradation of the Sakarya River. Additionally, the clay deposits placed across the Citadel Mound prior to the Late Phrygian occupation levels may have been used to raise the level of the city above the rising level of the floodplain (Marsh 1999; Voigt 1994).

Marsh hypothesizes that the aggradation of the Sakarya River, as well as that of the nearby Porsuk River, were the result of human-initiated landscape change in Central Anatolia (Marsh 1999:169–70, 2012). Widespread deforestation and grazing would have led to erosion of topsoil into nearby streams within the Sakarya basin, a process that continues today (Marsh 2005:170). Marsh (1999) notes that this phenomenon is similar to that observed during the contemporaneous BOP in southwestern Anatolia, although analogous stream responses have not been systematically identified in that area. Sedimentation rates at Bafa Gölü in western Anatolia, however, increased by a factor of 5–6 times during the late Hellenistic and Roman periods, corresponding to the period of prominence for the neighboring city of Herakleia and increased deforestation of the catchment basin for the Büyük Menderes River, which feeds Bafa Gölü (Knipping et al. 2008). Pollen data from the same area suggests this occurred somewhat later than the onset of the BOP, which began around 1200 BCE in Bafa Gölü cores (Knipping et al. 2008).

Evidence from central and southeastern Turkey shows that similar erosion patterns have occurred in that area during the last few thousand years. Rapid aggradation near Kurban Höyük began after 2000 BCE, corresponding to local population increase and landscape degradation (Miller 1990; Wilkinson 1999). Increased particle size after ca. 1000 BCE in the upper Euphrates River valley alluvial fills suggests intensified slope erosion; a similar pattern began in the Roman/Byzantine period on the Orontes near Antioch (Wilkinson 1999:567). These changes are more likely due to human intervention in ground cover and drainage systems than to climate change (Wilkinson 2003:30–31). Alluviation patterns also changed in the Konya basin during the Late Holocene with an abrupt discontinuity in deposition from earlier periods and increased sedimentation rates; it is not known, however, whether the origin of this change was climatic or anthropogenic (Roberts et al. 1999). In southwestern Turkey, near Sagalassos, sedimentation rates increased as a result of anthropogenic landscape disturbance during the first millennium BCE, followed by a second phase of increased sedimentation during the Medieval period (Dusar et al. 2012).

One new set of geomorphological data from a lake in eastern Central Anatolia comes from Tecer Gölü, where sedimentological and geochemical analyses of varved layers document climatic fluctuations since 4000 BCE (Kuzucuoğlu et al. 2011). Following the mid-Holocene wet period identified in most other studies, Tecer data record increasing aridity between 3000–2350 BCE. Following a 450-year hiatus, the period 1900–850 BCE is characterized by a shallow, mainly saline lake, suggesting climate variation within a generally dry period. During the period 850–70 BCE the lake was deep, indicating a humid climate with low evaporation (Kuzucuoğlu et al. 2011:183). Within this moist period, however, evidence for short periods of arid summers occurs three times: ca. 670–630 BCE, 450 BCE, and 300 BCE. The first millennium CE begins with an arid period (70 BCE–500 CE), followed by wetter winters (500–540 CE), increasing water inflow (540–630 CE), and decreasing evaporation (630–820 CE)—all indicative of a generally more humid climate. The following period of the Medieval Climate Anomaly (850–1150 CE) indicates generally high lake levels, and thus humidity, while the succeeding Little Ice Age (1150–1750 CE) begins more arid (ca. 1150–1450 CE) and ends humid (Kuzucuoğlu et al. 2011:186).

Data from the Levant include multiproxy alluvial data from coastal Syria and Levantine Turkey, as well as detailed studies of lake levels from the Dead Sea. Alluvial data indicate a period of abrupt aridification during the Late Bronze Age and Early Iron Age (1175–800 BCE), coinciding with a general depopulation of increasingly arid regions (Kaniewski et al. 2008; Kaniewski et al. 2010). Evidence for human intervention in soil transport systems is clear in Le-vantine Turkey, where intensive agricultural practices of uplands during the Hellenistic and Roman periods appear to have been the cause of observed patterns of severe soil erosion dating from 150–700 CE (Beach and Luzzadder-Beach 2008; Casana 2008).

Dead Sea lake level changes, reconstructed from sedimentary analysis, are the result of changing freshwater input levels to the basin and, thus, related to local humidity and climate change. Over the Late Holocene, lake levels dropped from about 1800–1300 BCE, gradually rose until about 100 CE, then fell abruptly to a low around 400 CE and rose again abruptly by 500–600 CE. Lake levels reached a new low at 800 CE and then gradually rose until extensive irrigation during the last century led to a drop of more than 20 m in the lake level (Bookman et al. 2004; Migowski et al. 2006). A study of modern precipitation and lake levels provides additional evidence that the two are strongly correlated, so that lake level changes in the past can serve as a proxy for climate changes within the catchment area of that lake (Enzel et al. 2003).

Paleoenvironmental Reconstruction of the Gordion Region

Paleoenvironmental conditions in Central Anatolia are not well documented in comparison to the coastal regions of Anatolia and the Levant, although recent efforts at Eski Acıgöl, Nar Gölü, and Tecer Gölü have produced rich datasets that provide new insights into environmental change in the eastern portion of Central Anatolia. The limited extent of data from most of the Plateau is due to the relative scarcity of permanent water bodies in which laminar sedimentary deposits accumulate; these records are especially absent in the dry northwestern portion of Central Anatolia near Gordion. The paleoenvironments of Gordion presented here are inferred based on local geomorphological data, regional proxy climate data, and knowledge of how present vegetation communities respond to climatic and anthropogenic pressures. I focus on paleoenvironments corresponding to the occupation period of the site known from excavation, from the Late Bronze Age to the Medieval Period, although settlement in the Gordion region has been identified as early as the Chalcolithic (Kealhofer 2005b; Marsh and Kealhofer 2014) and continues today.

Paleoclimate

The paleoclimatic data from many sites in the Levant and Anatolia are well calibrated to the timescale of occupation at Gordion. Tables 3.1 and 3.2 represent the most reliable and informative Holocene datasets from these two regions. I integrate these data in the discussion below by time period to hypothesize climatic change at Gordion during its occupation period. See also several recent synthetic reviews of Holocene climate change in the Eastern Mediterranean (Dusar et al. 2012; Finné et al. 2011; Roberts et al. 2011b).

Early and Middle Holocene

The Early and Middle Holocene in the Mediterranean are characterized by average moisture levels well above those of the present day (Roberts et al. 2011a; Roberts et al. 2011b). Almost every proxy paleoenvironmental record from Anatolia and the Levant shows a similar pattern, with many indicating continued humidity through the Chalcolithic, ca. 3000 BCE (e.g., Eski Acıgöl [Roberts et al. 2011b; Roberts et al. 2001] and Lake Van [Roberts et al. 2011b; Wick et al. 2003]), although in some records drying begins earlier. Nar Gölü was drier than present around 4500 BCE (Allcock 2013:175) and Gölhisar Gölü began drying shortly after 4000 BCE (Eastwood et al. 2007:338). Greater fluctuations are evident in Soreq speleothems, where aridification begins around 4500 BCE and 3500–2800 is a period of increasing humidity (Bar-Matthews and Ayalon 2011), and Dead Sea lake levels, which include at least two dry intervals in the Middle Holocene and a reduction in the lake level of the Dead Sea between ca. 6200–3600 BCE (Migowski et al. 2006:427).

The transition to the Early Bronze Age coincides with a period of regional aridification in the Eastern Mediterranean (Table 3.2; Roberts et al. 2011b). The timing of this event is less distinct in some proxy paleoclimate records than later aridification events, which are more abrupt. A short but intensely dry period recorded in the Soreq cores around 3300 BCE (Bar-Matthews and Ayalon 2004) correlates with a short arid period at Tecer Gölü dated ca. 3300–3000 BCE (Kuzucuoğlu et al. 2011), while a more gradual drying trend is apparent at Nar Gölü, Eski Acıgöl, Gölhisar, and Lake Van (Allcock 2013; Roberts et al. 2011b).

While Soreq then indicates a return to humidity until roughly 2500 BCE (Fig. 3.13), the Anatolia lake records agree on a drying trend continuing until about 2000 BCE (Table 3.2).

The period between 2200–2000 BCE is notable as the driest period recorded in Holocene records at Gölhisar Gölü (Eastwood et al. 2007) and very dry in the other Anatolian cores (Table 3.2), but also a period of abrupt aridification in the Levant (Bar-Matthews and Ayalon 2004) and the broader Near East (Cullen et al. 2000; Dalfes et al. 1997; Finné et al. 2011; Roberts et al. 2011b; Staubwasser et al. 2003; Weiss et al. 1993). This so-called "4.2 kya event" is the most distinct and abrupt climatic change in the Eastern Mediterranean during the Holocene and would have presented a challenge for inhabitants of the Gordion region at this time.

Well-dated archaeological evidence from Gordion, however, begins in the Middle Bronze Age, when paleoenvironmental evidence from Anatolia indicates a relatively moist climate. Humidity returns to Anatolia by 1900 BCE (Kuzucuoğlu et al. 2011; Roberts et al. 2011b), although records from Nar Gölü show little evidence for this shift (Allcock 2013). A general trend toward aridity follows, culminating in a notably arid period at the end of the Late Bronze Age ca. 1300 BCE.

Late Bronze Age to Early Iron Age Transition (YHSS 8/9–7)

The Late Bronze Age (YHSS 8/9, 1500–1200 BCE) is the first period at Gordion where substantial settlement is evident from archaeological remains and the first period from which substantial archaeobotanical remains have been recovered (Miller 2010; Voigt 2009, 2011). There is robust data for a significant regional climatic event near the end of the Late Bronze Age with a shift towards drier conditions, although the timing of this event varies between proxy data (Tables 3.1, 3.2). In the Levant, Mediterranean sea cores indicate rapid aridification between 1200–900 BCE (Schilman et al. 2002; Schilman et al. 2001), while Soreq speleothems date the same event to roughly 1100–900 BCE (Bar-Matthews and Ayalon 2004; Schilman et al. 2002). The Dead Sea level dropped abruptly to a low around 1400 BCE, but then increased gradually throughout the first millennium

Table 3.2. Summary table of paleoclimatic evidence from the Holocene of Anatolia matched up with Gordion archaeological phases; parenthetical periods are not documented in the Gordion region. Solid lines represent temporal limits of cores (Eski Acıgöl, Lake Van, and Gölhisar Gölü data from Roberts et al. 2011b:150; Nar Gölü data from Allcock 2013:175; Woodbridge and Roberts 2011:3387; Tecer data from Kuzucuoğlu et al. 2011:181; Gordion phasing after Voigt 2009 and Kealhofer 2005b).

Date	Gordion Archaeological Phase	Gölhisar Gölü	Eski Acıgöl	Nar Gölü	Tecer Gölü	Lake Van
2000						
	YHSS 1			fluctuating		
1000				arid / humid	fluctuating	fluctuating
	YHSS 2	stable	fluctuating	arid		
0 BCE/CE	YHSS 3			fluctuating		
	YHSS 4				humid	
	YHSS 5					
1000	YHSS 6-7	drying	arid	arid	arid	arid
	YHSS 8/9				fluctuating	
	YHSS 10	humid	humid			
2000		arid	arid		arid	
	Early Bronze Age		drying		fluctuating	drying
3000		drying		drying	arid	
					humid	
4000		humid				
	Chalcolithic					
5000		fluctuating	humid	humid		humid
6000		humid				
	(Pottery Neolithic)			fluctuating		
7000						
	(Pre-Pottery Neolithic)			humid		
8000						
9000						

Consensus humid periods · Consensus arid periods

BCE, implying increased regional humidity (Bookman et al. 2004; Migowski et al. 2006).

In Anatolia paleoenvironmental records do not record the same level of paleoclimatic detail as the Levantine data, primarily because unvarved lake cores cannot be dated as precisely as speleothem data. The Eski Acıgöl cores record signs of drought beginning by 1100 BCE (Roberts et al. 2011b; Roberts et al. 2001), as does Tecer Gölü, where there is a hiatus in lake laminae from 1300–1100 BCE, indicating that the lake became seasonally dry (Kuzucuoğlu et al. 2011). The Gölhisar Gölü core shows drying from the Late Holocene humid peak (around 1300 BCE) until 1000 BCE (Eastwood et al. 2007). Nar Gölü data indicate an arid period beginning much earlier during the Bronze Age (ca. 2500 BCE) and continuing until about 600 BCE (Allcock 2013:174). Lake Van shows a minor increase in aridity beginning around 1100 BCE and continuing until 600 BCE (Roberts et al. 2011b) within a trend of general aridification from the Middle Holocene through the Roman period (Wick et al. 2003).

Iron Age through Hellenistic (YHSS 6–3)

Data regarding paleoclimatic change during the first millennium BCE are the most discrepant among the time periods of occupation at Gordion. Many paleoenvironmental datasets record fluctuating wet and dry periods over these centuries, but these changes are not well correlated temporally among sites and within or between regions. Even Anatolian sites close together present conflicting records: Tecer Gölü was humid from 850–70 BCE (Kuzucuoğlu et al. 2011:185), while Nar Gölü was relatively dry (Allcock 2013:175) and Eski Acıgöl shows substantial wet-dry fluctuation, albeit within a relatively dry period in comparison to the Middle Holocene (Roberts et al. 2011b:150; Roberts et al. 2001:733). Lake Van similarly shows a variable climatic signal within a relatively dry period (Wick et al. 2003:670).

This is also one of several periods during which major inconsistencies exist between the Mediterranean sea cores and Soreq speleothems. From 900–600 BCE, the Soreq data show an increase in humidity, followed by a sharp drop, while the seafloor cores put the end of abrupt aridification at 900 BCE and show a plateau in humidity through 500 BCE (Fig. 3.15).

The consensus of these diagrams, however, is that the period between 900–100 BCE was one of aridity, although the first half of the millennium (900–500 BCE) was wetter than the second half (through 100 BCE). In contrast, the Dead Sea level increased gradually throughout the first millennium BCE, implying increased humidity and/or reduced evaporation during that period. One Dead Sea study shows a drop in lake levels from about 300–100 BCE (Migowski et al. 2006), but another does not (Bookman et al. 2004).

Taken together, these data primarily indicate a drier period during the Iron Age and Hellenistic periods across the eastern Mediterranean, albeit with substantial local climatic variation in the intensity and timing of climatic shifts. Further confounding the picture is increasing evidence for human intervention in the landscape, discussed further below. This period in particular emphasizes the limitations inherent in regional paleoclimatic synthesis and the need for local proxy data to make detailed assessments of climate change at Gordion.

Roman through Medieval (YHSS 2–1)

Roman control of Anatolia coincides with a significant shift towards a generally warmer period with a regionally variable rainfall regime across the eastern Mediterranean. The term Roman Warm Period (alternatively the Roman Climatic Optimum; Kuzucuoğlu et al. 2011) has been applied to these warmer climatic conditions, dated from the first century BCE through the first half of the first millennium CE, identified in multiple parts of the eastern Mediterranean and on a broader scale across the north Atlantic (Bianchi and McCave 1999; Finné et al. 2011; Patterson et al. 2010). During the first century BCE there is evidence for high levels of humidity in the Levant, with a rising Dead Sea lake level through roughly the BCE/CE transition (Bookman et al. 2004; Migowski et al. 2006), but aridification is already evident in Central Anatolia at Tecer Gölü (Kuzucuoğlu et al. 2011). The following centuries, until at least 400 CE, see increased aridity in a high-resolution Soreq cave speleothem (Orland et al. 2009), a decline in Dead Sea lake levels (Bookman et al. 2004; Migowski et al. 2006), increased evaporation at Tecer Gölü (Kuzucuoğlu et al. 2011), and decreased rainfall and rising salinity at Nar Gölü (Allcock 2013; Jones et al. 2006; Woodbridge and Roberts 2011). In contrast, Mediterranean sea cores show increased humidity between 100–700 CE (Schilman et al. 2002:186).

By 500 CE there is evidence for substantially increased humidity at Nar Gölü (Jones et al. 2006:362) and Tecer Gölü (Kuzucuoğlu et al. 2011:185), where increased winter precipitation is the initial driver of rising lake levels. Data from the Dead Sea present a conflicting picture indicating a possible sudden rise in lake volume around 400 (Bookman et al. 2004) or 700 CE (Migowski et al. 2006); both studies then agree on increased humidity resulting in higher lake levels following 1000 CE (Bookman et al. 2004). This signals the onset of the Medieval Climate Anomaly, which constituted a generally humid period across the eastern Mediterranean, although differing in intensity and timing between sites (Allcock 2013; Bakker et al. 2012a; Dean et al. 2013; Jones et al. 2006; Kuzucuoğlu et al. 2011; Roberts et al. 2012; Schilman et al. 2002). Records from Nar Gölü (Jones et al. 2006; Woodbridge and Roberts 2011) and Lake Van (Wick et al. 2003) suggest continued humid conditions through roughly 1400 CE, after occupation had ended at Gordion.

Vegetation and Soil Responses to Climate and Human Activity

Pollen and geomorphological studies give insight into how the vegetation in an area responded to climatic, ecological, and anthropomorphic changes in the environment, and to how those changes affect soil formation and transport. While no pollen studies are available from the immediate area of Gordion, coincident changes in the pollen profiles of neighboring regions give insight into the timing and severity of regional vegetation changes that may have also affected Gordion. There is general agreement that pollen records of the Late Holocene in the Near East are poor climate proxy sources because human activity intensified and became the driving force behind vegetation change (e.g., Dusar et al. 2012; Roberts et al. 2001; Rosen 2007; van Zeist and Bottema 1991). Geomorphological data from the arid Near East track a similar pattern of human-driven landscape change (Dusar et al. 2012; Rosen 2007; Wilkinson 2003). Despite climate change over the occupation history of Gordion, as described above, I argue that the regional patterns

of vegetation change described below are driven primarily by human activity. This perspective has ample support from detailed studies on Late Holocene settlement dynamics on broad regional (e.g., Dusar et al. 2012; Eastwood et al. 1998; Rosen 2007) and local (e.g., Allcock 2013; Bakker et al. 2012b; Marsh 2005, 2012; Vermoere et al. 2000) scales in the Eastern Mediterranean.

The only pollen cores from steppe areas of Central Anatolia are those from Tuzla Gölü and Seyfe Gölü (Bottema et al. 1994). Contemporary surface pollen samples from eight steppe and steppe-forest sites in the area show large contributions of Chenopodiaceae, *Artemisia*, and Poaceae pollen, which constitute major components of the local steppe vegetation, but also 30% or more arboreal pollen, which is not local (Bottema et al. 1994). The pollen is primarily pine, with some oak and juniper, and is blown into the Central Anatolian Plateau from neighboring forested mountains, suggesting that even open steppe sites should show a significant contribution of arboreal (especially pine) pollen. Both Tuzla Gölü and Seyfe Gölü include significant quantities of pine pollen. Seyfe Gölü appears to have been surrounded by a dry steppe for the entire duration covered by the core (unfortunately, not dated), with Asteraceae, Chenopodiaceae, and Poaceae pollen comprising 30–90% of the total pollen count at each sample point. The Tuzla Gölü core, estimated to begin around 1000 BCE, shows high proportions of Asteraceae, Chenopodiaceae, and Poaceae pollen in all periods. Pine pollen becomes more prevalent in the later half of the core, averaging around 50% of the total pollen count. This might be due to increased pine growth in neighboring areas or to shifting wind patterns, but is unlikely to represent pine growing locally (Bottema et al. 1994:46).

Although pollen has not been recovered from the immediate vicinity of Gordion, a similar pollen profile is likely to be produced by vegetation at Gordion. The major steppe taxa today include a number of grasses and members of the Chenopodiaceae and Asteraceae. Oak and juniper grow locally on hillsides, and pine stands still exist at higher elevations within 20 km of the site. Cyperaceae and other hydrophilic taxa grow along the river. The pollen profiles from Seyfe and Tuzla Gölü thus represent a typical Central Anatolian steppe vegetation community and suggest that steppe conditions have prevailed in Central

Anatolia for at least the last 3000 years, which is not surprising given the relatively stable and moderately arid climatic regime that has persisted throughout the Late Holocene.

The Beyşehir Occupation Phase (BOP) is the most visible sign of human impact on the environment during the period in which Gordion was inhabited. The BOP is most evident in pollen cores from southwestern Anatolia, but is also visible in pollen signatures from the Levant to Greece (Bottema and Woldring 1990). The onset of the BOP occurs at different times and with somewhat different signatures in different places, even within the same region. The date of onset for the BOP in southwestern Anatolia, for example, varies from 1700 BCE at Beyşehir to 1200 BCE at Söğüt (Eastwood et al. 1998), and at Gravgaz is subdivided into a deforestation phase (1520–1430 to 410–240 BCE) and a cultivation phase (410–240 BCE to 660–770 CE) (Vermoere et al. 2000). The duration of the BOP is even more variable, lasting only about 200 years at Köyceğiz but close to 2000 years in several other cores (Eastwood et al. 1998:75).

The BOP is often associated with archaeological evidence for increased human impact on the landscape. One of the major components of the BOP pollen signature is an increase in olive pollen, and large number of stone presses used for olive oil production have been found at Hellenistic and Roman settlements in southwestern Anatolia and contemporary settlements in the Levant (Baruch 1990; Roberts 1990; Rosen 2007; Vermoere et al. 2003). The end of the BOP at sites in southwestern Anatolia, including Beyşehir, Gölhisar, Pınarbaşı, Ağlasun, Bereket, and Gravgaz, is coincident with a general depopulation of the region and the abandonment of Sagalassos, a Roman city, in the seventh century CE (Bakker et al. 2012b; Kaniewski et al. 2007b; Roberts 1990; Vermoere et al. 2000).

The cause of the BOP is deliberate human intervention in the landscape on a scale previously unseen in the paleoenvironmental record of Anatolia. Forest areas were cleared for timber and agriculture, arboriculture increased, and vegetation succession was altered from previous patterns (Bottema and Woldring 1990; Eastwood et al. 1998; Roberts 1990). A similar phenomenon appears to have occurred at Gordion. A combination of deforestation and intensive grazing on hill slopes led to dramatic increases in erosion and alluviation of the Sakarya riverbed (Marsh 2005, 2012;

Marsh and Kealhofer 2014). Similar mechanisms for erosion have been inferred from geomorphological studies in western and southeastern Anatolia (Knipping et al. 2008; Wilkinson 1999), as well as the northern Levant (Beach and Luzzadder-Beach 2008; Casana 2008).

The pattern of erosion and alluviation at Gordion proceeded in two phases. The first began around 4000 BCE, coincident with the first settlement of the Gordion region (Kealhofer 2005b), when small upland streams began to aggrade, presumably due to human impacts on vegetation communities of hillslopes, with this aggradation rate reaching a peak around 2000 BCE, when Gordion itself was first established (Marsh 2012; Marsh and Kealhofer 2014; Voigt 2011). Although human settlement in the Sakarya valley began to increase during the Middle and Late Bronze Ages (Kealhofer 2005b; Marsh and Kealhofer 2014), Gordion grew dramatically in size during the Phrygian period, when it became the capital of a regional polity that controlled Central Anatolian trade routes. It was during the Middle and Late Phrygian periods, when the city reached its maximum areal extent and political power (Marsh and Kealhofer 2014; Voigt 2002), that the second, and more significant, erosional phase commenced, leading to the alluviation of the Sakarya River (Marsh 1999). This is substantially later than the onset of BOP in southwestern Anatolia (Eastwood et al. 1998; Marsh 1999).

Regional-scale erosion as evidenced in aggradation of the Sakarya accelerated during and following Roman occupation of Gordion (Marsh 2005:169). The Roman period is characterized by limited urban settlement at Gordion itself (Bennett and Goldman 2009; Goldman 2005; Voigt 2002) but a large regional population spread across the landscape (Kealhofer 2005b). It appears that this large rural population had a significant negative impact on regional vegetation communities leading to increased soil erosion rates (Marston and Miller 2014). Hellenistic and, to a greater extent, Roman settlement appears to have had similar effects across Anatolia and the northern Levant. The pattern of a delayed BOP is mirrored at Bafa Gölü in western Anatolia, where increased deforestation and alluviation coincide with a time of economic importance and population expansion at the neighboring city of Herakleia during the Hellenistic and Roman periods (Knipping et al. 2008). Data from Nar

Gölü suggest a similar timing for the BOP (Haldon et al. 2014:140–42). A similar pattern of economic and population expansion leading to landscape change can be seen in the Gravgaz core, where the cultivation phase identified in the pollen core is contemporaneous with increased settlement and economic activity in the countryside of Roman Sagalassos and at the site itself (Vermoere 2004; Vermoere et al. 2002a; Vermoere et al. 2000). Land use at Kinet Höyük (Beach and Luzzadder-Beach 2008) during the Roman period and in the Jebel al-Aqra (Casana 2008) during the Hellenistic and Roman eras also triggered large-scale regional erosion in the Northern Levant.

In total, these data suggest that the relationship between land-use patterns, vegetation change, and landscape change at Gordion is not atypical among settlements in Central Anatolia and the broader Near East. A period of landscape disturbance and erosion tracks early human settlement of the area ca. 4000 BCE, but a more significant landscape transformation appears to result from land-use practices of later eras, especially the Iron Age and Roman periods. Parallels for this pattern exist at sites across the broader region, where the timing and nature of human impacts broadly described as BOP are variable and appear dependent on the complex interplay among local settlement histories, land-use strategies, vegetation communities, geomorphology, and climate change.

From Environmental Reconstruction to Cultural Interpretation

Many attempts have been made to link climate change with human cultural development and downfall throughout the history of archaeological inquiry (Trigger 2006). The climatic changes in the Near East that occurred during the period of occupation at Gordion were substantial and may have played a role in the collapse of the Hittite Empire and the economic development of the Phrygian state (Neumann and Parpola 1987). There is little evidence, however, that climatic change played a major role in the development of vegetation communities in Anatolia in the Late Holocene. The extant pollen data support the

conclusion that Central Anatolia has been characterized by a semi-arid steppe-woodland since at least the Middle Holocene. The major factor in vegetation change has been the expansion of human population, economy, and activity in the landscape, especially since the expansion of complex polities across Anatolia beginning in the Late Bronze Age and significantly accelerating during the later Iron Age. Human impact on the environment of Gordion has been substantial over the past 6,000 years, as documented above.

This link between increased human population, wealth, and environmental change identified here should be of little surprise, given contemporary processes of global development. The purpose of providing a paleoenvironmental context for Gordion, however, is more specific. In order to model the effects of human subsistence strategies on pre-existing vegetation communities and on the biogeographic systems of the Gordion region, it is necessary to understand the baseline functioning of those systems and how they respond to specific human activities. Contextualizing landscape change within its environmental, social, and economic contexts, as a resilience perspective prescribes, also provides a mechanism for understanding how and why specific land-use strategies might have been chosen by the inhabitants of Gordion, allowing us to postulate directionality of cause and effect of linked changes in human and natural systems.

Ongoing climatic and environmental change constantly altered the parameters of human decision-making systems and led to shifts in land-use strategies, producing the amalgam of remains that comprise the environmental archaeological record. The cycle of change and response between cultural and environmental systems is complex, but such cycles can be approached through the use of both discrete ecological and behavioral models and integrated systems frameworks, such as those developed in Chapter 2 and investigated in the following chapters. Ultimately, such an integrated investigation provides a better understanding of how people adapted to environmental and social change throughout the history of Gordion and how the landscape of Gordion developed as a result.

4

Wood Use and Landscape Change

Gordion lies in the steppe-woodland of Central Anatolia, where landscape change over time can be measured using three datasets: proxies for steppe grassland change (in species composition and extent), proxies for woodland change (also in both species composition and extent), and geomorphological change. Evidence for paleoclimatic change, which may have driven changes in these three areas, and for local geomorphological change was discussed in Chapter 3. This chapter focuses on changes in regional woodland dynamics and attempts to ascertain the role of Gordion's inhabitants in affecting these transformations, using the analysis of wood charcoal remains from the ancient city as the primary proxy dataset. Chapter 5 addresses change in agricultural systems and grassland composition, drawing on carbonized seed remains from Gordion.

Questions and Approaches

Archaeological wood charcoal remains represent two distinct sets of information that are inextricably bound together within charcoal assemblages: woodland landscape composition and human behavioral choices in gathering, using, burning, and disposing of wood. Both availability of wood and human choices in acquiring wood for different uses leave traces in archaeological assemblages, and one challenge in analyzing those assemblages is disentangling those two factors. The longer history of paleoethnobotanical analysis of wood charcoal (often termed anthracology in European scholarship) has focused on ecological reconstruction. More recently, paleoethnobotanists have turned to the behavioral study of wood acquisition, including selection and deposition choices and the roles they play in assemblage formation. In addi-

tion, experimental research has been critical in shedding light on lingering taphonomic questions that inform methods used for the quantification and interpretation of charcoal fragments.

Here, I present a brief summary of the literatures on both the ecological and behavioral perspectives and present an argument for the use of a holistic framework for the analysis of the Gordion charcoal assemblage. In the analysis that follows, I classify tree taxa by ecological community to reconstruct both spatial patterns of wood foraging and change over time in wood availability. Through spatial analysis of wood charcoal by context across the site, I identify patterns of use and deposition that inform our understanding of wood disposal at Gordion. Finally, to explore behavioral choices, I distinguish charcoal from structure collapse, a proxy for construction materials, and charcoal from pyrotechnic features, where wood was deliberately burned as fuel. Drawing on foraging theory (see Ch. 2), I model key variables for wood selection for both construction and combustion and test them against the archaeological charcoal assemblage from Gordion to identify which variables structured wood acquisition patterns.

Reconstructing Woodland Ecology

The earliest publication dedicated to the synthetic analysis of charcoal from archaeological deposits is Salisbury and Jane's 1940 article in *Journal of Ecology* using wood charcoal excavated from Maiden Castle and Verulamium in England to interpret the composition of woody plants in the paleoenvironment of those sites and human use of those wood resources. Analysis of wood charcoal from archaeological sites expanded substantially in the late 1960s and 1970s with the advent of systematic flotation in the Americas and with the scientific underpinnings of the New Archaeology, which placed an emphasis on paleo-ecological interpretation (see historical overviews in Asouti and Austin 2005; Pearsall 2000; Smart and Hoffman 1988; Trigger 2006). The emphasis of char-coal analysis, both in Europe and the Americas, has usually been on paleoenvironmental reconstruction and an assessment of the comparative role of human activity and climatic change in producing diachronic variation in the composition of woody vegetation in a given area (e.g., Miller 1985; Pearsall 1983; Will-cox 1974). A separate research trajectory uses charred plant materials in the reconstruction of plant com-munity changes on a larger scale through the Quater-nary and Tertiary periods (reviewed in Delhon 2006; Figueiral and Mosbrugger 2000; Scott and Damblon 2010), but this literature is generally distinct from that of anthropological archaeology and paleoethno-botany published in English and German. Much of the literature combining paleoecological analysis of charcoal from archaeological and geological deposits was until recently published only in French, creating an independent research tradition that is only re-cently beginning to be referenced in the larger North American/British/German/Dutch body of literature (Asouti and Austin 2005; Chabal et al. 1999; Jacomet 2006b; Théry-Parisot et al. 2010).

Behavioral Studies of Wood Use

Scholars since the 1940s have also considered the role of human wood preference in creating archaeo-logical charcoal assemblages (Godwin and Tansley 1941; Salisbury and Jane 1940). Many see human preference as a source of variation that complicates charcoal analysis and renders charcoal less suitable for paleoenvironmental reconstruction than other botan-ical remains (Asouti and Austin 2005:1–2). Recent work has focused on taphonomic interpretation of charcoal assemblages to remove those that are poten-tially problematic (Asouti 2005; Asouti and Austin 2005; Chabal 1992; Chabal et al. 1999), with some scholars going so far as to advocate sampling only cer-tain types of deposits in the field (Asouti and Austin 2005:5). Most arguments for or against the validity of charcoal as a paleoenvironmental marker are based on behavioral assumptions (Godwin and Tansley 1941), experimental observations (Chabal 1992), or

ecological models for the comparison of archaeologi-cal and modern data sets (Smart and Hoffman 1988). When the focus remains solely on environmental re-construction, human preference is reduced to prob-lematic "noise" in the archaeological record. For these reasons, human behavior was rarely considered in ar-chaeological charcoal analysis during the earlier part of the 20th century.

A notable exception is the conceptual model of the relationship between environment and human activity proposed by Shackleton and Prins (1992). The authors identify the general assumption in char-coal studies as the "principle of least effort," namely that charcoal frequency directly tracks environmental presence with a one-to-one correspondence (Shack-leton and Prins 1992:632). They propose four differ-ent environmental circumstances reflecting different availability of preferred wood species and of dry ver-sus green wood and argue that under certain of these environments the assumption of the "principle of least effort" is valid, and in others it is not. Their conclu-sions are open-ended, and they suggest that no archae-ological mechanism currently exists to differentiate between environmental availability and human pref-erence in determining the composition of an archaeo-logical charcoal assemblage. Following Ford (1979) and their own prior research, Shackleton and Prins suggest that ethnographic analogy is the only method for inferring the likelihood that certain wood species would be preferred or avoided.

The work of Chabal (1988, 1992; Chabal et al. 1999) and the "Montpellier School" seeks to create an explicitly archaeological mechanism to differentiate between environmental availability and human pref-erence by rigorously assessing the depositional pro-cesses that result in charcoal assemblages and choosing specific contexts for analysis. While this solves the problem of what contexts to use for environmental reconstruction, it does so by ignoring behavioral inter-pretation that is possible with charcoal remains. The use of associated archaeological information to infer use context, coupled with ethnographic analogy, is an approach adopted by a number of scholars to attempt to understand wood use patterns, with some success (e.g., Asouti 2005:251–57; Henry and Théry-Parisot 2014; Smart and Hoffman 1988:168–70). While this can explain the archaeological distribution of some wood types and begins to relate human activity and decision making with environmental variation, it does so only on a case-by-case basis. Successful examples include studies of charcoal kiln sites in Western Europe (Bonhôte et al. 2002; Fabre and Auffray 2002; Ludemann 2002, 2006; Montanari et al. 2002; Nelle 2002), roof collapse in the Andes (Marconetto 2002), and elite burial furniture from Egypt and Anatolia (Asensi Amorós 2002; Simpson and Spirydowicz 1999).

An alternative approach to behavioral interpretation comes from the field of human behavioral ecology, which has been applied to archaeological problems with increasing frequency over the past 20 years (Bird and O'Connell 2006; Winterhalder and Smith 2000; see extensive discussion in Chapter 2). Behavioral ecology provides a common conceptual framework for the creation of models for human behavior that can be tested using archaeological plant remains (Gremillion 2014; Gremillion and Piperno 2009). Despite the evident utility of this approach, I was the first to apply behavioral ecology to archaeological wood charcoal assemblages in order to reconstruct patterns of wood acquisition behavior and shed light on the key variables behind wood selection, beginning with a study using data from Gordion (Marston 2009). The utility of behavioral ecology thinking and foraging theory models in particular, for exploring past human engagement with woodland landscapes has been acknowledged (Dufraisse 2012; Gremillion 2014; Salavert and Dufraisse 2014) and applied to other case studies from the Mediterranean (Rubiales et al. 2011) to the Maya world (Robinson and McKillop 2013). Such an approach, however, is cautioned against the potential to become mechanistic and isolated from the role of social and cultural systems (Asouti 2012).

In this chapter, I explore woodland use at Gordion by integrating all of the perspectives above: ecological, contextual, social, and behavioral. Following recent critiques of behavioral ecology modeling using foraging theory (Asouti 2012; Smith 2015; Zeder 2014, 2015), I apply elements of foraging theory to the careful contextual analysis of charcoal samples and directly test whether specific variables used in foraging models structure wood acquisition, rather than just assuming their importance. I argue that this integrated approach to wood charcoal analysis offers distinct advantages over case-by-case contextual approaches used in prior wood charcoal research and

permits better explanations of past human activity in a landscape than ecological interpretation alone.

Methods

This assemblage consists of all "hand-picked" wood charcoal fragments recovered from contexts excavated at Gordion between 1993 and 2002. It does not include charcoal from flotation samples, but comparative analysis of samples excavated in 1988 and 1989 indicates that hand-picked and flotation charcoal assemblages at Gordion are essentially identical in aggregate composition (Miller 2014). Consistent collection methods have been used since 1988, so methods for the recovery of this assemblage are consistent with those described by Miller (1991a, 2010) for the 1988–89 seasons. Although I introduced a few minor changes in laboratory methods, these data were collected to be directly comparable with Miller (1991a, 2007) and to be maximally comparable with other charcoal studies. In pursuit of that goal, I include sample-by-sample reporting of charcoal identifications (Online Appendix 1) in the electronic appendices to this volume.

Sample Collection

All charcoal analyzed here was collected by hand in the field. Excavators were instructed to recover all pieces of charcoal encountered during excavation, including those found during screening (1-cm screen was used for all deposits excavated). This creates a sampling bias in the assemblage, as large pieces of charcoal were most likely to be recovered, and different excavators were more and less conscientious in the amount of charcoal they collected. Flotation samples were also taken for any deposit with substantial amounts of ash or visible carbon, from surrounding control deposits, as well as from any context of interest to the excavator. This also affects the hand-picked charcoal assemblage, as some major charcoal deposits were removed as flotation samples and so are not represented here, but will be analyzed in future work. The advantage to using this encounter-contingent strategy for charcoal collection, as opposed to systematic flotation sample collection espoused by others (e.g., Asouti and Austin 2005; Chabal 1992; d'Alpoim Guedes and Spengler

2014), is that nominally all charcoal above a certain size threshold was collected, regardless of the type of feature (or lack thereof) with which it was associated. However, this strategy may be biased towards taxa that might be present in larger fragments and against those that might be preferentially represented by smaller fragments (<1 cm). A detailed study by Miller (2014), however, compared flotation charcoal to hand-picked charcoal assemblages and found no evidence for bias in taxon diversity or frequency as a result of the hand-picking method. Additionally, this approach was efficient, and the Gordion charcoal assemblage includes a representative diversity of context types. This allows contextual, ecological, and behavioral analyses to be conducted with the Gordion data, not all of which are possible with sampling strategies used by others.

Sample Analysis

Most charcoal samples included dirt from the field and very small pieces of charcoal (or even charcoal dust) resulting from post-depositional and/or post-excavation breakage of the charcoal pieces. To limit the sample to pieces of charcoal likely to be identifiable, all samples were screened through a 2 mm geological sieve and the resulting >2 mm fraction was weighed and retained for analysis. Only fragments with at least one growth ring visible were identified to decrease the identification bias associated with smaller pieces of charcoal; Chenopodiaceous taxa and other subshrubs, including monocots, that lack proper growth rings were also counted, but were extremely rare in the assemblage, perhaps in part due to a selection bias against smaller fragments of wood (although only three fragments of Chenopodiaceous charcoal were present among 1412 fragments recovered using flotation; [Miller 2014]). Each piece of charcoal was broken by hand to yield a transverse section and examined with a 7.5x-75x zoom stereomicroscope. Pieces requiring the examination of diagnostic characters in the radial and tangential sections were broken along those planes and examined at 100x–400x with an incident-light compound microscope. Each piece of charcoal was recorded and placed with like pieces from the same sample, which were weighed in aggregate by taxon at the conclusion of the analysis of the sample. Pieces of charcoal were not weighed individually, which differs from Miller's (1991a) methods of analysis.

At least ten pieces of charcoal were examined from each sample unless fewer than ten pieces were present, in which case all were examined. If the first ten pieces were the same taxon, and the rest of the sample appeared homogenous at a glance, then this sample size was deemed sufficient. If the samples pulled were of multiple taxa, if the sample was very large, or if the remainder of a sample appeared diverse, more pieces would be pulled until all were examined, no new types appeared (Fig. 4.1), or a maximum of 35 pieces was reached. This was an arbitrary limit used for the purpose of efficiency and departs from the number of charcoal fragments necessary for a fully representative sample of the taxonomic diversity of a context (at least 100 in temperate climates, 250 or more in Mediterranean and tropical zones), as prescribed by several scholars on the basis of statistical analysis of charcoal assemblages (Asouti and Austin 2005; Chabal et al. 1999; Delhon 2006; Scheel-Ybert 2002). The floristic diversity of woody plants, however, is much lower in arid regions. Therefore, a lower number of pieces should represent a statistically valid sample in a region such as Central Anatolia. Moreover, since the collection strategy included recovery of large charcoal pieces that fragmented further during and following collection, we expected many samples to be homogenous as the fragments resulted from a single original piece of wood. The number of pieces examined, and the percentage of the total sample weight to which that corresponds, are listed for every sample in Online Appendix 1. Although not more than 35 charcoal fragments were examined per sample, in the majority of cases more than 80 percent of the sample by weight was examined, and in only 3 percent of samples was less than 25 percent of the sample weight examined. I attribute the discrepancy between the large percentage of the assemblage identified and the small sample sizes (much less than the suggested minimum number of charcoal fragments) identified in my analysis as an effect of the hand-picked charcoal sampling strategy, which is not discussed by authors other than Miller (2010:22–23).

An attempt was made to pull a representative variety of sizes and shapes of charcoal fragments from each sample: so-called "grab-sampling" (Miller 1985). I disagree with Asouti and Austin (2005:7) that this introduces unacceptably subjective elements into the analysis. Their preferred method, dry sieving and sub-

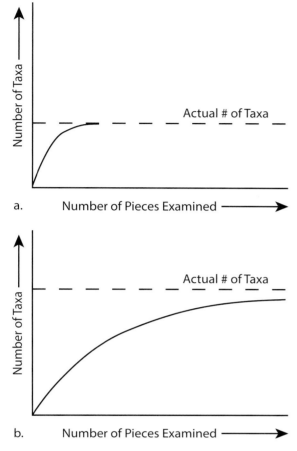

Fig. 4.1. Saturation curves representing the relationship between the number of pieces examined in a sample and the cumulative total number of taxa identified: (a) saturation curve for a small sample with few taxa; (b) saturation curve for a large sample with many taxa.

in a given sample. Instead, a flexible number is used to account for differences in diversity and sample size between samples. This follows the logic of the "saturation curve" phenomenon described by Asouti and Austin (2005:6–7) and others (e.g., Dufraisse 2008; Lepofsky and Lertzman 2005; Lyman and Ames 2004); Gini-Lorenz concentration curves have also been used to similar effect (Delhon 2006; Scheel-Ybert 2002; Scheel-Ybert and Dias 2007). If more fragments are identified in a sample, the number of taxa identified will increase according to a decelerating function with an asymptote at the actual number of taxa in the sample (Fig. 4.1). The shape of the curve is related to the sample size and the number of taxa present in the sample. Smart and Hoffman (1988:176) describe a formal way to plot this trend for a number of samples to ensure that subsamples are of an appropriate size to capture the majority of the diversity within each. In samples with one or two taxa in similar proportions, it will take only a few fragments to identify the full variation within the samples (Fig. 4.1a); note that this is the case for many, if not most, of the Gordion hand-picked samples. If many taxa exist, it will take more fragments to approach the asymptote (Fig. 4.1b). This feature of charcoal assemblages allows for a quick way of predicting sample variability. If pieces are chosen at random from a sample and identified and the first ten pieces out of a sample are all the same taxon, the probability is high that the saturation curve is steep and that few taxa are actually present within the sample. This efficient workflow allowed me to examine 3,748 charcoal fragments across 445 samples while identifying 83% by weight of the total charcoal assemblage >2 mm, giving a representative sample of the entire assemblage.

Identification

Identifications of the wood charcoal materials were based on comparisons with my own comparative collection gathered from Central Anatolia, currently housed at Boston University, and with Miller's comparative collection at the University of Pennsylvania, as well as with published wood atlases (Barefoot and Hankins 1982; Fahn et al. 1986; Panshin and de Zeeuw 1970; Schoch et al. 2004; Schweingruber 1990; Schweingruber et al. 2006). Most identifications were possible at the genus level. It is rarely possible to distinguish between the woods of different

sampling, with a preference for the >4 mm fraction, is essentially the same method as used here. A 4 mm cutoff for fragment size would limit analyses at Gordion, as many pieces of charcoal are in the 2–4 mm size class and would result in lower sample populations. Any subsampling method will reduce the sample size and increase the chance of sample size errors in analysis. Since the suggested Mediterranean sample size of 400–500 fragments for environmentally representative samples (Chabal et al. 1999) is not achievable and likely not necessary for Gordion hand-picked samples, this method was not adopted.

The methods used in this study do not require a certain number of charcoal fragments to be examined

species within a genus, although this is occasionally an option if particular species have distinctive characters; in some cases, species (or groups of species) can be distinguished within a genus, such as deciduous vs. evergreen oak (Schweingruber 1990). In this study, two identifications were made on phytogeographic grounds, as only one species of the genus grows in the Gordion region. This is the case for pine (*Pinus* cf. *nigra*) and hornbeam (*Carpinus* cf. *betulus*). In other cases, it is not possible to distinguish between the woods of related genera. The Maloideae, a subfamily of the Rosaceae, includes wild and domestic apple and pear trees, as well as several wild genera (*Crataegus, Cotoneaster, Sorbus, Pyracantha, Amelanchier*), none of which are possible to distinguish on the basis of wood anatomy, and many of which are indigenous to the Gordion region (Schweingruber 1990:617). The woods of willow and poplar, both within the family Salicaceae, are only distinguishable through one character in the radial section that is often ambiguous (Schweingruber 1990:673, 675; see also Appendix B). Analysis of the Salicaceae remains from Gordion show the presence of both willow and poplar, but relatively few charcoal fragments were distinguished and all Salicaceae wood is reported at the family level in Online Appendix 1.

Characters used for identification of different wood types are reported in Appendix B, following the format used by Miller (1991a: app. I, 2010: app. B) for previous work at Gordion. Identification rates for Gordion wood are high, with less than 2% of the total assemblage weight coded as unidentifiable or indeterminate types. Most of the wood is oak, pine, or juniper, consistent with Miller's previous assessment (1999b). New taxa identified in this analysis include willow (*Salix* sp., not definitively identified by Miller), hornbeam (*Carpinus* cf. *betulus*), and possible hackberry (cf. *Celtis* sp.). Miller did not identify shrub wood from the family Chenopodiaceae among the 1988–89 hand-picked samples, but did find Chenopodiaceae wood among charcoal recovered through flotation, identified as possible *Noaea mucronata* (Miller 2014). Not identified among these samples were alder (*Alnus* sp.) (Miller 1991a, 2010), dogwood or cornelian cherry (*Cornus* sp.) (Marston 2003), and grape (*Vitis vinifera*) (Miller 2014), which had been previously identified only in single deposits from Gordion (Table 4.1).

Quantification and Interpretation

Simple statistics, including the number of fragments identified per sample and the percentage of total sample weight, were calculated from these raw data. Based on the percentage of the total weight examined and the percentage composition of each taxon in the sample, estimated total weights for each taxon were calculated; these weights are reported in Online Appendix 1 for every sample. This method extrapolates the results from the subsample examined to the entire sample population. The accuracy of this transformation is related to the representativeness of the subsample, a point discussed above. For samples with multiple taxa, more pieces of charcoal were examined and a greater proportion of the total sample was analyzed in order to increase the relative subsample size in those situations. Fragment counts for each taxon as recorded during analysis are also included in Online Appendix 1, but total extrapolated charcoal weight is used in analyses due to the fact that the two measures of weight and count are closely correlated (Chabal et al. 1999; Delhon 2006:13–15; Théry-Parisot et al. 2010), including at Gordion (Miller 2010: fig. 4.3), and weight better represents the original mass of wood combusted. Data previously published by Miller (2007a: apps. 3 and 4, 2010: app. E2) have been transformed using this same methodology to yield comparable data; hence values for Miller's data presented in Table 4.2 and following differ slightly from those reported in summary statistics in her publications (i.e., Miller 2007a:4, 2010:26) and earlier versions of the combined dataset (i.e., Marston 2010:333; Marston and Miller 2014:768).

Analysis and interpretation of the charcoal assemblage required the collection of detailed stratigraphic data for as many of the charcoal samples as possible. This was done at Gordion in the summers of 2006 and 2007 through the use of published excavation reports (Kealhofer 2005a; Sams and Voigt 1995, 1996, 1997, 1998, 1999, 2003, 2004; Voigt 2002; Voigt et al. 1997; Voigt and Young 1999), unpublished excavation reports (including the preliminary stratigraphic analysis [Voigt 2004]), trench notebooks, original excavation photos and plans, and personal correspondence with Mary Voigt during the summer of 2006 and following years. Information recorded for each sample includes provenience data, occupation phase,

Table 4.1. List of archaeological wood taxa identified at Gordion (data from this study; Marston 2003; Miller 2007a, 2010, 2014; Simpson and Spirydowicz 1999).

Family	Taxon	English Name
Aceraceae	*Acer* sp.*†	Maple
Betulaceae	*Alnus viridis**	Alder
Betulaceae	*Carpinus betulus*	Hornbeam
Buxaceae	*Buxus* sp.*†	Boxwood
Chenopodiaceae	cf. *Noaea mucronata*¶	Possible thorny saltwort
Cornaceae	*Cornus* sp.*	Dogwood, cornelian cherry
Cupressaceae	*Juniperus* sp.†	Juniper
Fagaceae	*Quercus* sp.	Oak
Juglandaceae	*Juglans* sp.*†	Walnut
Oleaceae	*Fraxinus* sp.	Ash
Pinaceae	*Cedrus libani**†	Cedar
Pinaceae	*Pinus* sp.†	Pine
Poaceae	*Phragmites* sp.	Reed
Rhamnaceae	*Rhamnus* sp.	Buckthorn
Rosaceae	Maloideae	Hawthorn, pear, apple
Rosaceae	*Prunus* sp.	Apricot, peach, almond, plum
Salicaceae	*Populus* sp.	Poplar
Salicaceae	*Salix* sp.	Willow
Tamaricaceae	*Tamarix* sp.	Tamarisk
Ulmaceae	*Celtis* sp.	Hackberry
Ulmaceae	*Ulmus* sp.	Elm
Vitaceae	*Vitis* sp.*	Grape

*Taxa not identified in this study (from Marston 2003; Miller 1999b, 2010, 2014)
†Taxa identified from furniture in Phrygian tumuli (Simpson and Spirydowicz 1999)
¶Chenopodiaceous wood was identified in this study but not to genus; Miller (2014) identifies fragments of chenopodiaceous wood as possible *Noaea mucronata*.

depositional context, functional use context, and details of the specific excavation circumstances. I compiled a list of structures and activity areas from which substantial quantities of botanical remains (hand-picked charcoal and flotation samples) were recovered and coded each sample accordingly. Contextual analysis of the hand-picked charcoal assemblage includes consideration of four axes of variation: chronological phase, structure/activity area association (i.e., spatial patterning of structures and activities), functional use context (e.g., hearth, building collapse, etc.), and depositional context. Ecological analysis aims to interpret the natural plant associations represented by the species identified archaeologically, while behavioral analysis incorporates contextual data with these ecological data to test behavioral models relating to wood collection strategies, based on the theoretical frameworks described in Chapter 2. These analyses are presented sequentially below, and integrated with other botanical data (presented in Ch. 5) and biogeographic and paleoenvironmental data (from Ch. 3) in Chapter 6 of this work.

Table 4.2. Gordion hand-picked wood charcoal by chronological phase, as percent of total charcoal weight, based on extrapolated taxon representation by sample (see p. 66 for explanation of methodology). "Other" is all other hardwood taxa; "Indet." is unidentifiable fragments. Values of 0 represent presence but less than 0.5% by weight; blank values indicate absence. Sample size and total charcoal weight are given to allow comparison of sample sizes between the datasets. Miller's data (a) is drawn from Miller (2010: app. E2) for all periods save Medieval and Roman, and excludes assemblages from three burned buildings; Roman data come from Miller (2007a), while Medieval data combine samples from Miller (2007a, 2010: app. E2). These data are reduced by removing charcoal from structural collapse contexts and from mixed, indeterminate, or undated contexts to produce only well-dated occupation debris (b). Data from this study are presented for all contexts (c) and just occupation debris (d), following the same reduction method used in (b). Sample YH 35913 is also excluded from (d) as an outlier: it is a Roman pyrotechnic feature used for an unknown industrial purpose that contained a massive quantity of oak charcoal and appears to represent a specific combustion event rather than daily practice. Occupation debris analyzed in this study (d) is combined with Miller's data (b) to produce a complete dataset of all likely fuel remnants (e).

Percent of Total Weight							
a. Miller 2007a, 2010	Pine	Juniper	Oak	Other	Indet.	# of Samples	Total Charcoal Weight (g)
Medieval	74		13	7	1	34	101
Roman	85		15	0	0	16	200
Hellenistic	26	0	65	7	2	36	212
Late Phrygian	25	12	45	16	0	71	366
Middle Phrygian	22	0	74	3		8	173
Early Phrygian	71	14	10	2		16	43
Early Iron Age	14	70	14	0	1	52	363
Late Bronze Age	33	62	4	2		13	118

b. Miller 2007a, 2010, Occupation Debris Only	Pine	Juniper	Oak	Other	Indet.	# of Samples	Total Charcoal Weight (g)
Medieval	75		12	6	1	28	95
Roman	35		65	0		8	45
Hellenistic	18	1	70	9	2	31	144
Late Phrygian	21	11	47	20	0	46	273
Middle Phrygian	22	0	74	3		8	173
Early Phrygian	71	14	10	2		16	43
Early Iron Age	39	23	33	1	2	30	114
Late Bronze Age	14	79	5	2		10	82

c. This Study, Total	Pine	Juniper	Oak	Other	Indet.	# of Samples	Total Charcoal Weight (g)
Medieval	80	6	1	12	1	26	69
Roman	6	2	91	1	0	20	387
Hellenistic	35	4	41	18	1	117	1294
Late Phrygian	70	11	13	3	3	22	261
Middle Phrygian	39	17	41	2	0	16	386

Table 4.2. cont'd

d. This Study, Occupation Debris Only	Pine	Juniper	Oak	Other	Indet.	# of Samples	Total Charcoal Weight (g)
Medieval	83	4	1	11	1	22	60
Roman	39	17	36	7	1	15	49
Hellenistic	39	1	39	20	1	80	1117
Late Phrygian	26	35	22	8	9	18	84
Middle Phrygian	40	17	41	2	0	15	384

e. All Fuel Charcoal (b+d)	Pine	Juniper	Oak	Other	Indet.	# of Samples	Total Charcoal Weight (g)
Medieval	78	2	8	8	1	50	155
Roman	37	9	50	4	0	23	93
Hellenistic	36	1	43	18	1	111	1261
Late Phrygian	22	17	41	17	2	64	357
Middle Phrygian	34	12	51	2	0	23	557
Early Phrygian	71	14	10	2		16	43
Early Iron Age	39	23	33	1	2	30	114
Late Bronze Age	14	79	5	2		10	82

Woodland Ecology and Exploitation

Archaeological wood charcoal records the woodland ecology of the local region contemporary with those deposits, given the complicating factors of human selection and taphonomy. While human selectivity in resource acquisition can result in variation between the prevalence of taxa in the wood charcoal assemblage and in natural forests, examination of the wood species present at a site indicates which ecological zones were utilized as wood sources (Asouti and Austin 2005; Marston 2009; Théry-Parisot et al. 2010). Certain species serve as distinct markers for local plant communities and the presence of these key species has been used to interpret change or consistency in foraging patterns over time (e.g., Bishop et al. 2015; Gremillion 2015; Lentz and Hockaday 2009; Miller 1985, 1999b; Pearsall 1983; Willcox 2002; Wright et al. 2015).

The taxon list for woods identified archaeologically at Gordion (Table 4.1) includes identifications at the family (1 taxon, although possibly identified to species), subfamily (1 taxon), genus (17 taxa), and species level (3 taxa). This includes native and import-ed taxa, although almost all imported woods come from furniture found in the largest burial tumuli of the Middle Phrygian Period: MM, P, and W (Miller 2010; Simpson and Spirydowicz 1999). The wooded area surrounding Gordion contains several phytogeographic zones, as defined by Zohary (1973). This survey and more recent, though less formal, vegetation studies in Central Anatolia (Ch. 3; Miller 2010) provide a regional ecological context for the discussion of wood use at Gordion. I distinguish four woodland community types and place each wood type encountered archaeologically within one or more of these communities. This community ecology approach, based in part on regional vegetation survey, establishes a spatial framework for reconstructing change in wood exploitation over time at Gordion.

Woodland Structure in the Gordion Region

Zohary's *Geobotanical Foundations of the Middle East* (1973) remains the most comprehensive and detailed work describing the natural vegetation communities of Anatolia. Zohary synthesizes early publications and adds his own systematic field transects and other observations to create a detailed picture of Turkish landscapes in the middle decades of the twentieth century and to reconstruct the vegetation history of the area since the Pleistocene. He describes the vegetation of Central Anatolia as a mixture of Xero-Euxinian forest elements and Indo-Turanian steppe, and terms this association "steppe-forest" (Zohary 1973:579). I adopt Zohary's terminology here to discuss the woodland communities surrounding Gordion, although as he notes the name "steppe-forest" is often misleading, as trees are virtually absent in many areas of Central Anatolia. Still, the dominant trees in this area are Euxinian types (most prevalent along the northern Black Sea coast of Anatolia) and Zohary argues that remnant forest stands throughout the area evidence a broader distribution of these trees in a pre-agricultural landscape (Zohary 1973:124). Within the Xero-Euxinian steppe-forest vegetation type Zohary distinguishes specific plant associations, which other scholars might consider fully fledged plant communities in their own right (c.f. Kent 2012; Lortie et al. 2004) in contrast with the steppe-forest "metacommunity" (Leibold et al. 2004). The four primary plant community types found in the Gordion region are described in the following section and compared to evidence from modern botanical surveys of the area and archaeobotanical data from the site itself.

The major tree species of this vegetation type are Euxinian in origin and include three oaks (downy oak, *Quercus pubescens* ssp. *anatolica*; Turkey oak, *Q. cerris*; and pedunculate oak, *Q. pedunculiflora*), three conifers (black pine, *Pinus nigra* ssp. *pallasiana*; Greek juniper, *Juniperus excelsa*; and prickly juniper *J. oxycedrus*), several species in the Maloideae (a cotoneaster, *Cotoneaster racemiflora*; firethorn, *Pyracantha coccinea*; rowan, *Sorbus aria*; hawthorns, *Crataegus laciniata* and *C. monogyna*; pears *Pyrus elaeagnifolia* and *P. syriaca*; and a serviceberry, *Amelanchier integrifolia*) and in the Prunoideae (a wild plum, *Prunus spinosa*, and wild almonds, *Amygdalus orientalis* and *A. webbii*), as well as a hackberry (*Celtis tournefortii*), elm (*Ulmus campestris*), viburnum (*Viburnum lantana*), spindle tree (*Euonymus verrucosa*), and privet (*Ligustrum vulgare*) (Zohary 1973:124). This contrasts with the Indo-Turanian steppe vegetation dominant in the treeless area south of the Gordion region

surrounding and to the west of Tuz Gölü (Zohary 1973:177). Zohary (1973:180) describes three types of steppe: those dominated by broad-leaved plants (maloccophyllous steppe), grasses (grass steppe), or gummy milkvetch (*Astragalus*) plants (tragacanthic steppe). Grasses naturally dominate the steppe-forest areas around Gordion, although most remnant steppe is now highly degraded and characterized by antipastoral plants, those resistant to ruminant grazing due to physical or chemical defenses that render them unpalatable.

In the Gordion region, Miller (2010) and I have recognized four woodland types. I term these montane forest, scrub forest, open steppe woodland, and riparian thicket. The first three fall within Zohary's Xero-Euxinian steppe-forest vegetation community, within which he identifies two "alliances" based around *Quercus pubescens* and *Pinus nigra*, and a number of plant associations within those types based on observations in various areas of Central Anatolia. The riparian thicket type instead corresponds with Zohary's hydrophytic vegetation association of Populetea, dominated by poplar and willow trees (Zohary 1973:602, 607). The following sections describe each of the four woodland types and compare existing published phytogeographical data with observations from my own fieldwork to interpret the ecological and phytosocial implications of the archaeological wood charcoal assemblage from Gordion.

Montane Forest

Montane forest can be best observed today 30 km northwest of Gordion on Hamam Dağı in the vicinity of Mihalıççık and the village of Hamidiye (Fig. 4.2, see color insert; see also Fig. 3.1). It is dominated by black pine (*Pinus nigra*) with an understory that includes downy oak (*Quercus pubescens*), both juniper species (*Juniperus oxycedrus* and *J. excelsa*), and a wild pear (*Pyrus* sp.). This forest is open above 1100 meters above sea level (masl) and dense above 1300 masl. This plant community closely matches Zohary's description of the "Pinion nigrae xero-euxinium association of *Pinus nigra*–*Quercus pubescens* ssp. *anatolica*" (Zohary 1973:581). In this association, identified 30 km south of Ankara (about 80 km east of Gordion) at 1300 masl, Zohary lists the dominant types as *Pinus nigra* and *Quercus pubescens* ssp. *anatolica*, with *Juniperus oxycedrus, Pyracantha coccinea, Sorbus umbellata,* and *Amelanchier integrifolia* as secondary types. Atalay (2001:32) considers this to be the climax forest of Central Anatolia.

Evidence for the use of montane forest wood resources at Gordion is substantial. The archaeological charcoal assemblage at Gordion is dominated by pine and oak, with juniper and Maloideae (potentially including *Pyrus, Sorbus, Pyracantha,* and/or *Amelanchier*) as significant secondary components found following the Early Phrygian period (Table 4.2). As previously argued by Miller (1991a, 1999b, 2010), I believe that the inhabitants of Gordion exploited the high-altitude montane forest extensively throughout the occupation history of the site, as it is the only major source for pine wood. The most proximate source of this wood was likely to have been the same area where montane forest is found today, on Hamam Dağı and, to a lesser extent, Çile Dağ. The inhabitants of Gordion may have selectively cut only pine from the montane forest and procured juniper, oak, and Maloideae trees from scrub forest areas closer to the city, as described below.

Scrub Forest

Dominated by shrubby downy oak (*Q. pubescens*), this forest type is much less dense than montane forest, with open grass steppe areas between stands of trees (Fig. 4.3, see color insert). Today the closest scrub forest can be found around the village of Avşar on the lower slopes of Çile Dağ, about 15 km northeast of Gordion at an altitude of 1000 m. Aside from scrub oak and occasional prickly juniper (*J. oxycedrus*), trees in the area include elm (*Ulmus minor*), wild plum (*Prunus divaricata* ssp. *divaricata*), almond (*Amygdalus* cf. *orientalis*), and hawthorn (*Crataegus* sp.). About 20 km northwest of Gordion, on the lower slopes of Hamam Dağı, lies a different type of scrub forest dominated by *Quercus pubescens* and *Juniperus oxycedrus,* with *J. excelsa,* Russian olive (*Elaeagnus angustifolia*), a hawthorn, a wild pear, and barberry (*Berberis* sp.) also present. Most trees are 3 m in height or shorter; some oak remains less than 0.5 m in height due to grazing pressure.

This scrub forest plant community contains elements of two associations described by Zohary within the "Quercion pubescentis xero-euxinium association of *Quercus pubescens*–*Quercus cerris* stepposum" (Zo-

hary 1973:580). The dominant trees of this association are *Quercus pubescens* and *Q. cerris* in a shrubby habit, along with *Juniperus oxycedrus.* Another association, identified within 10 km of Ankara, includes *Quercus pedunculiflora, Celtis australis, Crataegus laciniata, Prunus spinosa, P. divaricata, Amygdalus balansae, Pyrus elaeagnifolia, P. boissieriana,* and *Colutea cilicica* (Zohary 1973:580). The association of scrub oak and juniper with a few trees of the Ulmaceae, Maloideae, and Prunoideae is common in the region. This oak-juniper association may have expanded as a result of the degradation of climax pine forest (Atalay 2001:32), indicating that forest structure at the time of Gordion's foundation likely included less scrub forest and more montane forest.

Miller has suggested (2010:31) that this scrub oak-juniper association grew closer to Gordion in the past and was, therefore, more accessible to the inhabitants of the ancient city. Oak was the primary fuel wood from the Middle Phrygian through the Roman period (Fig. 4.1), and good quality fuel wood could have been acquired by harvesting deadwood or cutting scrub oak. Juniper is another frequent fuel wood, and scrub *J. oxycedrus* would have been a more accessible fuel source than larger *J. excelsa* trees, which grow more than 20 km from Gordion today. The presence of Maloideae, Prunoideae, hackberry, and elm wood in the archaeological assemblage likely indicates the harvesting of less frequent trees within the oak-juniper scrub forest association. *Colutea* is the only taxon attributed to this association by Zohary that has not been identified archaeologically at Gordion, which may be due to chance, as it is uncommon in the landscape today. The closest I have seen it grow to Gordion today is in the Sakarya River gorge about 20 km north of Gordion, especially in a wash by the village of Oğuzlar close to the confluence of the Sakarya and Ankara Rivers (Fig. 3.1; see Chapter 3 for other forested areas where *Colutea* has been identified).

Open Steppe Woodland

Much of the area surrounding Gordion today consists of open agricultural fields and degraded (overgrazed) steppe. Trees do appear in this landscape, however, as isolated stands or individuals along field edges or in the open steppe (Fig. 4.4, see color insert). These isolated trees are often fruit trees, including hawthorn (*Crataegus pontica*), almond (*Prunus amygdalus*), wild plum (*Prunus divaricata*), and wild pear (*Pyrus elaeagnifolia*), as well as Russian olive (*Elaeagnus angustifolia*), elm (*Ulmus glabra*) and, near irrigated gardens and fields, willow (*Salix*) and poplar (*Populus*) species (Miller 2010:12). Isolated junipers (*Juniperus excelsa*) and tamarisk (*Tamarix* sp.) also appear occasionally in this habitat. In areas not under cultivation, a variety of steppe grasses, antipastoral dicots, and field weeds grow between these scattered trees. Two plant associations described by Zohary within the *Q. pubescens*-dominated alliance are what he terms "wild orchards" (Zohary 1973:580–81), in which scattered fruit trees survive in an otherwise deforested environment. He cites *Pyrus elaeagnifolia*, *Crataegus laciniata*, *Prunus spinosa*, *Cotoneaster racemiflora*, *Amygdalus balansae*, *Rhamnus rhodopeus*, and *Jasminum fruticans* from the area south of Ankara and in the southwest of inner Anatolia, as well as *Noaea mucronata*, a chenopodiaceous shrub (Zohary 1973:580). In all cases, these vegetation associations appear on relatively flat land, often at lower elevations, that is suitable for farming and grazing. These trees are woodland relicts preserved as useful for fruit, shade, or as field boundary markers, characteristic of a heavily modified anthropogenic landscape.

Archaeological wood from Gordion includes many of these taxa as secondary components of the assemblage. Maloideae, *Prunus*, and willow are relatively frequent, while elm and buckthorn are less common archaeologically. Fruit-bearing trees are useful for more than fuel, so their selective preservation within heavily used grazing and agricultural areas is understandable, perhaps accounting for their relative frequency in the landscape compared with their archaeological underrepresentation. This vegetation community provides the closest source of elm and buckthorn to the site of Gordion today. The rare appearance of chenopodiaceous wood in the archaeological assemblage also represents use of this vegetation community, as non-agricultural areas are dominated by chenopodiaceous shrubs today, including *Noaea*, *Kochia*, and *Krascheninnikovia* (see Ch. 3). The lack of significant use of chenopodiaceaceous shrubs for fuel suggests that alternative fuel sources, including wood and animal dung, were sufficiently available and preferred over shrubs such as *Noaea* (Marston 2009; Miller 1999b; Miller and Marston 2012).

Riparian Thicket

Unlike the other plant associations described by Zohary, the riparian thicket is not part of the Xero-Euxinian steppe-forest community. Hydrophilic trees that grow preferentially near water sources are widespread across large areas of the Near East, although tightly concentrated along permanent or seasonal waterways. Zohary describes the Populetea class as dominated by willow (*Salix*) and poplar (*Populus*) species, as well as several species of tamarisk (*Tamarix*) and plane tree (*Platanus orientalis*), ash (*Fraxinus rotundifolia*), oleander (*Nerium oleander*), chastetrees (*Vitex* spp.), Russian olives (*Elaeagnus hortensis* and *E. angustifolius*), and elm (*Ulmus canescens*) (Zohary 1973:602). Zohary also describes a particular plant association along the northern part of the Sakarya River (at 400–500 masl) as including *Tamarix smyrnensis*, *Salix alba*, *Phrgamites communis*, *Clematis vitalba*, and *Vitis vinifera* as dominant trees and shrubs (Zohary 1973:607). Along the Sakarya near Gordion today (Fig. 4.5, see color insert), dominant trees are willow (*Salix* sp.), poplar (*Populus alba*), and tamarisk (*Tamarix* sp.), as well as reeds that grow in the river itself (*Phragmites* sp.). Miller (2010:16) cites informants who observed dense stands of willow, poplar, wild pear, wild apple, elm, reeds, and cattails (*Typha* sp.) along the river before it was channelized in the 1950s. In that period, the riverine thicket was a source of wood for the village of Yassıhöyük (Miller 2010:17).

Willow and poplar provided an important source of wood to the inhabitants of Gordion, and this wood was most likely collected along the banks of the Sakarya and Porsuk Rivers. Although willow occasionally grows as isolated trees in open steppe woodland, it is not a primary component of any of the steppe-forest associations described by Zohary (1973). Today, poplar is often planted for timber harvesting in many parts of Turkey (Hopkins 2003:52) and can be seen as such in Yassıhöyük and Çekerdeksiz, but this practice may be a relatively recent phenomenon. Reeds were used in construction at Gordion from the Early Iron Age (Miller 1999b, 2010) and appear in collapse deposits from the Hellenistic period examined in this study. The riparian thicket may also be the source of the ash, elm, and Maloideae wood from the archaeological assemblage, especially the ash, which is archaeologically rare at Gordion and has not been observed growing in the Gordion region recently.

Imported Woods

Several woods from archaeological contexts at Gordion do not grow locally today and are likely to have been imported. This includes many of the woods used in the furniture of the major Phrygian tumuli, which integrate local juniper, poplar, and pine with imported cedar, boxwood, maple, and walnut (Table 4.1; Simpson and Spirydowicz 1999). Miller (1991a, 2010) tentatively identified *Alnus viridis* in one object, apparently a wooden tray, found on the floor of a burned building of the Early Iron Age. This species is not reported in Turkey, so the tray could be an import from Europe. Two other woods identified archaeologically, *Cornus* sp. and *Carpinus betulus*, are not described by Zohary as part of the Xero-Euxinian steppe-forest community and have not been observed growing locally by myself or Miller (2010). There is no reason to believe these are necessarily imports, however, as both are components of the mixed coniferous Euxinian forest and, thus, do appear in Central Anatolia as well (Zohary 1973:575–77). Davis cites a collection of *Cornus mas* near Ankara (Davis 1972:451) and Ellen Kohler remembered dogwood growing along the Sakarya in the 1950s or 1960s (personal communication 2003). *Carpinus betulus* is more distant today, as a rather mesic Euxinian species (Zohary 1973:365), growing about 60 km north of Ankara in the Karagöl region (Davis 1982:684).

Walnut is another wood that was possibly grown locally, although it is not native to the Central Anatolian steppe-forest. Davis (1982:654) presumes that *Juglans regia* grows natively in northeastern Turkey and cites naturalized distribution of the tree throughout the Euxinian forest but not in Central Anatolia. It is grown for nuts and timber throughout Turkey today and is grown near Gordion as a garden tree. Maple also could have grown in the northern Euxinian forest within 100 km of Gordion, as *Acer campestre* is an important forest component in northwest Turkey (Davis 1967:513). Boxwood also grows in the northern Euxinian forest and cedar (*Cedrus libani*) can be found in the southern Taurus mountains (Davis 1965:71, 1982:631).

There is a clear negative correlation between the frequency of a wood type in the archaeological assemblage and the distance at which it grows from Gordion. The rarest types found archaeologically are either rare in the surrounding woodlands today or absent entirely. Fully half of the taxa identified within the elite tomb furniture are imported from at least 100 km away, suggesting that the Phrygian kings of Gordion had regional trade connections at significant distances. Within the archaeological deposits investigated in this study, the most common wood types were pine, oak, juniper, willow, poplar, Maloideae, tamarisk, and *Prunus*, which are the most common trees and woody shrubs within 30 km of Gordion today. Wood acquisition was an important activity at Gordion, and trees were likely managed for fuel, fruit, fodder, and construction. Detailed examination of wood anatomy, use, and deposition patterns of archaeological charcoal can provide evidence for such management.

Human Impacts on Woodland Structure

Modern woodland structure provides two important data points for considering past use of wood by inhabitants of Gordion. First, ecological relationships between plant species and biogeographic patterns of species distribution can only be observed in the present, but these ecological relationships are rooted in the evolutionary history of the region. The plant communities observed today are likely to have existed in the past, although their extent and particular species membership may have changed due to human intervention and both climatic and geomorphological changes. Second, present plant communities are the endpoints of a long history of human activity in the landscape of Central Anatolia and, thus, anthropogenic processes observed to affect plant communities today are likely to have had analogous effects in the past. Present woodland ecology and ethnographic observations of human impacts on woodlands provide the perspective needed to interpret primary archaeological data on wood use from Gordion.

Two types of archaeological evidence provide indications of woodland ecology in the past. The more straightforward is taxon representation in the archaeological wood assemblage. Changing frequencies of different wood types over time, when both selection and deposition are controlled as variables, should indicate differences in availability of specific trees. Such inferences have limited interpretive value, however, without an ecological framework for interpretation that connects coincident changes in multiple tree taxa to exploitation of different woodland communities over time. The second line of evidence comes from individual charcoal fragments, where observations of modifications to their wood anatomy disclose the growth and environmental history of the tree from which they came and the wood state at the time of collection. Such details, termed dendrological analyses, provide additional context for understanding environmental pressures, both climatic and anthropogenic, on past woodlands (Dufraisse 2006a; Schweingruber et al. 2006; Théry-Parisot and Henry 2012). Here I report briefly the results of dendrological analysis on select wood fragments from Gordion before turning to broader evidence for diachronic change in woodland structure.

Dendrological Observations

A dendrological approach gives insight into aspects of wood collection (Dufraisse 2006b; Ludemann 2006), management (Bernard et al. 2006; Thiébault 2006), and the "life history" of individual timbers (Asouti 2005; Dufraisse 2006b). As climate, parasites, fungal decay processes, and human forest management practices affect the development of anatomical structures within wood, these traces can often be correlated to the original source of stress based on comparison with modern specimens.

Specific methods for the identification of parasitic deformation on wood charcoal samples (Schweingruber et al. 2006:204–7) including fungal hyphae (Asouti 2005; Carrión Marco 2006), cell collapse from brown rot (Marston 2003), and aphid scars (Dufraisse 2006b) can indicate the burning of weathered or decayed wood. Wood diameter can be estimated through the quantification of annual ring diameters of larger wood charcoal fragments and comparison of the resulting distribution with those obtained through experimental study (Ludemann 2006) or mathematical calculation (Dufraisse 2006b); systematic comparison of large numbers of fragments allows the identification of wood management practices, including various types of trimming, coppicing, and pollarding. The presence of pollarding or other trimming regimes can indicate management of forest resources for fuel wood or fodder (Forbes

1998; Halstead 1998; Halstead and Tierney 1998; Rackham 1980; Thiébault 2005), and this practice has been identified archaeologically from a number of areas and periods in Europe (Haas et al. 1998; Haas and Schweingruber 1994; Thiébault 2005).

I did not practice systematic measurement and recording of all traces of wood anatomy modification during this study but did notice two variable attributes that are markers of stress during the growth period of the tree. The first is compression wood in several pine samples, and the other is narrow growth rings, which suggest human or animal impact on oak growth.

Compression Wood

Various lateral stresses on developing wood can result in the asymmetric effect of growth hormones within a growing stem, resulting in differential wood densities across a growth ring (Panshin and de Zeeuw 1970:288; Schweingruber et al. 2006:52–55). This phenomenon appears in branch wood as a marker of straightening a leaning stem and results in the formation of reaction wood, which comprises tension wood in angiosperms and compression wood in gymnosperms (Panshin and de Zeeuw 1970:288). Compression wood occurs on the underside of a branch and "is usually indicated by eccentric growth rings which appear to contain an abnormally large proportion of latewood in the region of fastest growth" (Panshin and de Zeeuw 1970:289). This effect of compression wood is visible macroscopically in the transverse section, but other seasonal factors can also result in the formation of large quantities of latewood. Microscopic characteristics of compression wood include the modification of the S2 cell wall, resulting in a pattern of helical cavities within the tracheids. These are visible in the transverse section at high magnification as longitudinal cracks and in the radial and tangential section as what appear to be dense spiral thickenings (Panshin and de Zeeuw 1970:292–94). These false spiral thickenings are the most readily visible microscopic feature of compression wood and provide confirmation to macroscopic observations.

Within this study, compression wood of pine was observed in at least five samples (YH 44875, YH 44863, YH 44883, YH 45340, YH 50775), three of which are from deposits in the same area of Operation (Op) 35 that date to the same period. Multiple fragments of compression wood were identified in each sample. Due to the rarity of compression wood in the overall assemblage, I infer that these fragments within each sample result from the degradation of a single piece of burned wood. I also believe that the three samples from Op 35 are likely to have resulted from the combustion of a single branch or tree. There is not likely to be any effect on the fuel value of these pieces, although reaction wood is slightly denser than normal wood, but the formation of compression wood has substantial negative effects on the suitability of timber for construction, as compression wood shows differential longitudinal shrinkage leading to splitting within the timber (Panshin and de Zeeuw 1970:296–97). Panshin and de Zeeuw (1970:297) note that compression wood forms preferentially in trees growing rapidly. Perhaps this explains why compression wood of pine but not juniper was observed in this study, as juniper is a much slower growing tree. Additionally, the frequency of compression wood in a species of pine varies from 6% in relatively straight trees up to more than 65% in very crooked trees (Haught 1958). Perhaps the presence of compression wood in these few contexts indicates the use of crooked pine branches or trunks as fuel, since they would be unsuitable for construction due to their twisting and poor durability. Alternately, if the wood were used in construction and failed, it may have been subsequently burned. Only one of the five samples containing substantial quantities of compression wood can be related to a specific use context: YH 45340 represents the roof collapse of an industrial building from the Late Hellenistic period. One could imagine that differential shrinkage of this beam resulted in the collapse of the roof, but there is a lack of additional information to support such an interpretation.

Growth Ring Reduction

Many oak fragments observed in this study showed a number of tightly packed, narrow growth rings in a small fragment of charcoal. Large earlywood pores, often with only a single row of large pores forming the entire growth ring, dominated such rings. I did not quantify the frequency of these pieces, due to the variable nature of wood growth, but they were not uncommon among the oak pieces examined. Miller noted a similar phenomenon: "usually rings were so narrow

that only large, early-wood pores were present" (Miller 2010:78). Several processes can result in the production of such narrow rings, including natural branch growth, systematic trimming by humans, and grazing by animals. I consider these three options in turn.

Some of these fragments with narrow rings also had small ring diameters, so represent the wood of small branches and twigs. Natural growth results in variations in latewood width (Panshin and de Zeeuw 1970:262) and often produces a general decrease in ring size with age from "juvenile wood" to "adult wood" (Schweingruber et al. 2006:102). In many Gordion fragments, narrow rings were visible on specimens with wide ring diameters, i.e., large branch or trunk wood. Although a gradual decrease in ring width can be the result of this aging process, more sudden and substantial changes in ring width and early- to latewood proportions are caused by environmental stresses, whether climatic or physical.

Climatic change can result in the formation of abruptly narrower or wider growth rings, and it is this variation that permits dendrochronological analysis (Speer 2010). More substantial changes, however, are produced by physical damage to a tree, especially when the crown is removed (Schweingruber et al. 2006:113; Thiébault 2006). Grazing, trimming, coppicing, pollarding, insect or fungal infestations, and natural traumas can result in growth ring diameter reduction and an increase in the proportion of earlywood to latewood within a ring. Experimental and ethnoarchaeological studies have demonstrated the effects of regular trimming (Bernard et al. 2006; Haas et al. 1998; Haas and Schweingruber 1994; Halstead and Tierney 1998; Thiébault 2006) or grazing (Schweingruber et al. 2006:166–67) on patterns of growth ring formation.

Without systematic measurement of the frequency and size of narrow, earlywood-dominated growth rings in oak charcoal fragments from Gordion, it is not possible to distinguish conclusively between the various options described above. I suggest, however, that trimming and grazing are most likely to have affected oak growth in the region of Gordion. Modern oak scrub within 15 km of the site has been heavily affected by grazing in the last century, to the point where some mature trees stand less than 50 cm in height (Fig. 4.3). Recent shifts away from sheep and goat herding in the area, combined with a restriction on fuel cutting, have allowed the trees to regrow to larger heights in much of this region. Alternately, the cutting of oak branches for fodder (Halstead 1998; Thiébault 2005) or fuel on a fixed multi-year schedule produces predominantly narrow, earlywood-dominated growth rings (Bernard et al. 2006). Systematic management of scarce and valuable wood resources close to Gordion could have led to the adoption of such a system and produced oak charcoal with narrow, earlywood-dominated growth rings such as those identified in this study and by Miller (2010). The trend towards increasing use of oak fuel at Gordion, detailed below, may be related to woodland management strategies that rendered oak a more sustainable fuel source than other local woods.

Diachronic Woodland Change

Previous analysis of Gordion wood charcoal by Miller (1991a, 1999b, 2007a, 2010) focused on diachronic changes in wood use at the site and their ecological implications. Miller (1999b, 2010) noticed several trends in the use of wood at Gordion from these samples, excavated in the deep sounding operations of 1988–89 (Table 4.2.a). Miller distinguished occupation debris contexts, where she argues that charcoal represents the cumulative record of daily practice (cf. Asouti and Austin 2005; Théry-Parisot et al. 2010; van der Veen 2007), from burned buildings, in which charcoal represents construction debris and burned furnishings (Miller 2010:25). Full data from the wood charcoal samples analyzed in this study are presented in Table 4.2.c. To limit this analysis to only occupational debris contexts, I follow Miller's methodology by excluding all contexts identified as resulting from structure construction and collapse, as well as mixed deposits of uncertain provenience (Table 4.2.d). To ensure data consistency, I followed the same procedure with Miller's data (Table 4.2.b) and then combined these two restricted datasets to produce a methodologically coherent dataset for occupation debris at Gordion (Table 4.2.e). The "other" category includes all hardwoods that are not oak; the majority of this type is willow/poplar (Salicaceae; Table 4.3). The two occupation debris datasets (from Table 4.2.b and 4.2.d) are compared in Figure 4.6. Diachronic change in wood fuel use is presented using the combined occupation debris dataset (from Table 4.2.e) in Figure 4.7.

Two major trends are apparent in this data. The first is the intense use of juniper as fuel during the Late Bronze Age, followed by a sharp decline in juniper use to about 20% of the assemblage by weight from the Early Iron Age through the Late Phrygian period, and then near absence through the rest of the sequence. The second trend is the general rise in hardwood use from the Late Bronze Age through the Roman period, both a result of increasing oak and other hardwood use, although oak is always a more significant fuel source. Drawing solely on samples excavated in 1988 and 1989, Miller observed a similar trend, albeit with the decline in juniper occurring more gradually from the Late Bronze Age to the Middle Phrygian period, rather than suddenly following the Late Bronze Age (Miller 2010:30). Miller suggests an ecological interpretation of this pattern, arguing that juniper was originally a preferred wood for combustion due to its proximity to the site, but was replaced by oak and other hardwoods as juniper became less available after the Early Iron Age period (Miller 1999b:21, 2010:31–32).

Other evidence from Anatolia supports an ecological interpretation of these coincident trends. Juniper and pine forests of Anatolia typically have an understory of oak; if the conifers are preferentially removed, oak often regrows more quickly (Bottema and Woldring 1984:139; Zohary 1973:280, 556, 581). Construction data, presented in detail below, also support this trend, with juniper used in earlier periods but essentially absent from both domestic and public building collapse contexts following the Early Phrygian period (Kuniholm et al. 2011; Miller 2010:31, 33). In prior work with preliminary versions of this integrated dataset, I have agreed with Miller that ecological succession explains this change in wood fuel use over time (Marston 2012a:391–92; Marston and Miller 2014:768); the complete dataset presented here (Table 4.2.e; Fig. 4.7) continues to support this explanation. Such an explanation is consistent with regional patterns of forest modification associated with the Beyşehir Occupation Phase phenomenon observed across Anatolia between 1500–700 CE (Bottema and Woldring 1990; Bottema et al. 1986; Eastwood et al. 1998; Roberts 1990; see also Ch. 3).

Further investigation of secondary hardwood exploitation provides additional context on this hypothesized ecological transition between early and

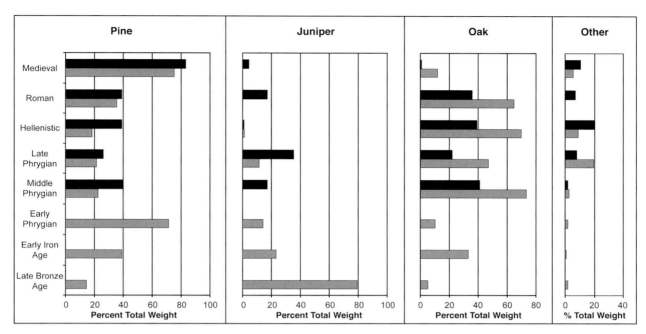

Fig. 4.6. Comparison of Gordion wood charcoal from occupation debris contexts excavated in 1988–1989 and 2004–2005 analyzed by Miller (grey) and 1993–2002 analyzed in this study (black) by occupational phase (data from Table 4.2). Charcoal data are quantified by percent of total assemblage weight for each phase. Note that excavations from 1993 to 2005 did not extend below Middle Phrygian deposits.

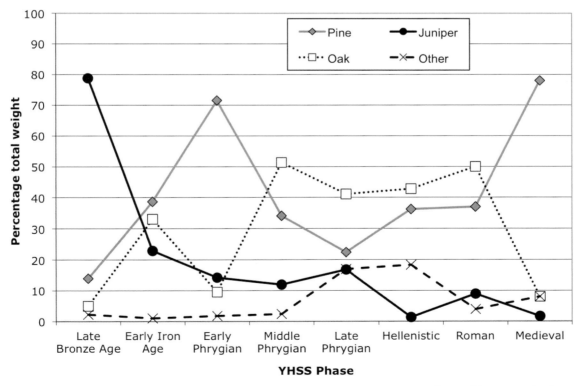

Fig. 4.7. Diachronic change in wood charcoal from occupation debris contexts. Data from Table 4.2.e; "other" includes all non-oak hardwoods.

later periods of occupation at Gordion. If primary forest dominated by oak, pine, and/or juniper is cut down, we expect increased reliance on less numerous "wild orchard" trees (Zohary 1973:363, 376) described in the open steppe woodland plant community above, including both scrub oak (*Quercus pubescens*) and juniper (*Juniperus oxycedrus*) but also elm, buckthorn, hornbeam, the genus *Prunus* (wild plum, almond), and the subfamily Maloideae (wild pear, hawthorn, cotoneaster, rowan, serviceberry). Additionally, we would expect increased and, perhaps, more sustained harvesting of riparian thicket species, which include the fast-growing Salicaceae (willow and poplar), as well as ash and tamarisk. Table 4.3 presents the weights of charcoal from these rare taxa found in occupation debris contexts—presumably burned as fuel—over the occupation sequence at Gordion; these comprise the "other hardwood" category in Table 4.2. By combining the taxa characteristic of open steppe woodland and those from riparian thicket communities, and comparing their frequency over time by both total weight (presented logarithmically to account for large differences in sample weight) and ubiquity (percentage of samples in which any of the taxa from each plant community appear), it is possible to assess whether inhabitants of Gordion relied more on these communities for firewood over time (Fig. 4.8). These two measures show excellent correlation, indicating these are robust proxies for change in wood acquisition patterns. Both the weight and ubiquity datasets

indicate an increase in utilization of both open steppe woodland and riparian thicket from the Late Bronze Age to the Late Phrygian period. Use of riparian thicket continues to increase in both the Hellenistic and Medieval, with an apparent dip during the Roman period, while trees of the open steppe woodland decline, or at most plateau, from the Late Phrygian through the Medieval period (Fig. 4.8). While these data are based on relatively few charcoal fragments for several periods (Table 4.3), their very scarcity prior to the Middle Phrygian period provides evidence for increasing importance over time. These results are consistent with the oak and juniper data that indicate anthropogenic modification of woodland structure and decreased availability of wood becoming more significant determining factors in fuel wood acquisition over time.

Pine wood use differs from the overall pattern described above. Following the Late Bronze Age, pine comprises 20–40% of the total fuel wood assemblage, except during the Early Phrygian and Medieval periods, when it comprises more than 70% of the total fuel assemblage. Miller suggests that pine may have been systematically harvested and imported from more distant montane forests, perhaps in the form of charcoal, as fuel for Early Phrygian Gordion (Miller 2010:32). Miller (1999a, 2010) does not discuss the sudden drop in oak and corresponding increase in pine use in the Medieval period, as this trend is less noticeable in her original 1988–89 data, but is more

Table 4.3. Minor hardwoods from Gordion hand-picked wood charcoal fuel contexts, by ecological community and chronological phase, in grams, based on extrapolated taxon representation by samples analyzed by both Miller (2007a, 2010) and Marston (this study). Data comprise "other" category from Table 4.2.e.

| | Open Steppe Woodland | | | | | Riparian Thicket | | | # of |
	Prunus	Maloideae	Elm	Buckthorn	Hornbeam	Salicaceae	Ash	Tamarisk	Samples
Medieval		0.96	2.98		0.62	8.43			15
Roman			0.62				0.55	2.30	3
Hellenistic	2.90	6.42	0.62			155.49	0.54	64.75	27
Late Phrygian		19.48	6.57	8.72	0.75	11.39	1.24		14
Middle Phrygian	1.62	7.86						3.97	3
Early Phrygian	0.56								2
Early Iron Age							1.13		1
Late Bronze Age			1.86						1

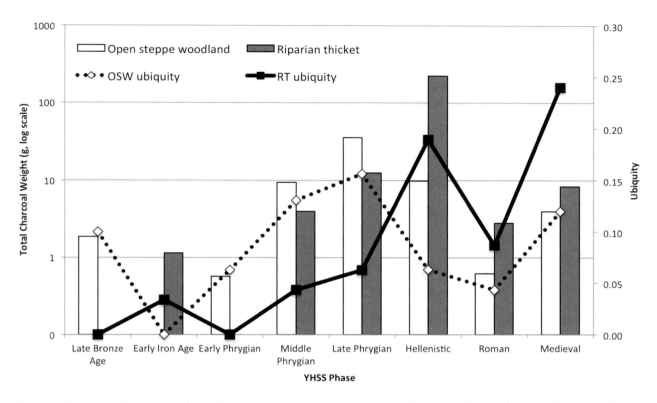

Fig. 4.8. Diachronic change in weights (bars, left axis on logarithmic scale) and ubiquity (lines; right axis) of minor hard-wood species that are components of open steppe woodland communities (*Ulmus*, *Prunus*, Maloideae, *Carpinus*, *Rhamnus*) and riparian thicket communities (Salicaceae, *Tamarix*, *Fraxinus*). Weight data from Table 4.3 and ubiquity data calculated from Marston (2009:2197) and Online Appendix 1.

evident both in her Roman/Medieval report (Miller 2007) and in data published in this study (Table 4.2; Fig. 4.7). In light of the substantially increased sample size reported here (50 occupation debris samples), an ecological explanation for Medieval, and Early Phrygian, pine use patterns is not readily tenable. Instead, there appears to be a selective force that influenced pine use during these two periods, perhaps reflective of differential regional land-use patterns (Miller et al. 2009:921–23). Miller's (2010:32) hypothesis that pine use peaks in the Early Phrygian period due to large-scale fuel wood acquisition in the montane forest is plausible for that period of growing economic and political power at Gordion, but much less likely for the smaller Medieval town. We are left with an interesting question: if wood resources were increasingly hard to come by in the immediate surroundings of Gordion, and dung was used as a substitute fuel, why

would the Medieval remains reflect the use of pine, a distant wood, for the majority of fuel needs? I suggest two possible interpretations.

If local juniper and oak scrub forest had become depleted by the Medieval period, it is possible that adequate supplies of fuel wood were simply unavailable in the near environment of Gordion. Oak and other hardwood use peaks in the Hellenistic and Roman periods, although use of both open steppe woodland and riparian thicket trees increase from the Roman to the Medieval periods (Fig. 4.7). The use of dung as an alternative fuel source is discussed at length in Chapter 5, but dung fuel use appears to peak early in the sequence and decline well before the Medieval period. Perhaps the montane forest became the easiest fuel source to access in the Medieval period, and the abundance of pine charcoal in those levels reflects the need to travel longer distances for wood.

An alternate explanation is that the presence of pine in fuel use contexts is related to easier access to pine than to other, locally abundant, fuel resources. Pinewood may have been readily accessible from collapsed buildings and other contexts at Gordion, and was reused by the Medieval inhabitants for fuel. It is also possible that Medieval trade networks brought pine wood or prepared charcoal to Gordion, making the acquisition of pine possible locally. Either interpretation could explain preference for pine over other fuel sources.

Overall, these sitewide diachronic trends suggest a predominantly ecological explanation for changing wood fuel use over time, driven by sustained human intervention in local woodland structure, albeit with some evidence for preferential wood selection during a few occupation periods (notably, pine in the Early Phrygian and Medieval periods). These data, however, do little to explain the patterning of wood remains within phases and across different areas of the site. In Miller's initial study of the 1988–89 samples (1999), she distinguished between structural collapse and occupational debris, arguing that differential use needs led to the selection of different wood types for specific purposes. Accordingly, I distinguished those context types in the analyses above to separate construction and fuel wood. It is possible, however, to go even further and interpret the 1993–2002 charcoal samples by individual structure or activity area and by patterns of functional use. This detailed level of analysis permits further interpretation of the use context of wood types at Gordion and enables the application of behavioral models to the Gordion charcoal assemblage to further investigate the nature of wood acquisition and use at the site over time.

Spatial and Depositional Patterns of Wood Use

Careful stratigraphic analysis of the context of use and deposition for wood charcoal fragments allows distinction of how wood types may have been selected for specific purposes, allowing the distinction of different human activities and taphonomic processes that are otherwise conflated, obscuring behavioral patterns in the archaeological record (Asouti and Austin 2005; Marston 2009; Théry-Parisot et al. 2010).

Careful stratigraphic work allows the distinction of likely fuel remnants in occupational debris (discussed above) from construction wood remains in building collapse (discussed below), but also allows different feature types, use areas (e.g., domestic vs. industrial), and depositional contexts to be distinguished.

Stratigraphic analysis of portions of the area excavated at Gordion between 1993 and 2002 was completed by Voigt (2004) prior to the initiation of this study. In order to give context to the greatest number of botanical samples possible, I identified a total of 50 structures and activity areas with associated botanical remains. Some of these have been published and described in detail by the excavators (Dandoy et al. 2002; Sams and Voigt 1995, 1996, 1997, 1998, 1999; Voigt 2002; Voigt et al. 1997; Voigt and Young 1999), while others were reconstructed solely from the original excavation notebooks and plans. Voigt assisted me in this contextual analysis, although it should still be regarded as preliminary and subject to future revision upon definitive publication of the 1993–2002 excavation seasons.

In this section, I examine spatial trends in the deposition of wood charcoal: both by use and depositional context across deposit and feature types sitewide and within fifteen discrete structures and activity areas. In the discussion below I distinguish between "use context" and "depositional context" for each sample. "Use context" here refers to the spatial aspects of use as well as the specific behavioral activities of the users, and can be correlated with the "systemic context" of Schiffer (1972, 1976). I differentiate between domestic and industrial use contexts, although the specific activity for which the wood was procured (combustion) may be the same in both areas, in order to assess possible distinctions in fuel use: if the goal is to bake bread or smelt metal, technical aspects of the wood type chosen may differ, leading to different selective practices. "Depositional context," on the other hand, is a result of both wood use and disposal choices and, for this study, is based on feature class and stratigraphic analysis. For example, I regard wood charcoal recovered in and around a pyrotechnic feature to be most likely remains of spent fuel from that feature, and wood from building collapse deposits as construction material from the structure in question, while pits and midden deposits are likely to contain a palimpsest of activities and disposal episodes from the surrounding area of the site.

Analysis by Depositional and Use Context

All samples with sufficient stratigraphic provenience were divided among domestic, industrial, and public use contexts. The identification of context type was made based on published excavation reports and unpublished excavation records and stratigraphic analysis relating these structures and activity areas to specific functions, as described above. Samples were also assigned to categories of depositional context when possible. Frequent categories identified include structure collapse, surfaces, pyrofeatures (hearths, ovens, furnaces), pits, stratified trash deposits, and wash deposits. These data are included with each sample entry in Online Appendix 1.

Preliminary analysis of distinctions between these contexts indicated that depositional context explained the patterning of charcoal within specific contexts more than use context (Marston 2010). Depositional factors appear to outweigh the context of use due to secondary deposition and mixing of originally discrete use contexts. This pattern is best demonstrated by contrasting collapse deposits and pyrofeatures (Table 4.4; Fig. 4.9)—which appear to contain primary deposition of charcoal within their use context—with pits, surfaces, stratified trash deposits, and other secondary deposits (Fig. 4.10). There is noticeably less variation in species composition among collapse and pyrofeature contexts than among other contexts, which supports the interpretation that the former better represent primary wood deposition than the latter.

Another approach to disentangling contexts of use and deposition for wood is to explore spatially and stratigraphically defined structures and activity areas, and to compare charcoal assemblages from areas with similar depositional patterns to reconstruct potential differences in original wood use. Although this limits the dataset from the sitewide data analyzed above, it allows more detailed contextual study of stratigraphy leading to a better understanding of patterns of wood use and disposal at Gordion.

A total of 50 activity areas and structures from which plant remains were collected were identified from three areas of the Citadel Mound (Northwest, Southwest, and Southeast Zones; see Fig. 1.5), from the Lower Town, and from the Outer Town (Fig. 1.4). Thirty-five of these areas have hand-picked charcoal samples securely associated with them. Of these, I

Table 4.4. Hand-picked charcoal samples by depositional and use context, for all periods and structures. "Other" is all other hardwood taxa; "Indet." is unidentifiable fragments. Values of 0 represent presence but less than 0.5% by weight; blank values indicate absence.

Depositional Context	Use Context	# of Samples	Percent of Total Weight							Total Weight (g)
			Pine	Juniper	Oak	Willow/ Poplar	Tamarisk	Other	Indet.	
Collapse	Domestic	5	17	10	54			19		8.3
Collapse	Industrial	9	20	16	59			1	3	36.5
Collapse	Public	5	100							162.6
Pyrofeature	Domestic	16	9	2	68	1	19			315.4
Pyrofeature	Industrial	12	2	1	97					349.1
Pit	Domestic	9	9	14	74			1	2	51.6
Pit	Industrial	6	41	6	28		17	3	4	22.1
Surface	Domestic	20	29	3	65			1	2	89.9
Surface	Industrial	6	44	8	8	36				20.6
Surface	Public	4			80				20	7.4
Trash	Domestic	47	61	15	18	3		2		845.1
Wash	Industrial	4	29	20	9			18	18	16.3
Deposit	Domestic	15	64		11	24				584.8

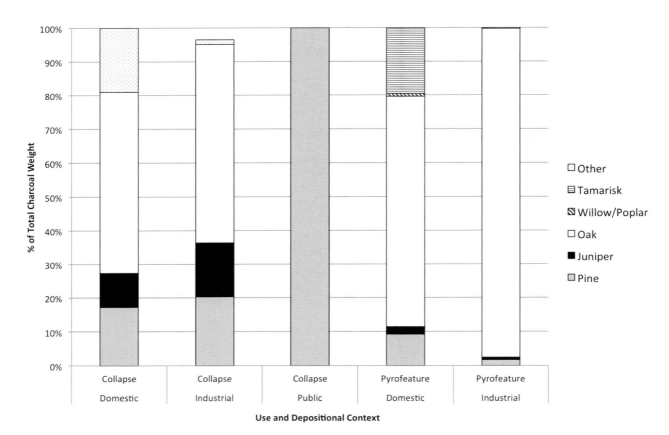

Fig. 4.9. Hand-picked wood charcoal from collapse and pyrofeature deposits, by use context. Wood types are displayed as percent of total charcoal weight for each contextual assemblage; sums below 100% result from the presence of indeterminate charcoal fragments. Data from Table 4.4.

chose 15 areas for detailed analysis: all those with four or more samples per area. This includes 1 context from the Lower Town, 1 from the Outer Town, and the 13 from the Citadel Mound: 2 from the Southeast Zone, 3 from the Southwest, and 8 from the Northwest. For ease of reference, each area was given a code consisting of two letters to designate area and two digits to identify the structure; these codes are particular to this study and are not comparable with other Gordion publications. The association of context, phase, and location for each sample is given in Online Appendix 1, while a brief summary of the structures and activity areas appears in Table 4.5. Summary results of charcoal analysis for these areas, grouped by depositional context, are given in Table 4.6. Detailed stratigraphic analysis of these areas is given in the following section, but first comparative trends in areas with similar depositional patterns are considered.

Pyrotechnic Features

The most noticeable trend in charcoal assemblages from pyrotechnic features is the homogeneity of charcoal from individual features (Fig. 4.11) in contrast to the diversity of wood types found across all such features sitewide (Fig. 4.9). Of the seven discrete pyrotechnic contexts considered here, four contain only one such feature (NW15, SW02, NW02, NW09) and each of those contains a single wood that totals 90–100% of each charcoal assemblage, although the wood type burned varies (pine, oak, or tamarisk). Contexts with multiple pyrotechnic features have more diversity in their charcoal assemblages (NW01, NW02, NW03). I suggest that most fuel loads contained a single wood type and the majority of charcoal excavated from a pyrotechnic feature comes from the most recent burning episode. If these two suppositions hold in the majority of cases,

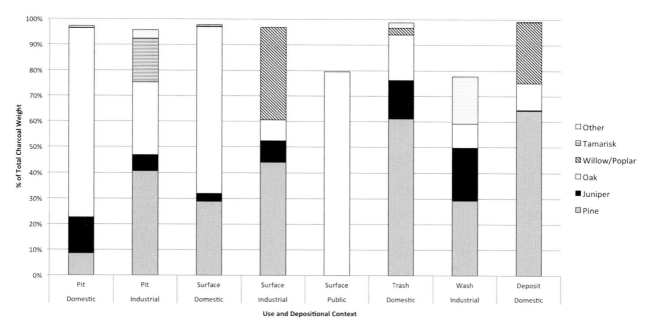

Fig. 4.10. Hand-picked wood charcoal from pits, surfaces, stratified trash, wash, and deposits of mixed origin ("deposit"), by use context. Wood types are displayed as percent of total charcoal weight for each contextual assemblage; sums below 100% result from the presence of indeterminate charcoal fragments. Data from Table 4.4.

Table 4.5. Structure codes for areas with four or more hand-picked charcoal samples. Structure names come from Voigt (2004) and unpublished excavation notebooks.

Structure Code	Name of Area	Contexts Present	YHSS Phase
LT06	RAKS SW Structure	Domestic structure	YHSS 4
NW01	Medieval Ovens	Domestic ovens	YHSS 1
NW02	Medieval Furnace and Hearths	Industrial furnace and domestic hearths	YHSS 1
NW03	Roman Building 1	Domestic structure and courtyards	YHSS 2
NW07	Galatian Building 5 Exterior	Poorly preserved structure exterior	YHSS 3A
NW08	Galatian Building 4	Figurine workshop and collapse	YHSS 3A
NW09	Galatian Building 3	Unroofed metallurgy workshop collapse	YHSS 3A
NW15	Hellenistic Courtyard Complex	Exterior activity area and features	YHSS 3B
NW25	Galatian Building 1	Public building	YHSS 3A
OT02	Domestic Pits	Refuse deposits	YHSS 4?
SE01	Hellenistic House Complex	Domestic structure exterior	YHSS 3A
SE02	Mosaic Building Porch	Public structure collapse	YHSS 4
SE02	Hellenistic Accumulation	Refuse deposits	YHSS 3B
SW02	Abandoned Village House	Domestic structure	YHSS 3A
SW05	Pithouse Accumulation	Refuse deposits	YHSS 3/4
SW08	Middle Phrygian Dump	Refuse deposits	YHSS 5

Table 4.6. Charcoal from each analyzed structure or activity area with four or more samples, as percentage of total weight. Some activity areas are separated into multiple depositional contexts, which may not be synchronic. "Other" is all other hardwood taxa; "Indet." is unidentifiable fragments. Values of 0 represent presence but less than 0.5% by weight; blank values indicate absence.

Percent of Total Weight

Depositional Context	Use Context	Structure Code	Area and Context	Period	n=	Pine	Juniper	Oak	Willow/ Poplar	Tamarisk	Other	Indet.	Total Weight
Pyrofeature	Domestic	NW01	Medieval Ovens	YHSS 1	8	70	12	3	8		6	2	20.7
Pyrofeature	Domestic	NW02	Medieval Hearths	YHSS 1	7	93			4		1	2	4.5
Pyrofeature	Domestic	NW03	Roman Building 1 Hearths	YHSS 2	5	10	19	60		11			21.3
Pyrofeature	Domestic	NW15	Hellenistic Courtyard Complex Hearth	YHSS 3B	1					100			54.9
Pyrofeature	Domestic	SW02	Abandoned Village House Hearth	YHSS 3A	1			100					41.0
Pyrofeature	Industrial	NW02	Medieval Furnace	YHSS 1	5	91			9				27.6
Pyrofeature	Industrial	NW09	Galatian Building 3 Furnace	YHSS 3A	1	100							0.7
Trash	Domestic	OT02	Domestic Pits Trash	YHSS 4?	5	23	30	42	1		4	1	10.6
Trash	Domestic	SE01	Hellenistic House Complex Fill	YHSS 3B	5	25	11	64					49.2
Trash	Domestic	SW05	Pithouse Accumulation	YHSS 4/3	19	72	11	7	4		3	1	479.5
Trash	Domestic	SW08	Middle Phrygian Dump	YHSS 5	11	64	29	5			2	0	226.0
Trash	Industrial	NW07	Galatian Building 5 Exterior Trash	YHSS 3A	7	54	16	9			10	10	30.1
Collapse	Public	NW25	Galatian Building 1 Collapse	YHSS 3A	4	100							5.1
Collapse	Public	SE02	Mosaic Building Porch Collapse	YHSS 4	1	100							157.5
Use and Collapse	Domestic	LT06	RAKS SW Structure Collapse/ Accumulation	YHSS 4	4	4		96					8.4
Use and Collapse	Domestic	NW15	Hellenistic Courtyard Complex Surfaces and Pits	YHSS 3B	12	3		96				1	94.7
Use and Collapse	Domestic	SE02	Hellenistic Accumulation	YHSS 3B	7	5		95					88.3
Use and Collapse	Domestic	SW02	Abandoned Village House Collapse/ Accumulation	YHSS 3A	8	66	0	9	24		0	0	574.4
Use and Collapse	Industrial	NW08	Galatian Building 4 Collapse and Floor	YHSS 3A	7	7	4	71			8	11	6.5
Use and Collapse	Industrial	NW09	Galatian Building 3 Collapse and Floor	YHSS 3A	3	10	22	38	30				24.7
Use and Collapse	Public	NW25	Galatian Building 1 Floor Deposits	YHSS 3A	4			77		2		21	6.4

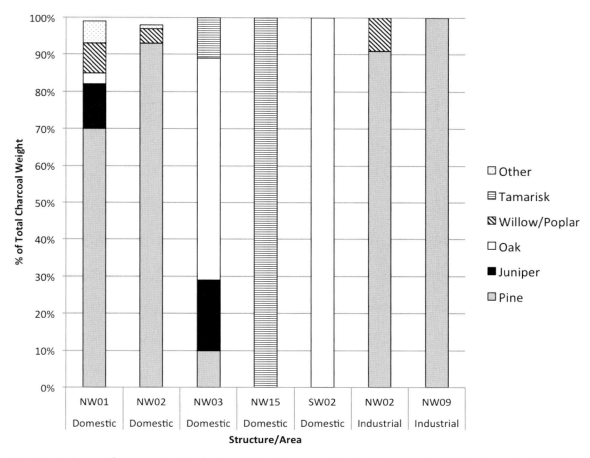

Fig. 4.11. Wood charcoal from pyrotechnic features. Wood types are displayed as percent of total charcoal weight for each contextual assemblage; sums below 100% result from the presence of indeterminate charcoal fragments. Structure codes and data are from Table 4.6.

each individual hearth would contain a vast majority of one charcoal type. However, if different fuels are used in separate burning episodes over a span of time, the accumulated charcoal assemblage would become more diverse. That is especially noticeable in the Roman Building 1 hearths (NW03).

Trash Deposits

The five trash deposits reported here served a variety of functions, as interpreted by the excavators and described in detail in the following section. What these deposits have in common, however, is that they are composed of deliberately deposited waste in secondary depositional context. Despite the disparate origins, both spatial and behavioral, of their contents, they have remarkably similar charcoal composition profiles (Fig. 4.12). Although pine is often the most

common wood type present in these trash deposits, they also all contain juniper, an uncommon wood in the periods studied (Fig. 4.7). At a minimum, each trash deposit contains three wood types (pine, oak, and juniper) and most contain five or more. This pattern holds true across three time periods and four spatially distinct areas of the site, an excellent indication that depositional patterning, rather than similarity in wood use, is the primary factor determining the composition of charcoal deposits at Gordion. Trash deposits include the outcome of multiple wood use episodes and combine fuel, structural, and crafted wood into one archaeological deposit. I agree with Miller (1999b:18) that assemblages of burnt wood produced through repeated action over time are more representative of the current state of wood use and environmental availability than are individual construction beams or pieces of furniture, and my data support

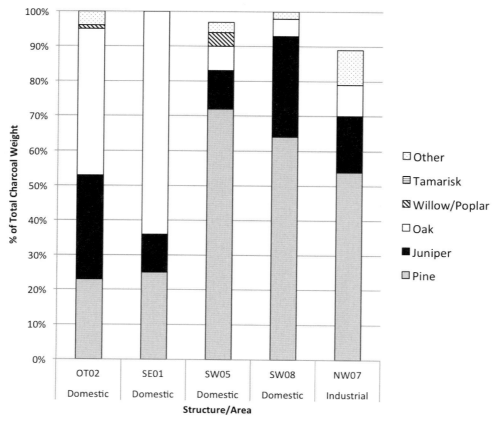

Fig. 4.12. Wood charcoal from stratified trash deposits. Wood types are displayed as percent of total charcoal weight for each contextual assemblage; sums below 100% result from the presence of indeterminate charcoal fragments. Structure codes and data are from Table 4.6.

Chabal's (1992; Chabal et al. 1999) and Asouti and Austin's (2005) arguments that stratified trash deposits are the best source of data for assessment of any diachronic change in wood use or availability.

Structure Construction and Collapse

The interpretation of structure construction and collapse contexts is challenging, in part because many of these contexts contain charcoal from deposits that include structure collapse, but are also potentially mixed with interior deposits. The two collapse contexts that are certain to contain only remains of burned structural members (Fig. 4.13) contain only pine wood, but other contexts with collapse material mixed in with potential occupation debris contain oak, willow/poplar, and some juniper as well (Fig. 4.14). Different wood types appear to have been used in the construction of different structures. The two large public buildings

with definitive collapse contexts, Galatian Building 1 (NW25) and the Mosaic Building (SE02), both used pine exclusively, probably as roof beams (Fig. 4.13). The Abandoned Village House (SW02) also used pine in construction, as evidenced by its direct association with reeds embedded in clay matrix (described below), and more pine was present in this structure than any other domestic or industrial collapse context (Fig. 4.14). The RAKS SW domestic structure (LT06), as well as the industrial Galatian Buildings 3 (NW09) and 4 (NW08), had a majority of oak in their collapse deposits, making oak the most likely structural wood. In the case of the RAKS SW structure and Galatian Building 4, oak comprises nearly the entire assemblage and so appears to have been the only structural wood. Galatian Building 3 has more variety and also a more complicated construction history involving the reuse of several walls from pre-existing structures as described below, perhaps explaining its heterogene-

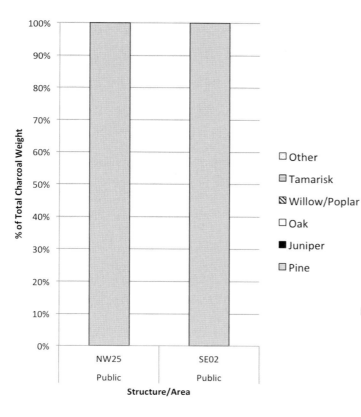

Fig. 4.13. *left* Wood charcoal from structure collapse. Wood types are displayed as percent of total charcoal weight for each contextual assemblage. Structure codes and data are from Table 4.6.

Fig. 4.14. *below* Wood charcoal from mixed structure use and collapse contexts. Wood types are displayed as percent of total charcoal weight for each contextual assemblage; sums below 100% result from the presence of indeterminate charcoal fragments. Structure codes and data are from Table 4.6.

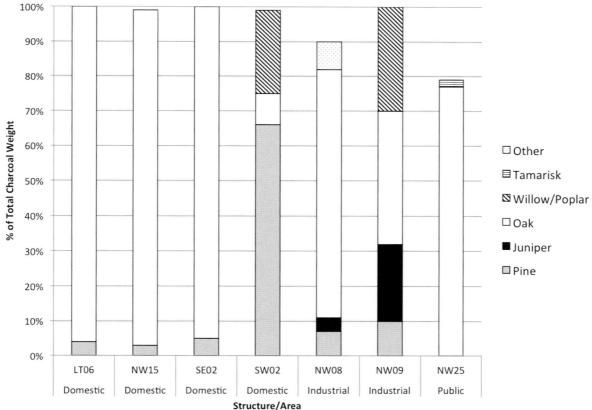

ity. Additionally, this diversity of wood may include non-structural wood use inside the building as well. It is interesting to note that the floor and surface deposits of the Galatian Building 1 (NW25) and Hellenistic Courtyard Complex (NW15), as well as Hellenistic accumulation atop the Mosaic Building Porch (SE02), are nearly entirely oak. This indicates that oak was used on a regular basis in both domestic and public contexts for non-structural purposes, both for fuel (as discussed earlier) and potentially for furniture or tools as well.

Analysis by Structure/Activity Area

Additional perspective on the structures and activity areas used in the contextual analysis above is provided by detailed stratigraphic analysis of each area. Here charcoal deposition within each of the 15 areas listed in Tables 4.5 and 4.6 are described in detail, with deposits listed in alphabetical order.

LT06–RAKS SW Structure

This Late Phrygian semi-subterranean structure was excavated in 1995 in the Lower Town Area B operations (Sams and Voigt 1997:479, plan 473). It measures approximately 4 x 4 m square and had a small hearth set in the floor surface. Four charcoal samples come from this structure, three of which were removed from a hard, brown deposit with white flecks that overlay the structure's foundations. The depositional event that this layer represents is unclear, but it may represent melted mudbrick collapse. The fourth sample (YH 47829) comes from a hard, bricky collapse deposit lying just outside the NW wall of the pit structure. All four samples contain oak charcoal, and one includes several pieces of pine charcoal as well. The sample from unambiguous collapse is entirely oak, suggesting that oak was used as a support for the wall. The other samples are likely to represent collapsed structural members, but may instead be fuel remains deposited secondarily within and around the structure; their context remains ambiguous (see also further discussion in Marston [2012b:52–53]).

NW01–Medieval Ovens

This area was excavated in 1995 and 1997 in Op 35 and 44 and includes material from two occupation

phases, YHSS 1B and YHSS 1A. According to preliminary stratigraphic analysis, "these two occupation phases are closely related architecturally and represent a short period of time" (Voigt 2004:2). Although architecture in this area is not well preserved, several recognizable pyrotechnic features were excavated, including domed ovens and a type of grill. Excavation reports describe these as a "well-preserved oven or tanour [tandır] lined with clay coils and vented to the surface. More unusual was a round griddle heated by a firing chamber beneath its slate floor" (Sams and Voigt 1999:565). These features are dated to the Medieval period based on associated coins and pottery (Sams and Voigt 1999:565).

All eight charcoal samples from this activity area are the preserved contents of these pyrotechnic features (YH 57387, the tandır oven, and YH 44599, another oven) or from the ashy deposits surrounding those features, covering contemporary adjacent surfaces. I interpret these samples as all representing spent fuel from these pyrotechnic features, which appear to be domestic ovens. Support for this interpretation comes from the shape and size of the tandır and coil-built domed ovens, which are common Medieval features at Gordion and comparable to household ovens from the 1200–1400 CE citadel of Taşkun Kale (Goldman 2000:69–70; McNicholl 1983) and are also ethnographically documented from villages in Anatolia (Yakar 2000:153, 164). All eight charcoal samples contain pine, willow/poplar is present in three, and juniper, oak, and Maloideae charcoal appear in one sample each. This variety is consistent with the results of a number of burning episodes over a period of time in which different woods were used to fire the ovens. The large quantity of pine, more than 70% of the samples by weight, suggests that this was the most available wood fuel during the period in which these ovens were used. Another possibility is that the last few firings of the ovens used more pine, so it is more prevalent, but due to its ubiquity through both use phases of this area, it is likely that it was most commonly used wood over the entire use period of these pyrotechnic features.

NW02–Medieval Furnace and Hearths

In contrast to the Medieval ovens in Op 35 and 44 discussed above, the major pyrotechnic feature uncovered in Medieval levels of Op 49 and 51 is clearly industrial in function. This round structure has a stone foundation and remains of a fired tile dome. Iron slag was found within the furnace suggesting that it served as a kiln for roasting ore or a furnace (Sams and Voigt 2004:198). Two smaller hearths lie close to the furnace in what appears to have been a domestic courtyard.

Of the twelve charcoal samples from this area, seven were removed from the interior fill of the furnace and the other five from the exterior hearths and surrounding domestic surfaces. As such, I analyze the charcoal assemblages from these two areas separately. Over 90% of the charcoal from the furnace is pine, although single pieces of willow and hornbeam wood were present in the furnace. Surprisingly for a feature of its size, the furnace contained only 5 g of charcoal; most of the charcoal was found in the adjacent hearths and ash lenses. This is likely due to the quantity of furnace fill removed as 21 separate flotation samples and is not illustrative of the true quantity of wood burned within the furnace. Analysis of the charcoal from the flotation samples will allow re-examination of this feature. The domestic hearths also contain more than 90% pine charcoal by weight, with the remainder consisting of willow charcoal. While it is possible to argue that pine was the preferred fuel for this furnace—at least in the last few firings that the contents of the feature represent—the similarity between industrial and domestic charcoal assemblages in this area make such a conclusion tenuous at best. Pine appears to have been the most used wood type at Medieval Gordion, regardless of context.

NW03–Roman Building 1

Excavations in this area of the Citadel Mound by Rodney Young in 1950 uncovered Roman remains, including the plans of a multi-room house with a central courtyard (Young 1950). Excavations in 1997 and 2001 revealed earlier building phases of this structure, discussed at length by Goldman (2000:71–76). Charcoal remains from this structure were recovered from two small interior hearths and two adjacent surface deposits. One hearth contained oak charcoal only, but the other had substantial quantities of juniper and tamarisk without the presence of oak. Both surface deposits include small quantities of pine, and the ashy deposit adjacent to the second hearth also includes

oak. It is likely that all of this charcoal was produced by combustion of wood in these two hearths. Interpretation of these deposits suggests that a variety of wood types were burned in these domestic hearths, the specific function of which may have been heat or cooking.

NW07–Galatian Building 5 Exterior

This building dates to the third and latest construction phase of the Late Hellenistic Galatian period in the northwest quadrant of the Citadel Mound and was heavily damaged by later Roman construction (Sams and Voigt 1999:564). The function of this building is unknown, although its predecessors were involved in specialty craft production, and the series of trash deposits, hearths, and surfaces to the west and south of the structure appear to be industrial features as well. Charcoal in this area comes from trash deposits to the south of the structure. Charcoal from these trash deposits includes pine, oak, juniper, and Maloideae wood. Of these seven samples, six contain pine, making it the most common and most ubiquitous wood type in those deposits. The use context of the wood is difficult to infer from their depositional context of secondarily deposited trash, but it is most likely that the charcoal was produced in the immediate area and, thus, is the result of industrial burning. The diversity of wood types in this area suggests that the charcoal remains represent an accumulation of spent fuel from repeated combustion events.

NW08–Galatian Building 4

This structure dates to the second construction phase of Late Hellenistic Galatian settlement at Gordion (Sams and Voigt 1999:563). This building was used contemporaneously with Building 1 and Building 3, and incorporates part of the earlier Monumental Hellenistic Wall into its foundation course (Voigt 2004:18, 20). This building appears to have been destroyed at the same time as those two neighboring structures and large quantities of broken pottery covered the floor, lying directly below roof collapse. The final depositional pattern shows unusual disturbance, suggesting that "someone must have systematically searched and perhaps looted the building after its occupants had left" (Sams and Voigt 1999:563). Based on the contemporaneity of this archaeological evidence with a Roman military expedition to Galatia described by Livy, the excavators have hypothesized that this building was ransacked and destroyed by a Roman army under the command of Manlius Vulso in 189 BCE (Dandoy et al. 2002; Goldman 2000:23–24; Sams and Voigt 1998:683, 1999, 2004:197). The function of Building 4 was most likely a specialized terracotta production facility based on fragments of terracotta figurines, a mold for a lion head figurine, small containers of colored pigments and paints, and a set of iron and bone tools with shaped ends that were found on the floor of the building directly underneath collapse deposits (Sams and Voigt 1998:683, 1999).

Seven samples of charcoal come from this building, all of which date to the destruction of the structure: three from building collapse, two from floor surfaces, and two from the fill of pits within the structure. The weight of charcoal from the collapse contexts totals less than 3 g, mainly oak and ash, with some juniper. It is possible that all three woods were used in the construction and/or roofing of Building 4, but due to the small overall sample size, it is also possible that wooden objects burned in the collapse of the building became mixed with wall and roof collapse. Charcoal from the pit deposits is entirely oak; pine and oak are present on the surface. Contextual information is not sufficient to identify the construction material of this building unequivocally.

NW09–Galatian Building 3

This building, contemporaneous with Building 4 described above, dates to the second construction phase of the Late Hellenistic Galatian period at Gordion (Sams and Voigt 1999:563). Building 3 is a reuse of Building 2 (described below) as an unroofed enclosure or courtyard in which metallurgical activity appears to have occurred (Sams and Voigt 1999:563). A pile of iron slag in one corner of the area is associated with a furnace that produced a small quantity of pine charcoal (YH 56773). The other charcoal samples within Building 3 were from wall collapse or rested on the floor surface. These three samples total nearly 25 g of charcoal, and include substantial amounts of pine, juniper, oak, and poplar wood. It is likely that the collapse context samples represent support beams from the collapsed walls, and the large piece of poplar resting on the floor may be remains of the same.

The unusual design of this unroofed building, using some pre-existing walls and some newly constructed, perhaps explains this diversity of support elements in use at the time when the building burned.

NW15–Hellenistic Courtyard Complex

The Early Hellenistic Period occupation at Gordion is characterized by residential settlement across the Citadel Mound (Voigt et al. 1997:12). In the western half of the Citadel Mound, a number of houses were uncovered with a large exterior space between them, within which a number of ash-filled pits and a basin hearth were built into successive use surfaces (Sams and Voigt 1998:683; Voigt 2004:28; Voigt et al. 1997:12). The thirteen samples from this area yielded large quantities of charcoal, almost 150 g in total. The vast majority of the charcoal from the surfaces, pits, and fill deposits is oak (96%), with a small amount of pine in one sample. The basin hearth contents, however, include 55 g of only tamarisk charcoal, likely representing the burnt remains of one firing episode. The difference between the hearth contents and the surrounding deposits may represent the difference between repeated burning episodes of oak and one large fire of tamarisk. Taking the charcoal from this area as an aggregate would be a mistake, as the tamarisk is not illustrative of routine activities in the area, which generally used oak (cf. van der Veen 2007:986–87).

NW25–Galatian Building 1

This structure is the largest and most impressive of the Late Hellenistic Galatian buildings in the northwest area of the Citadel Mound. This monumental public building was constructed with multicolored ashlar stone foundations and divided from the adjacent residential area by an ashlar-faced wall more than 2 m thick (Sams and Voigt 1999). Within this compound were a number of specialized structures for craft production, including pottery (Building 2) and figurine manufacture (Building 4) and metalworking (Building 3). Building 1 is not only the most carefully constructed of these, but also the largest, over 12 m in length (Sams and Voigt 1999:570). Also notable is its tile roof, unique among buildings from the Late Hellenistic Galatian period (Sams and Voigt 1997:481). Building 1 was built in the initial building phase of this area and refloored in the second building phase (Sams and Voigt 1999:563). It was burned at the end of this building phase, when the roof collapsed. This is contemporary with the destruction of Buildings 3 and 4, and appears to have occurred in 189 BCE when the Roman army of Manlius Vulso visited the site (Dandoy et al. 2002:45).

While this structure is inferred to have a public function due to its size and location, little is known about the activities that occurred within it. Charcoal samples were recovered from primary roof collapse deposits, which contained or directly underlay roof tiles, and floor deposits, which underlay the roof collapse. Four samples from roof collapse contain only pine charcoal. Three samples underlie roof collapse, and these consist of oak and unidentifiable charcoal. A final sample comes from a deposit above the roof collapse dating to the third construction phase of Building 1, which consists of oak and tamarisk charcoal. Based on this analysis, it appears that pine was used in the support beams for the roof, while oak was used within the building. Oak may have been used in vertical structural elements, for tools or furniture, or as fuel wood on the floor of the building when it burned. The use of wood in the third construction phase is stratigraphically unclear.

OT02–Domestic Pits

Several small excavation units were placed in the Outer Town in the 1993, 1994, and 1995 excavation seasons. Operation 43, on a low ridge above the alluvial plain of the Sakarya, was excavated in 1995 and revealed a number of pits with domestic refuse, probably dating to the Late Phrygian period (Voigt et al. 1997:11). One large pit with a mudbrick lining on the top contained large quantities of ash and charcoal at various levels. The contents of this pit included five charcoal samples representing a diversity of taxa. Oak is the most common charcoal type and comprises 42% of the total charcoal weight, but pine and juniper also appear in significant quantities (23% and 30% by weight, respectively). Small amounts of Maloideae and willow/poplar were also present in the pit fill. The fill of this pit appears to be secondarily deposited refuse from multiple use contexts, so contextual analysis of the charcoal function is not appropriate. The variety of wood types present in this context suggests a

number of distinct burning episodes contributed to the fill of this pit (see also further discussion in Marston [2012b:53]).

SE01–Hellenistic House Complex

In the southeastern area of the Citadel Mound, excavations in 1995 uncovered part of a domestic complex including parts of several rooms and an open courtyard dating to the Late Hellenistic period. Several charcoal samples were recovered from trash deposits, in both the construction fill of the house complex and a plastered bin within it. The final sample comes from a wash layer above one of the floor surfaces. The construction fill contained nearly equal amounts of pine, oak, and juniper charcoal, while the bin contents and wash were almost exclusively oak. It is not possible to reconstruct the use contexts of the wood that produced this charcoal, but the variety of taxa in these trash deposits again suggests that they are the product of multiple discrete burning episodes.

SE02–Mosaic Building Porch

The Late Phrygian Mosaic Building was originally excavated by Young in the early 1950s (Young 1953), but excavations in 1995 uncovered a previously unknown porch along the north side of the west end of the structure (Voigt et al. 1997:9; Voigt and Young 1999:223). While the construction date and function of this space relative to the main building are uncertain, the Mosaic Building itself was an administrative building used during the Late Phrygian period as a part of Achaemenid control of Gordion (Voigt et al. 1997). Charcoal samples from this area include one massive charcoal deposit just outside the colonnade of the porch. This sample (YH 50730) appears to represent roof collapse from the porch and consists of more than 150 g of pine charcoal. One fragment of pine charcoal still bears tool marks and was shaped into a notched beam 4.1 cm in depth and more than 8 cm in width (the edge is broken). I believe that all the pinewood from this context is remnants of the same or similar beams. The other seven samples from this area are all wash deposits overlying the porch fragments, likely trash from the Early Hellenistic period. These deposits are almost entirely oak, with one piece of pine charcoal amounting to 5% of the total charcoal weight for these trash layers (see also further discussion in Marston [2012b:51–52]).

SW02–Abandoned Village House

Excavations at the southwestern end of the Citadel Mound uncovered a well-stratified sequence of domestic occupation levels, including buildings dating to the Late Hellenistic Period (Sams and Voigt 1995:374–75). Nine charcoal samples were taken from the best-preserved structure and include the contents of one hearth within the structure and a number of deposits overlaying the floor surface. These samples likely represent both building collapse and secondarily deposited trash that accumulated in the abandoned structure. The contents of the hearth are entirely oak charcoal, while the accumulated deposits include a surprising diversity of charred wood remains.

Within the accumulated deposits, over 500 g of charcoal were recovered, including pine, juniper, oak, willow/poplar, elm, and ash, as well as carbonized reeds. Pine is the largest component of the charcoal assemblage, above 65%, followed by willow/poplar at 24% and oak at 9%. Juniper, elm, and ash are each present at less than 1% of the total charcoal weight. Four of the eight samples have carbonized reeds present and in two cases multiple parallel reeds were embedded in an inorganic clay or plaster matrix, likely fragments of wall or roof construction. In one case, a piece of pine wood backs the reed-clay matrix at a perpendicular orientation, supporting the theory that structural elements of wood supported lighter walls or, more likely, roofs. It is unclear whether these elements come from the collapse of the building in which they were located, although that is the closest potential source for collapsed roofing. These samples likely represent a combination of collapsed structural elements (certainly the pine and reeds, possibly the willow/poplar and oak) and spent fuel discarded within the abandoned structure.

SW05–Pithouse Accumulation

Excavations in Op 17 at the southwest end of the Citadel Mound identified a semi-subterranean structure that has been reconstructed as a two-roomed pithouse (Sams and Voigt 1997:479; Voigt and Young 1999:224–25). The walls of the superstructure include a base course of mud-plastered stone and packed mud above (Voigt and Young 1999:224–25); the floor is 1.2 m below grade, so the height of the walls was probably less than one meter more. All 19 charcoal samples from this area come from the accumulated collapse and trash within the pithouse, so their specific use context is not possible to determine. Pine is the most ubiquitous wood, appearing in 95% of the samples, while oak, juniper, and willow/poplar each appear in 42% of the samples. Maloideae wood appears in just two samples. By weight, pine is the most common charcoal (74% of total assemblage weight), followed by juniper (11%), oak (8%), willow/poplar (4%), and Maloideae (3%). This diversity again suggests a mixed origin of the samples, although it is possible that pine was used in wall support of this or another nearby collapsed structure (see also further discussion in Marston [2012b:53]).

SW08–Middle Phrygian Dump

Directly below the earliest Late Phrygian structure in Op 17 is a deep pit cut into the existing Middle Phrygian architecture. It was filled with layers of reconstructable pottery, animal bone (some articulated), and metal objects, including iron and bronze arrowheads, lead and gold sheet, bronze horse fittings, a silver ring, and bronze pins (Sams and Voigt 1998:684; Voigt and Young 1999:211). Voigt has suggested that this deposit may represent the remains of a wealthy household that opposed the Achaemenid conquest of Gordion at the end of the Middle Phrygian period (Voigt and Young 1999:211). Charcoal remains from this pit include 11 samples totaling 225 g of charcoal. Pine comprises the majority of the wood charcoal by weight (64%), although juniper is also a significant component (nearly 30%). Oak is rare, appearing in four samples and totaling 5% by weight. Small amounts of *Prunus* and Maloideae charcoal complete the assemblage. The significant presence of juniper is surprising, as juniper use for fuel at Gordion was severely diminished after the Late Bronze Age. This unusual assemblage may include burned furniture or tools that rarely appear in trash deposits elsewhere at the site. Regardless, this charcoal assemblage contains the mixed debris from several use contexts, thus, its contextual interpretation remains elusive (see also further discussion in Marston [2012b:53]).

Wood Selection and Foraging

Detailed contextual analyses of the type presented above allow the identification of wood use in the construction and maintenance of individual structures, while sitewide depositional analyses and diachronic perspectives on the disposal of different wood types also shed light into broader patterns of wood selection behavior in the past. As discussed in the initial section of this chapter, a theoretical basis for understanding patterns of wood use and deposition that create archaeological charcoal assemblages has been only developed to a limited extent, with the "principle of least effort" (Shackleton and Prins 1992) and the approach of the Montpellier school (Asouti and Austin 2005; Chabal 1992; Chabal et al. 1999; Théry-Parisot et al. 2010) serving as the only core conceptual models for the field. In an article (Marston 2009), I suggested that human behavioral ecology provided an alternative framework for assessing patterns of human behavior in wood selection through taphonomically informed analysis of wood charcoal remains from archaeological sites. I also argued that such a model provided a mechanism for testing the validity of the principle of least effort—typically just used as an assumption—for different behaviors. This article presented a case study drawing on wood charcoal data from Gordion to discern patterns of wood selection and foraging for two wood uses that could be distinguished archaeologically: fuel and construction.

Here, I focus on the analogic process by which core elements of behavioral ecology were applied to create the wood acquisition model and discussion of its implications for understanding wood use at Gordion. I briefly summarize that research and present its core findings related to wood acquisition at Gordion, but readers should consult the article for further discussion of the theoretical implications for wood charcoal studies (Marston 2009).

A Foraging Model for Wood Collection

Assumptions

The foraging model for wood acquisition is based on four assumptions. Some are inherited from previous work in behavioral ecology, while some are unique to the conditions presented here. The accuracy of testing a model is directly related to the ability of the modeler to ensure that the model tests the antecedent theory, that the theory is sound, and that the assumptions of the model accurately relate to the assumptions of the theory and are explicitly stated (Winterhalder 2002:210). The following are the four assumptions of the wood acquisition model proposed here:

1) Humans behave as do other animals in terms of foraging behavior (*generality of behavioral models assumption*)

2) Humans assess costs and benefits of foraging options and generally prefer those outcomes with the greatest benefit/cost ratios (*optimization assumption*)

3) Foraging for non-food resources has the same degree of evolutionary relevance as does foraging for food resources (*search strategy assumption*)

4) Archaeological context can be used to infer the behavioral activities that result in archaeological deposits with a relatively high degree of confidence (*archaeological analogy assumption*)

The *assumption of the generality of behavioral models* to humans as well as animals is the core assumption of human behavioral ecology and is supported by nearly 30 years of work in behavioral ecology (Bird and O'Connell 2006; Winterhalder and Smith 2000). The *optimization assumption* is a component of all foraging models within behavioral ecology. Even costly signaling models, which predict non-optimal behavior, are predicated upon superior reproductive success conferred by that costly behavior (Grafen 1990; Zahavi 1975). Although recent critiques of behavioral ecology challenge the optimization principle as a general explanatory logic for human behavior (Smith 2009a, 2015; Zeder 2014, 2015), advocates still support the broad use of optimization (e.g., Gremillion et al. 2014; Mohlenhoff et al. 2015), and optimization is less controversial as an explanatory assumption in well-defined cases of tradeoffs (see further discussion in Ch. 2).

The *search strategy assumption* is unique to the study presented here. I argue that this is a valid assumption based on analogic reasoning, since the variables involved in the formal diet breadth model are the same as those involved in the acquisition of any natural resource, as discussed in Chapter 2. The evolutionary relevance of applying behavioral ecology models to resources that do not directly measure the currency of reproductive success has been questioned for the currency of caloric intake (e.g., Bamforth 2002) but this is not universally accepted as relevant. Cultural transmission models, such as those of Boyd and Richerson (1985, 2005; Richerson and Boyd 2005), predict that "natural selection can act upon all types of phenotypic variability, whether genetically or culturally transmitted...differential reproductive success ensures that genetic variants associated with successful phenotypes will be passed on to future generation" (Gremillion 1996:184). The currency of calories does not have to be the only proxy for reproductive fitness, all that is required is that the currency be measurable in terms of material benefit (Gremillion 1996:185, 199). Some recent work has moved towards a broader interpretation of the relationship between foraging activities and reproductive fitness in which the caloric value of the resource is not the currency of measure (e.g., Bliege Bird and Smith 2005; Hawkes 1991; Hawkes and Bliege Bird 2002). Selection remains appropriately at the level of individual behavior (Smith 2006; Winterhalder and Goland 1997).

The final assumption, termed here the *archaeological analogy assumption*, is central to the discipline of archaeology, and is met to varying degrees by different data sets (Binford 1967; Fogelin 2007; Schiffer 1976, 1987; Wylie 1985). Certain archaeological contexts are more reliably identified than others, and some regional archaeological traditions are better understood than others. Meeting this assumption requires careful consideration of archaeological context and the selection of only reliably documented archaeological material for analysis. Analogy also requires judicious use of comparative material (archaeological, ethnographic, or historical) and is most reliable when the relevant analogy explains more details better than competing theories, following the principles of inference to the best explanation (Fogelin 2007).

Variables

Foraging logic requires the measurement of several variables related to the availability of a resource. The most important is the currency value of the resource itself, typically defined by caloric value for food resources, although other currencies have been modeled, including protein (Keegan 1986), fats (Binford 1978; Morin 2007), and micronutrients (Belovsky 1978). Other important variables include handling time, which incorporates pursuit and processing, and encounter rate. The relationship among these variables allows the construction of a rank-order list of prey preference (Krebs and Davies 1981; Stephens and Krebs 1986). Archaeologists have used ethnographic data to create such prey lists for human foragers and then tested them against the archaeological record (e.g., Broughton 1997; Diehl and Waters 2006; Zeanah 2004). In creating this foraging model, I distinguished between the three variables of value, encounter rate, and handling time, substituting a quantifiable proxy measure for each. I tested each proxy measure against archaeological data to assess the relative contribution of each variable to the behavioral patterns leading to wood gathering at Gordion.

Wood had a variety of uses at Gordion and any behavioral model must distinguish between the remains of fuel and construction, which are the activities most reconstructed from archaeological context and appear to have been the most common uses of wood at the site (Marston 2009, 2012b; Miller 1999b, 2010). Different currencies are associated with the two activities of construction and combustion. For construction, people need long, straight, durable trunks. For combustion, most important is the energy content of the wood, which is a factor of the density, water content, and resin/oil content of the wood (Hall and Dickerman 1942). Given these parameters, different tree species produce woods that are more or less useful for the two activities. I constructed rank ordered lists of the fifteen potentially local tree taxa identified archaeologically at Gordion (Tables 4.7 and 4.8), based on data from botanical survey, regional biogeography, and experimental analysis (Betts 1913; Campbell 1918; Davis 1965–2000; Fisher 1908; Graves 1919; Hale 1933; Hall and Dickerman 1942; Krishna and Ramaswami 1932; Marsh 2005; Miller 1991a, 1999b; Panshin and de Zeeuw 1970; Parr and Davidson 1922; Reynolds and Pierson 1942; Zohary 1973).

The rank order for fuel woods and construction woods is based on experimental data from American and European species (Table 4.7). The rank order for

Table 4.7. Rank orders for wood value: for fuel, by volume (density-dependent) and by weight (density-independent), based on mean published caloric values of wood taxa (Betts 1913; Campbell 1918; Fisher 1908; Graves 1919; Hale 1933; Hall and Dickerman 1942; Krishna and Ramaswami 1932; Parr and Davidson 1922; Reynolds and Pierson 1942; The Heat Shed n.d.); for construction, based on published durability values (Panshin and de Zeeuw 1970:504–5, 627–29). See Marston (2009) for additional methodological details. Dashes indicate that taxon does not appear in these publications. Height data represents the maximum heights for Turkish species growing in the Gordion area (Davis 1965–2000).

Taxon	Fuel Value Rank (by Volume)	Fuel Value Rank (by Weight)	Construction Value Rank	Maximum Height (m)
Acer	5	3	5	25
Carpinus	1	2	10	25
Celtis	8	5	9	5
Fraxinus	6	-	6	30
Juglans	7	-	1	25
Juniperus	9	1	2	15
Maloideae	2	-	-	10
Pinus	11	7	4	30
Populus	13	6	8	30
Prunus	10	9	3	10
Quercus	3	8	3	10
Rhamnus	12	-	8	10
Salix	14	-	10	30
Tamarix	-	4	-	5
Ulmus	4	-	7	30

fuel woods is an average of heat values given by various sources; the construction rank series is based on published information on resistance to disease, compression, bending, and warping (Panshin and de Zeeuw 1970). The height of the main trunk is also of importance, especially in roofing, and I give the maximum height of the taxon here as a proxy. The straightness of the main trunk is another important variable, but one that I was unable to quantify.

Tree taxa were also ranked by their relative frequency in the landscape and the distance from Gordion at which they grow today (Table 4.8). Two species of oak and juniper grow in the Gordion region, and those suitable for construction (*Quercus cerris* and *Juniperus excelsa*) grow further from Gordion today. As a result, I have differentiated between distance ranks for fuel and for construction woods, as this spatial pattern of tree growth was likely similar in the past, with some minor caveats discussed below (see also Ch. 3).

Testing Wood Foraging Behavior

The charcoal data from Gordion analyzed in this study are grouped into three categories by deposition context: construction collapse, pyrotechnic feature contents, and occupation debris. Occupation debris contexts are defined as discussed earlier, with all contexts coded as "collapse," "construction," "burial," or "mixed," as well as those with unclear provenience, removed from the assemblage. This follows the expectations of other scholars (e.g., Asouti and Austin 2005; Chabal et al. 1999; Miller 1999b) that generally scattered charcoal at a site is representative of fuel use. All areas of the site and occupation periods have been combined for this analysis. Table 4.9 presents these three data sets by total charcoal weight; they are ranked similarly by ubiquity as well (Marston 2009:2197). From these data a rank order of importance was created for each depositional context, which

Table 4.8. Rank orders for modern frequency and distance from the site of Gordion. Data are based on my own survey (Ch. 3) and those of Miller (2010) and Zohary (1973). The difference between the distance ranking for fuel and construction species is based on the location of *Quercus pubescens* vs. *Q. cerris* and *Juniperus oxycedrus* vs. *J. excelsa*, as the latter of each pair, which are suitable for construction, appear at higher elevations further from Gordion.

Taxon	Frequency	Freq. Rank	Fuel Species		Construction Species	
			Distance (km)	Dist. Rank	Distance (km)	Dist. Rank
Acer	Uncommon	3	100	4	100	5
Carpinus	Rare	4	100	4	100	5
Celtis	Rare	4	10	2	10	2
Fraxinus	Rare	4	1	1	1	1
Juglans	Rare	4	100	4	100	5
Juniperus	Common	2	10	2	20	3
Maloideae	Uncommon	3	1	1	1	1
Pinus	Dominant	1	30	3	30	4
Populus	Common	2	1	1	1	1
Prunus	Uncommon	3	1	1	1	1
Quercus	Dominant	1	10	2	20	3
Rhamnus	Rare	4	1	1	1	1
Salix	Common	2	1	1	1	1
Tamarix	Common	2	1	1	1	1
Ulmus	Uncommon	3	1	1	1	1

were compared to ranks predicted by the variables of interest (Table 4.9). Correspondence between the expected and observed rank orders was calculated using a bivariate correlation test and quantified with Kendall's tau-b coefficient, which is an appropriate correlation coefficient for ordinal rank data on a variable scale with ties present in the ranks (Kendall 1938). The values of the tau coefficient range from 1.0 (perfect positive correlation) to -1.0 (perfect negative correlation). Values of Kendall's tau-b and the two-tailed significance of the tests are given in Table 4.9; calculations of Kendall's tau-b and its estimated significance were made using SPSS 15.0 software for Windows.

In all three tests tree frequency was significantly strongly positively correlated (0.788–0.906) to observed charcoal composition at the 0.05 level. Surprisingly, the only other statistically significant rank order correlation is a strong negative correlation between construction collapse woods and distance. While foraging theory suggests that closer woods should be used more frequently and, thus, comprise a greater proportion of the archaeological charcoal assemblage,

this result indicates the opposite, as a result of pine being the top ranked construction wood taxon. I attribute this correlation to the fact that, for construction woods only, frequency and distance ranks were significantly negatively correlated (-0.836, $p < .05$). As a result, the negative correspondence of observed and distance rank orders is the result of the significant positive correlation of observed and frequency rank orders and the significant negative correlation of frequency and distance rank orders.

With the goal of identifying which of the variables tested may have structured wood acquisition practices, I did not attempt to weight these variables as do holistic foraging models, such as the diet breadth model. Based on these tests, I concluded that the inhabitants of Gordion, for both fuel and construction uses, utilized frequent wood types to a far greater extent than rarer wood taxa, regardless of distance to those resources. This finding matches the expectations of the "principle of least effort" (Shackleton and Prins 1992) and the Montpellier school (Chabal 1992; Chabal et al. 1999), as well as the implicit as-

Table 4.9. Bivariate rank order correlation test for wood from (a) construction collapse contexts, (b) pyrotechnic features, and (c) all occupation debris contexts. Archaeological charcoal is ranked by weight; values based on ubiquity rankings are similar (Marston 2009). Values of Kendall's tau-b are given for the pairwise correlation of the observed rank order with each of the modeled rank orders. Starred values are statistically significant at the 0.05 level. Dashes indicate that rank values are not available for that taxon (see Table 4.7).

a. Taxon	Weight (g)	Observed Rank	Construction Value Rank		Frequency Rank	Distance Rank
Pinus	174.90	1	4		1	4
Quercus	25.85	2	3		1	3
Juniperus	6.71	3	2		2	3
Ulmus	0.92	4	7		3	1
Fraxinus	0.84	5	6		4	1
Maloideae	0.30	6	-		3	1
	Kendall's τ-b		0.200		*0.788	*-0.856
	Significance		0.624		0.032	0.024

b. Taxon	Weight (g)	Observed Rank	Density Value Rank	Weight Value Rank	Frequency Rank	Distance Rank
Quercus	562.43	1	3	8	1	2
Pinus	69.04	2	11	7	1	3
Tamarix	61.17	3	-	4	2	1
Juniperus	9.67	4	9	1	2	2
Ulmus	7.29	5	4	-	3	1
Salicaceae	2.91	6	13	6	3	1
Maloideae	1.32	7	2	-	3	1
Carpinus	0.06	8	1	2	4	4
	Kendall's τ-b		-0.333	-0.600	*0.906	-0.124
	Significance		0.293	0.091	0.003	0.688

c. Taxon	Weight (g)	Observed Rank	Density Value Rank	Weight Value Rank	Frequency Rank	Distance Rank
Quercus	1144.54	1	3	8	1	2
Pinus	1068.74	2	11	7	1	3
Salicaceae	191.44	3	13	6	3	1
Juniperus	179.29	4	9	1	2	2
Tamarix	73.35	5	-	4	2	1
Maloideae	22.82	6	2	-	3	1
Ulmus	7.54	7	4	-	3	1
Prunus	4.52	8	10	9	3	1
Carpinus	1.22	9	1	2	4	4
Fraxinus	1.08	10	6	-	4	1
	Kendall's τ-b		-0.222	-0.333	*0.795	-0.249
	Significance		0.404	0.293	0.003	0.357

sumptions of other authors (e.g., Asouti and Hather 2001; Pearsall 1983; Willcox 1974), that archaeological charcoal assemblages mirror (perhaps imperfectly) the frequency of tree species in an environment, since people generally use the most available resources preferentially. These results initially suggest that wood selection was not practiced, which is why a coarse-grained modeling approach alone is not sufficient to understand wood use. Contextual analysis provides another dimension to this analysis.

The two identified roof collapse contexts from this study (Fig. 4.13), as well as those reported by Miller (1999b) and Kuniholm (1977) from earlier excavations on the Citadel Mound and by Liebhart and Johnson (2005) from the Tumulus MM tomb structure, show a clear preference for two woods in the construction of large buildings: juniper and pine. Juniper is used as a major fuel source only during the Late Bronze Age and then abruptly drops off in significance (Fig. 4.7) except in elite burial contexts (Liebhart and Johnson 2005; Simpson and Spirydowicz 1999), where it persists through the Middle Phrygian period. Miller has suggested that this is the result of over-harvesting of the local juniper species for fuel (Miller 2010:31). The negative correlation between distance and construction wood value is a result of a landscape in which people need to go further to find trees of a height suitable for roofing large buildings, a distance of 30 km to the montane forest where pine trees are abundant. Additional insight into the value of pine beams for construction is that many were reused on the Citadel Mound, some hundreds of years after their initial cutting (Kuniholm 1977:48).

Although selection of wood for fuel was not demonstrated on a sitewide basis, many trees infrequent in the present landscape (e.g., *Colutea*, *Pistacia*, and the spiny *Elaeagnus* and *Berberis*) are entirely absent from the archaeological assemblage, as would be predicted by the diet breadth model for low ranked resources outside of the optimal diet breadth. It appears that for fuel use in general, whether assessed through the contents of pyrotechnic features or through general occupation debris contexts across the site, the overriding criterion for wood choice was availability. Distance appears to have been less of an impediment than we might expect, as pine remained a primary fuel source through the Medieval period (Fig. 4.7). The continuing role of pine suggests that the combina-

tion of environmental and economic factors rendered pine a useful fuel source, perhaps as a by-product of construction or as part of a dedicated system of fuel procurement in the montane forest that could potentially have included charcoal fuel production (Miller 2010:32). Ethnoarchaeological research has shown selection of wood for specific uses based on a number of criteria, including taxon but also branch size, moisture content, and fungal alteration (e.g., Henry and Théry-Parisot 2014; Picornell Gelabert et al. 2011; Théry-Parisot and Henry 2012); perhaps pine wood may have been preferred for some of these reasons despite its distance from the site. The importance of cultural preference as determining which species are preferentially adopted or avoided may also play a role in wood selection (e.g., Asouti 2012; Wright et al. 2015), but evidence for such selection is not evident at Gordion for fuel.

Modeling behavioral variables derived from foraging theory allows for a preliminary assessment of how the ancient inhabitants of Gordion utilized their landscape for wood technology needs. As demonstrated here, behavioral models are rendered more interpretable when detailed contextual and ecological data are available. The combination of these approaches can lead to a more comprehensive understanding of past behavioral patterns than any one alone. As foraging models have been applied to other archaeological assemblages of wood charcoal (e.g., Dufraisse 2012; Robinson and McKillop 2013; Rubiales et al. 2011; Salavert and Dufraisse 2014), it has become evident that general models of human behavior aid in explaining the patterning of archaeological charcoal remains. In the case of Gordion, behavioral ecology gives insight into the importance of spatial characteristics of woodland ecology in determining fuel and construction wood selection.

Conclusions on Wood Use at Gordion

Ecological survey in the Gordion region permits the distinction of four woodland types, each characterized by distinct plant communities, that are likely similar in composition, if not location and extent, to those that surrounded Gordion during its occupation. Ecological analysis of charcoal evidence indicates that inhabitants of Gordion structured their woodland environment through selective use of preferred woods for construction and proximate woods for fuel. A reduction in the amount of juniper used as fuel following the Late Bronze Age and subsequent increase in the use of trees from open steppe woodland and riparian thicket communities indicate changes in local forest structure driven by wood use. The spatial analysis of charcoal assemblages across the site indicates that depositional history, rather than use pattern, most distinguishes feature types, but detailed analysis of coherent activity areas and structures still provides a perspective on construction and fuel use at specific times and places. Juniper, and later pine, was preferred for construction of large public and elite buildings, while oak was commonly used in the construction of smaller domestic structures. Evidence for thatching, likely of roofs, is present within structural collapse. Although there is no sitewide evidence for the selection of wood for fuel use, and the principle of least effort appears a valid supposition for fuel acquisition at Gordion, differences in wood use between specific pyrotechnic features do not rule out fuel selection at the scale of the individual fire.

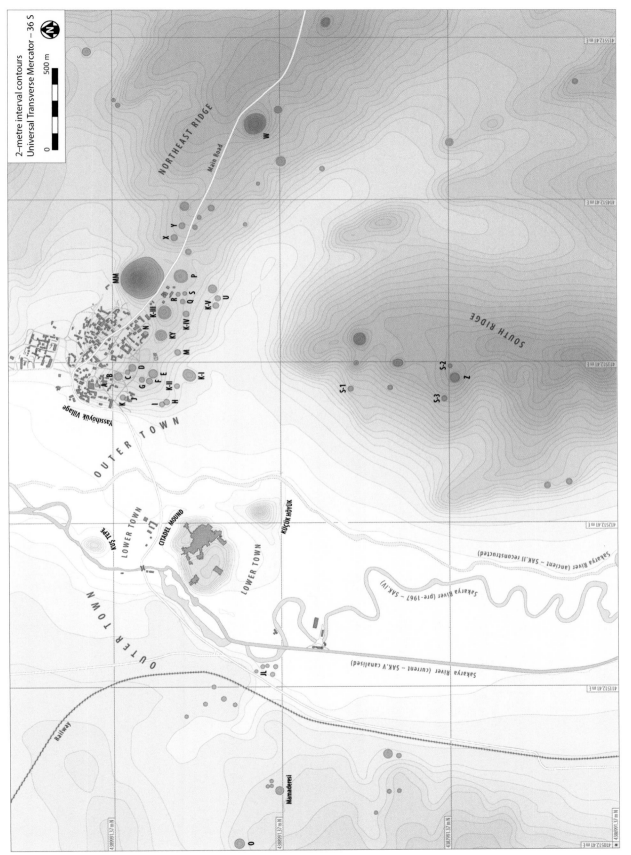

Fig. 1.3. Map of burial tumuli. KI–KV were excavated by the Körte brothers, Tumuli A–Z were named (though not all excavated by Young, and unlabeled tumuli have been surveyed since that date. Image created by Gabriel H. Pizzorno and Gareth Darbyshire, courtesy of Gordion Project Archive, Penn Museum.

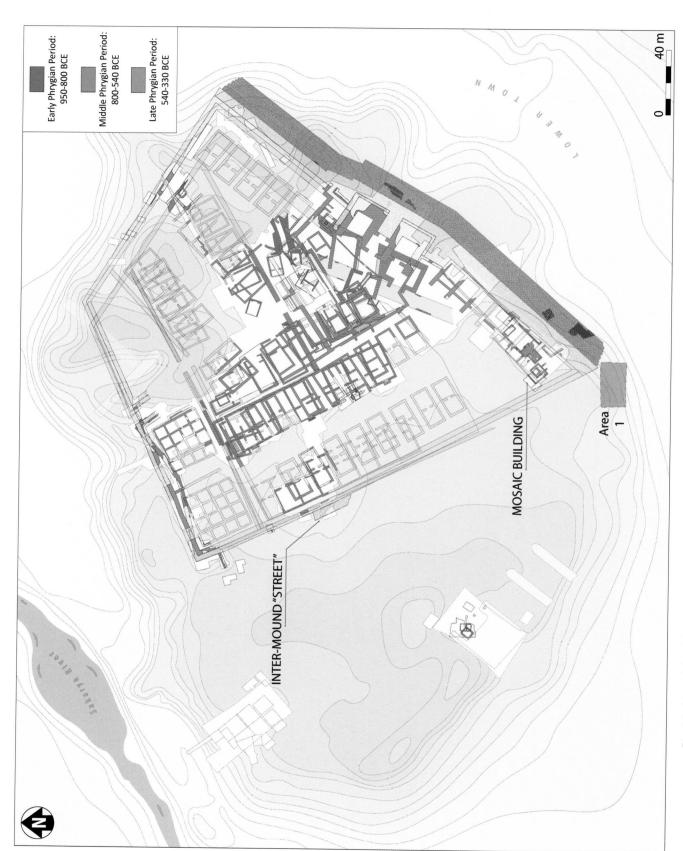

Fig. 1.5. Map of Citadel Mound as excavated in 2015. Image created by Gabriel H. Pizzorno and Gareth Darbyshire, courtesy of Gordion Project Archive, Penn Museum.

Fig. 3.5. Map of areas surveyed by foot within 5 km of Gordion (satellite imagery from Quickbird). Red line indicates informal transect path.

Kizlarkayası

Sakarya River

Porsuk River

Kuş Tepe

Citadel Mound

Conglomerate Ridge

Küçük Höyük

Sakarya River

Tumulus MM

Tumulus P   Tumulus T7

South Gypsum Ridge

N

0   1   2   3   4
km

Fig. 3.8. Mixed forest along river gorge above Beypazarı.

Fig. 3.10. Sakarya riverbank vegetation. Note tamarisk in foreground and willow in distance.

a

b

Fig. 3.11. (a, at left) Protected and (a, at right, and b) unprotected steppe.

a

b

Fig. 3.12. (a) Unirrigated bread wheat and (b) irrigated onion fields. Image (a) was taken in 2007, a particularly dry year.

Fig. 4.2. View north towards Sakarya River valley from Hamam Dağı, dominated by *Pinus nigra*.

Figure 4.3. Scrub forest on the southern lower slopes of Hamam Dağı, dominated by *Juniperus oxycedrus* and *Quercus pubescens*.

Fig. 4.4. Open steppe woodland near Gordion. Trees are elm (*Ulmus glabra*), the ground cover is cultivated wheat.

Fig. 4.5. Riparian thicket along the east bank of the Sakarya River at Gordion, dominated by *Salix* sp. and *Populus alba*.

5

Agriculture, Risk, and Environmental Change

The primary evidence for agricultural practices and environmental change at Gordion comes from animal and plant remains systematically collected during excavations between 1988 and 2005. Paleoethnobotanical remains from the site offer one of the most comprehensive datasets in the Near East in terms of number of samples analyzed and the time depth of deposits sampled, allowing detailed examination of agricultural and environmental change over the course of occupation in the region. In this chapter, I present the analysis of 484 flotation samples excavated between 1993 and 2005, including 220 flotation samples previously published only in summary form (Marston 2012a). I integrate this dataset with zooarchaeological data to explore diachronic trends in diet, agricultural systems, land-use strategies, and environmental transformation.

Questions and Approaches

Archaeological remains of Near Eastern agriculture include its primary durable products—charred plant fragments and animal bones—and secondary records, including cooking features and utensils, food storage installations, and, in some cases, historical documentation of production, trade, and consumption of foodstuffs. Going beyond the simple identification of food production to understanding agricultural systems within their economic and environmental setting, however, usually requires the application of structured modes of inference to connect patterns of behavior observed in the present with those possibly adopted in the past. As a result, the analysis of agricultural systems requires robust models that combine theoretical expectations with empirical datasets analyzed within a framework that accounts for the specific cultural and

environmental settings of the past. In this chapter, I draw on more than a decade of local ecological investigation and models derived from behavioral ecology and resilience thinking, with a particular focus on risk management, to develop metrics for identifying the archaeological signatures of specific agricultural strategies, which are then tested using data from Gordion.

Approaches to Reconstructing Agricultural Strategies

There are multiple lines of evidence that indicate how the inhabitants of Gordion may have practiced agriculture over the millennia during which the site was occupied: archaeological, historical, and ethnographic. The most direct evidence for agricultural practices at Gordion comes from results of those agropastoral production efforts: plant remains of agricultural crops and domestic animal bones. A related line of evidence is the seeds of field weeds that infest agricultural fields under certain conditions. The presence or absence of these seeds may indicate the way in which fields were planted and tended, including whether or not the fields were irrigated. The final lines of evidence are historical and ethnographic accounts of agriculture in the region of Gordion and other semi-arid landscapes of the Near East; these provide concrete data on traditional and historical agricultural adaptations to regional landscapes that may or may not leave archaeological traces. These ethnographic and historical accounts, combined with appropriate theoretical frameworks, generate expectations about the patterning of archaeological remains that can then be tested using the available paleoethnobotanical data, the approach taken in this chapter.

Wild seeds include both ruderal plants, which colonize areas of anthropogenic disturbance, and field weeds. Drawing on regional ecological studies, experimental archaeology, and archaeobotanical research, it is possible to connect groups of wild species with specific agricultural strategies, using them as proxies for agricultural practices. Such an approach can be based on the presence of specific field weed taxa that occupy distinct ecological niches (e.g., Miller 2010; Miller and Smart 1984; Riehl 2014) or on functional ecological characteristics of those weeds, such as maturity season (Bogaard et al. 1999; Charles et al. 1997; Jones et al. 2010).

Historical accounts of agriculture in Central Anatolia begin with Middle Bronze Age records from the Assyrian trading colony of Kültepe/Kanesh (Dercksen 2008a, 2008b). More copious records exist from the Hittite records of the Late Bronze Age, which depict an agropastoral strategy remarkably similar to that practiced in the region into the 20th century (Hoffner 1974; Yakar 2000). Although we generally lack texts from the Early Iron Age of the Eastern Mediterranean (the so-called "Dark Age"), agriculture in the Greco-Roman world is much better understood because we have both literary sources that describe agricultural practices (e.g., Hesiod's *Works and Days*, Cato the Elder's *On Agriculture*, Columella's *On Agriculture* and *On Trees*, and Pliny the Elder's *Natural History*) and numerous documentary records of agropastoral production, trade, and taxation, which enable detailed reconstruction of agricultural practices (Erdkamp 2005; Gallant 1991; Garnsey 1988, 1999; Kron 2000, 2012a, 2012b). Historical records from the Medieval Islamic caliphates describe the introduction of an agricultural system that combined new agricultural techniques with crops introduced from the tropical and subtropical lands of south Asia (Decker 2009; Watson 1974, 1983); innovative ways of cooking these and traditional staple foods were introduced to the region and documented in Seljuk cookbooks (Oral 2002).

The ethnography of agriculture in Central Anatolia is well studied in comparison to many parts of the world. Several monographs document agricultural and land-use practices in the region (Makal 1954; Yakar 2000) and two series of concentrated ethnographic studies have been conducted in the village of Yassıhöyük itself (Gürsan-Salzmann 1997, 2005; METU 1965). In addition to these local accounts, ethnographic studies from other areas of the semi-arid Near East provide valuable insight into how villagers adapt their subsistence strategies to account for environmental, economic, and political changes on both local and regional scales. These ethnographies include rich data on the impact of social and economic change on agricultural practices and on how farmers and families make tradeoffs between farming, herding, and wage labor (e.g., Horne 1982; Martin 1980; METU 1965).

Integration of these multiple lines of evidence allows a much more complete perspective on past agricultural practices, but remains a challenge, as establishing the relevance of each dataset to a particular time and place requires distinct modes of analysis and inference. Recent research on paleoclimate reconstruction that incorporates both paleoenvironmental proxy records and historical documentation (e.g., Haldon et al. 2014; Izdebski et al. 2016) provides one example of such successful integration, as do more focused regional studies that integrate archaeological data with paleoclimate records (e.g., Allcock 2013; Rosen 2007). The key is finding ways to operationalize expectations, whether derived from theory, historical references, or ethnographic analogy, into measurable metrics for archaeological data that allow discrimination among alternative agricultural strategies. This mechanism for identifying agricultural decision making is facilitated through careful selection of theoretical models appropriate to the data types available from a given archaeological site. In prior work, I identified risk management as a powerful perspective for testing theoretical expectations with paleoethnobotanical and zooarchaeological data (Marston 2011, 2015); a risk-management perspective engages both behavioral ecology and resilience thinking, and allows broader application of archaeological cases to contemporary concerns of anthropogenic environmental change (cf. Butzer 1996; Chase and Scarborough 2014; Cooper and Sheets 2012; Jackson et al. 2001; Redman et al. 2004; van der Leeuw and Redman 2002).

Identifying Agricultural Decision Making in the Paleoethnobotanical Record

Paleoethnobotanists have approached agricultural decision making from a variety of perspectives, drawing on bodies of theory as varied as practice theory (Morehart and Morell-Hart 2015; Morell-Hart

2015), entanglement theory (van der Veen 2014), political economy (Fuller and Stevens 2009), human behavioral ecology (Gremillion 2014; Piperno and Pearsall 1998; VanDerwarker et al. 2013), resilience thinking (Marston 2015), niche construction (Smith 2014; Zeder 2012), and co-evolutionary relationships between plants, environments, and social practices (Asouti 2013; Fuller et al. 2010). The key challenge in each of these cases, however, is operationalizing theoretical expectations to produce testable metrics that can be applied to archaeological data to identify the presence or absence of specific agricultural practices. Some theoretical frameworks lend themselves more readily to such applications than others, which require a more qualitative analysis of data.

As discussed earlier in Chapter 2, risk management is an effective concept for operationalizing theoretical expectations for agricultural decision making. It is a central concern of farmers (Cancian 1980; Fleisher 1990; Goland 1993; Halstead and Jones 1989; Walker and Jodha 1986), as well as foragers (Bliege Bird et al. 2002; Kaplan et al. 1990; Smith and Boyd 1990), and theoretically compatible with both foraging models (Gremillion 2014; Marston 2011; VanDerwarker et al. 2013; Winterhalder 1986; Winterhalder and Goland 1997) and resilience thinking (Marston 2015). Critically, however, risk can be conceptually rendered with ready metrics for paleoethnobotanical investigation. Diversification and intensification are the two primary categories of risk-management strategies for agropastoralists (Marston 2011). General concepts such as crop diversification can be tailored to the plants and climates of specific regions. For example, within the semi-arid Near East, diversification metrics include ratios of barley to naked wheat as a measure of drought sensitivity among cereal crops (Marston 2011); bitter vetch to lentil serves as an equivalent among pulses. Diversification among animal species is also drought sensitive, with the ratio of sheep and goats to cattle one useful metric (Miller et al. 2009). Diversity indices, such as the Shannon-Weaver Diversity Index (Shannon and Weaver 1949; Shannon 1948a, 1948b), prove a useful metric for quantifying diversity among animal and plant species. Spatial diversification of agriculture can be assessed using crop weed species that serve as ecological indicators; stable carbon and nitrogen isotopes (Fiorentino et al. 2012; Fiorentino et al. 2015), stron-

tium isotopes (Balasse et al. 2002; Bogaard et al. 2014; Giblin 2009; West et al. 2009), and trace elemental analyses (Ethier et al. 2014) can also tie plants and animals to specific locations of growth. Temporal diversification in resources across the seasons can be detected through age-at-death estimations of herd animal populations (Payne 1973; Reitz and Wing 2008) and, for plants, by the presence of crops planted and harvested at different times of year (McCorriston 2006). Markers of agricultural intensification include the presence of weeds of wet areas as a proxy for irrigation intensity, especially Cyperaceae seeds (Marston 2012a; Miller 2010; Miller and Marston 2012), and ratios of cereals to wild seeds from contexts arising from the burning of animal dung, as a measure of the degree to which domestic animals were fed agricultural products as fodder (Miller 1996, 1997; Miller and Marston 2012; Miller and Smart 1984).

The environmental impacts of these agricultural practices can also be identified through specific metrics for archaeological plant remains, as well as associated geoarchaeological and paleoenvironmental proxy measures. The presence of seeds of specific wild taxa harvested together with those crops is one measure of the level and type of ecological disturbance brought by agriculture (Jones 1988; van der Veen 2007). Ratios comparing the presence of seeds of taxa characteristic of healthy grasslands to seeds of taxa found in severely overgrazed areas also serve as a proxy measure of grassland health and the environmental impact of overgrazing, when applied to contexts resulting from the burning of ruminant dung as fuel (Marston 2012a). Finally, erosion resulting from human land use can be identified using a variety of geoarchaeological and geomorphological methods (e.g., Casana 2008; Dusar et al. 2012; Marsh and Kealhofer 2014; Stinchcomb et al. 2011; Wilkinson 2003).

In the analyses presented below, I develop many of these metrics to achieve three goals: 1) to identify agricultural practices utilized at Gordion, 2) to quantify their relative importance during each period of occupation, and 3) to assess their environmental impact through proxy measures of ecological and landscape change. In these analyses, I focus on the botanical assemblage of seeds and associated plant parts recovered from systematic flotation at Gordion, but also incorporate previously published results of zooarchaeological and geoarchaeological investigations at Gordion and in its surrounding region, many of which are summarized from the prior two chapters.

Methods

The data presented here derive from the analysis of 220 flotation samples excavated at Gordion between 1993 and 2002. Consistent collection methods have been used since 1988, so methods for the recovery of this assemblage are consistent with those described by Miller (2010) for the 1988–89 seasons. I introduced a few minor changes in laboratory methods, mainly relating to how the identified seeds were segregated after initial analysis, but these data were collected to be directly comparable with Miller's and to be maximally comparable with other archaeobotanical studies in Central Anatolia (e.g., Asouti and Fairbairn 2002; Fairbairn 2006; Nesbitt and Samuel 1996). To ensure the maximum comparative value of this study, I include sample-by-sample reporting of flotation samples in the electronic appendices to this volume (Online Appendix 2).

Sample Collection

Excavators at Gordion were instructed to take soil samples for flotation of approximately 10–15 liters (one bucketful of soil) from all archaeological contexts of interest. This included contexts with evidence of burning (e.g., ashy soils, hearth contents), other well-defined contexts (including bins, pits, burials, surfaces, and jars), and control samples from surrounding matrices. Sampling was, however, uneven between trenches and excavation units depending on the nature of deposits uncovered and the complexity of the stratigraphy encountered. All flotation samples were bagged in large, heavy plastic bags.

Nearly all samples were floated during the year of excavation; a few of the latest samples each year were saved over the winter and floated at the beginning of the subsequent season. Nearly all of the samples were floated by Miller (1993–2001); I floated samples excavated in 2002. Sample volume was measured using a graduated container and recorded prior to flotation. Flotation was initially accomplished using a Siraf-style device (Nesbitt 1995:128; Pearsall 2000:44–52) borrowed from the British Institute in Ankara, later

replaced with a new device built for the Gordion project in 1993. The new Gordion device was built on the design of the British Institute one, both of which pumped water from the Sakarya River using a gasoline-powered pump. The heavy fraction was caught in a synthetic window screen mesh (mesh size 1 mm) placed into the device before each sample was added, while the light fraction was caught in a fine (mesh size <0.1 mm) polyester cloth supported in an agricultural sieve. Each sample was floated for 2–20 minutes, depending on size and soil consistency, with an average time of 10–15 minutes. The heavy and light fraction cloths were removed after each sample was floated and placed to dry: the heavy fractions on a plastic tarp in the sun and the light fractions hung indoors. This system was efficient, allowing the processing of up to 30 samples during a full workday with the help of one assistant, and provided excellent recovery: few carbonized plant remains were found in the heavy fractions. Once dried, all heavy fractions were sorted in the field and all light fractions were packaged in plastic bags and shipped to the Penn Museum with the permission of the Museum of Anatolian Civilizations in Ankara and the Turkish Ministry of Culture and Tourism.

Sample Analysis

Of the 1266 flotation samples collected at Gordion between 1993 and 2002, 432 were requested from the Penn Museum for this project. Those samples were selected on the basis of stratigraphic reconstruction, with the aim of acquiring a representative sample of all chronological periods, use and deposition contexts, and excavation areas of the site. Stratigraphic reconstruction was done as for the hand-picked charcoal fragments described in Chapter 4 (see pp. 66–67). Of the 432 samples loaned, 220 were analyzed, representing all chronological phases from the Middle Phrygian (YHSS 5) to the Medieval (YHSS 1) periods and 29 distinct structures and use contexts.

Laboratory protocols for each sample followed those used by Miller (2010). A maximum of 100 ml of light fraction volume was sorted for each sample, so samples that were larger were split using a riffle box sample splitter. The sample was split in half continuously until one fraction was less than 100 ml, and only that portion of the sample was sorted and identified. Context information, including the volume of the original soil sample, was recorded on a data sheet, as was the fraction of the total sample analyzed and the volume and weight of the sample (or subsample, if split). The sample was then poured through a series of geological screens to size-fractionate the sample: a 2-mm screen, 1-mm screen, and 0.5-mm screen. This resulted in four fractions, including the <0.5 mm fraction.

To ensure efficiency and maximum collection of useful data, each fraction was sorted differently. The 2-mm fraction was sorted entirely into seeds, charcoal, other plant parts, dung, other charred material, and residue (dirt, rocks, rootlets, leaves, modern insects, etc.). Bone and shell were also removed at this stage and placed in labeled containers for zooarchaeological analysis. Seeds, charcoal, plant parts, dung, and indistinct charred material were weighed separately at this stage. The presence of modern botanical and other non-botanical materials from each sample are recorded alongside carbonized remains in Online Appendix 2. Seeds and plant parts were then identified as described below; charcoal was reserved for subsequent analysis. In the 1-mm fraction, only seeds (whole and broken fragments) and identifiable plant parts were removed and the remainder (including charcoal) was left as residue. Below 1 mm, only whole seeds (i.e., seeds with the complete shape more or less intact, not small fragments of seeds) were pulled from the residue. Identifiable plant parts, mainly cereal rachis fragments, were pulled from all fractions. Many samples were initially sorted by undergraduate volunteers in the UCLA Paleoethnobotany Laboratory, but I checked every fraction of every sample that was sorted by volunteers to ensure that all identifiable seeds and plant parts had been removed from the residue. Seeds were then identified and recorded at three size fractions: >2 mm, >1 mm, <1 mm. These distinctions between fractions remain on the original data sheets but were summed for the final data reported here.

Identification

The identification of seeds from Gordion flotation samples relied primarily on three resources: the previous work of Miller (1999a, 1999b, 2007b, 2010), which established a baseline for taxa likely to be found in this study; my modern comparative collection of more than 700 Near Eastern and Mediterranean taxa (now housed at Boston University [Boston University 2015]); and illustrated paleoethnobotanical reports from the region (Cappers 2006; Fairbairn et al. 2007; Goddard and Nesbitt 1997; Kroll 1983; Miller 1998, 2010; Nesbitt and Summers 1988; van Zeist and Bakker-Heeres 1982, 1984a, 1984b, 1985; van Zeist et al. 1984), including the *Flora of Turkey* (Davis 1965–2000) and seed identification guides (Cappers et al. 2006; Nesbitt 2006; Renfrew 1973; Schoch et al. 1988). Seeds were examined microscopically at magnifications up to 70x using a stereoscopic light microscope and selected type specimens were examined using a variable-pressure scanning electron microscope (VPSEM) at up to 10,000x magnification when resolution of minute details aided identification. Additionally, I brought unknown seeds to the Ethnobotanical Laboratory at the Penn Museum to refine identifications with the help of Miller and her reference collections.

Most seeds were identified to genus, with some taxa identified to species based on morphology (e.g., domesticates) or phytogeography/monospecificity in Turkey (e.g., *Peganum harmala*, *Cardaria draba*). In many cases, archaeological types fail to match a modern type precisely; such uncertain identifications are marked with a "cf." before the genus or species, depending on the level of certainty.

Quantification and Interpretation

Several basic quantitative measures were calculated for each sample to enhance comparability of the samples with one another and with the results of other paleoethnobotanical studies. The first is the standardized measure *charred density*, the weight of charred remains (charcoal, seeds, and plant parts) >2 mm divided by the volume of soil floated and corrected by the proportion of the sample analyzed if subsampled (Miller 1988:74; Pearsall 2000:196). The charred density measure corrects for initial sample size variation and permits comparison between large and small soil samples (Marston 2014:166–67).

A second type of quantitative analysis is based on relative ratios, with unlike numerators and denominators; these can be used to standardize the numerator of interest against a variable background material, such as charcoal (Marston 2014:168–69; Miller 1988:75; Pearsall 2000:203). I calculate *seed to charcoal* ratios to address the different contributions of agricultural products and wild seeds to the archaeobotanical assemblage (Klinge and Fall 2010; Miller 1996, 1997, 2010; Miller and Marston 2012). The *total seed to charcoal* ratio permits comparison between samples with variable burning of wood: the relative proportion of seeds to charcoal standardizes seed count to burning intensity in the formation history of the sample (Fritz 2005:794; Miller 1999b; Pearsall 2000:203; Wright 2010:54–55). The *cereal to charcoal* ratio similarly standardizes cereal counts (or weights) to charcoal density to compare the amount of agricultural product between samples (Miller 2010:53). The *wild seed to charcoal* ratio in areas where animal dung was burned as fuel is a measure of the relative contribution of dung and wood to the fuel assemblage (Klinge and Fall 2010; Miller 1996, 2010:54; Miller and Marston 2012; Miller and Smart 1984; Miller et al. 2009). The *wild seed to cereal* ratio is a derivative of the latter calculation and, when applied to samples that represent burned animal dung, directly compares the proportion of wild seeds (proxy for grazing) to that of cereals (proxy for fodder) to measure the relative contribution of those two food sources to animal diet (Klinge and Fall 2010; Miller 1997, 2010; Miller and Marston 2012; Miller et al. 2009). All these ratios are density-independent measures, as they are standardized against a common component of flotation samples, so do not need to be further adjusted to account for differences in soil volume or for subsampling, and can be directly compared between samples within and between archaeological sites (Marston 2014).

The quantitative measure of *ubiquity* calculates the frequency of a particular taxon (or group of taxa) across a number of samples, up to an entire chronological phase or site (Marston 2014; Pearsall 2000:212–16; Popper 1988:60–64; Wright 2010). Ubiquity is calculated by dividing the number of samples that contain the taxon of interest by the total number of samples analyzed. This measure allows for the identification of differential intensity of use (or disposal) between different taxa at a site. As cautioned by other authors, however, this measure is only useful when samples are numerous and diverse, and only within a site, not for inter-site comparison (Pearsall 2000:214; Popper 1988:63). Both criteria are met in this analysis and although some samples analyzed here were taken from the same archaeological deposits, the impact of this repetition on taxon ubiquity measures is insig-

nificant due to the large number of samples included in the study. I use ubiquity between taxa or groups of taxa as the most straightforward measure of differences in taxon frequency by area, functional and depositional context, and chronological period.

An additional mechanism for relative measurement of taxon frequency between groups of samples is *diversity indices*, including the Shannon-Weaver index (Shannon and Weaver 1949) and Simpson's diversity index (Simpson 1949). These indices indicate the homogeneity or heterogeneity of a sample or group of samples, incorporating measures of species evenness and richness to calculate diversity between samples. The notable difference between the Shannon-Weaver and Simpson's indices is that Simpson's index is less sensitive to the presence of few rare taxa than the former, so may be more appropriate for the analysis of botanical assemblages that are numerically dominated by a few ubiquitous taxa, but presents problems for the analysis of widely varying diversity between groups of samples. As such, the Shannon-Weaver index is more appropriate for the Gordion dataset. This index calculates sample diversity (H') on a scale of 0 (only one taxon present, no diversity) to a maximum relative to the number of taxa present (multiple taxa, evenly distributed) and can be calculated through a variety of equivalent equations (Peres 2010:29; Popper 1988:67; Reitz and Wing 1999:105; Simpson 1949). Because the index is sensitive to both the number of taxa present and their relative degree of evenness, it is not appropriate for comparisons between sites or other groups of samples that have widely varying numbers of identified taxa. This measure is also sensitive to the level of identification employed, so sites with a higher degree of identification specificity (i.e., those that identify to the genus rather than family) will show higher values of H'. Within a single site, however, the Shannon-Weaver index can indicate differences in evenness between groups of samples (e.g., chronological phases) and show varying trends of resource utilization or specialization across spatial or chronological axes.

Diet and Economy

In order to identify diet and agricultural practices at Gordion, I integrate the results of my study with those of previous paleoethnobotanical (Miller 2007b, 2010), zooarchaeological (Miller et al. 2009; Zeder and Arter 1994), chemical (McGovern et al. 1999), and historical (DeVries 1990; Voigt and Young 1999; Watson 1983) analyses to provide a complete picture of both agricultural production and subsistence at Gordion during its occupation sequence. The major components of the paleoethnobotanical and zooarchaeological assemblages at Gordion are the primary staples of Near Eastern agriculture: cereals, pulses, and domesticated bovids and pigs (Colledge et al. 2004; Miller 1991b, 2010, 2013; Miller et al. 2009; Zeder 1991, 2008, 2011; Zeder and Arter 1994). Although other animals and plants, both domestic and wild, were eaten in varying frequencies at different points in time, the staples of wheat, barley, pulses, sheep, goats, cattle, and pigs formed the basis of agricultural production at Gordion across the entire occupation period of the site. Variations in the percentages of these major taxa and more substantial changes in the contributions of minor taxa, however, give evidence for shifting agricultural and agropastoral strategies related to environmental and cultural changes in the region.

The Gordion Diet

Although agricultural production is not equivalent to diet, botanical remains represent the range of plant foods brought on site, at least in part for human consumption, and, thus (though imperfectly), reflect local diet. Here I combine paleoethnobotanical data from Miller's (2007b, 2010) studies and my own to produce a dataset that spans the entire period of occupation at Gordion as investigated during excavations directed by Voigt (described in Voigt 1994, 2004, 2009, 2011, 2013). This is possible because all flotation samples were collected with a single protocol, Miller and I adopted functionally identical laboratory methods, and our standards for the identification of major crop taxa were consistent. As a result, our studies can be used as a single dataset on economic activity at Gordion. These data have been published in summary form in several articles (Marston 2011, 2012a; Marston and Miller 2014; Miller and Marston 2012), but are presented in full detail for the first time here, and on a sample-by-sample basis in Online Appendix 2.

Cereals

At Gordion, the primary cereals farmed during all phases of occupation were the free-threshing, or naked, bread and hard wheats (*Triticum aestivum/durum*, indistinguishable based on seed morphology), and hulled barley (*Hordeum vulgare*); these seeds are ubiquitous (Table 5.1) and make up the majority of the total identifiable cereal assemblage in every phase (Tables 5.2 and 5.3; Fig. 5.1). Relatively pure concentrations of both of these cereals were found in the Early Phrygian Destruction Level, YHSS 6 (Miller 2010; Nesbitt 1989), and the Burnt Reed House, YHSS 7 (Miller 2010); note that samples from these contexts, treated extensively by Miller (2010:59–61), are excluded from many of the analyses below as pure crop caches such as these distort most cereal-based metrics (Table 5.2). The ratio of barley to naked wheat fluctuates by period and by measure used (seed or rachis fragment), but similar trends are indicated by both measures (Fig. 5.2). This shifting ratio is a marker of changing risk-return tradeoffs made by farmers, as described further below.

Both two-row and six-row barley are present in the Gordion seed assemblage. Their relative presence can be measured by the proportion of barley grains that show twisting (present in the side grains of six-row barley, thus pure six-row barley should have a 2:1 twisted:straight grain ratio) and rachis fragment morphology, as documented for 1988–89 samples by Miller (2010: table 5.6). She notes (Miller 2010:43) that two-row and six-row barley have different starch content, making the two-row variety more suitable for beer-making, while six-row is more valued for fodder but has greater water needs, so any significant shift between the two types would have environmental implications. In Table 5.4, I estimate the relative proportion of six- to two-row barley by period based on Miller's data (2010: table 5.6), both by grain and rachis morphology. These estimates vary widely based on method of estimation (i.e., how indeterminate grains are treated in the calculation) and whether grains or rachis fragments are used to estimate the frequency of six-row barley—estimates for a single period can diverge by more than 75%. Due to this imprecision, although two-row versus six-row barley agriculture has potentially significant risk implications, I do not include these data in later analyses of risk adaptation.

Table 5.1. Ubiquity of all economic plants by period; dashes represent taxa not present (data from Online Appendix 2; Marston and Miller 2014: electronic supplementary material 1; Miller 2010: app. F).

YHSS Phase	Late Bronze Age	Early Iron Age	Early Phrygian	Middle Phrygian	Late Phrygian	Hellenistic	Roman	Medieval
# of samples	32	78	21	43	108	118	27	19
Cereals								
Naked wheat	0.94	0.92	0.62	0.74	0.95	0.88	0.52	0.84
Barley	1.00	0.97	0.67	0.79	0.97	0.85	0.48	0.84
Einkorn	0.63	0.67	0.24	0.05	0.13	0.03		0.05
Emmer	0.09	0.17	0.19	0.02	0.04	0.05		0.11
Triticum sp.	0.50	0.65	0.29	0.16	0.33	0.15		0.21
Rye	0.06	0.03			0.04	0.05		0.11
Oat						0.01		
Broomcorn millet				0.05	0.05	0.02		0.11
Foxtail millet		0.13	0.05	0.07	0.16	0.17	0.11	0.32
Rice								0.37
Unidentified cereal	1.00	0.97	0.62	0.95	1.00	0.95	0.74	0.63
Pulses								
Bitter vetch	0.72	0.47	0.19	0.16	0.43	0.39	0.15	0.32
Lentil	0.19	0.12	0.29	0.23	0.20	0.23	0.11	0.16
Chickpea			0.14			0.01		
Unidentified pulse	0.41	0.27	0.19	0.37	0.37	0.39	0.04	0.32
Fruits and Nuts								
Grape	0.06	0.04		0.07	0.05	0.10	0.07	0.16
Cherry					0.01			
Fig				0.05	0.02			0.05
Pistachio shell	0.25	0.05			0.06			
Almond shell		0.04	0.05	0.02	0.01			
Acorn shell	0.06	0.03				0.02		
Oil and Fiber Plants								
Flax			0.05					
Cotton								0.16
Safflower					0.03	0.03		0.05

Regardless of type, however, barley was brewed into beer at Gordion. Distinctive "beer strainer" vessels are considered a core component of the Early and Middle Phrygian pottery tradition, reflecting beer consumption as a distinctive marker of Phrygian identity, as attested by contemporary Greek sources (Sams 1977). The use of barley for beer was confirmed by chemical analysis of drinking vessels from the royal tomb within Tumulus MM, which indicated that an alcoholic mixture of barley beer, honey mead, and grape wine was drunk from these vessels during the funerary rites for the king (McGovern et al. 1999). The presence of calcium oxalate, which has been identified as a distinctive marker for beer in other archaeological contexts (Michel et al. 1992), confirms the contribution of barley to this beverage. Beer consumption appears to have fallen

Table 5.2. Sample numbers, key statistics, and summary ratios for flotation samples by period. Sample YH 52724 (Hellenistic) is excluded as a far outlier (contains a cache of more than 2000 *Alhagi* seeds); burned crop stores from the Early Iron Age and Early Phrygian are excluded from analyses involving cereal and seed weights (top half of table). "nc" = not calculable (data from Online Appendix 2; Marston and Miller 2014: electronic supplementary material 1; Miller 2010: app. F).

YHSS Phase	Late Bronze Age	Early Iron Age	Early Phrygian	Middle Phrygian	Late Phrygian	Hellenistic	Roman	Medieval
Excluding burned crop stores								
Number of flotation samples	32	66	8	43	108	117	27	19
Barley seed weight (g)	7.90	12.42	0.14	2.17	25.32	11.95	0.37	1.36
Naked wheat seed weight (g)	5.36	7.97	0.08	1.28	15.23	8.85	0.44	1.44
Minor cereal total seed weight (g)	0.23	2.74	0.38	0.01	0.25	0.25	0.00	0.26
Barley to naked wheat ratio (by seed weight)	1.47	1.56	1.75	1.69	1.66	1.35	0.84	0.94
Barley rachis fragment count	264	486		23	1599	735	10	674
Naked wheat rachis fragment count	356	544		33	4366	1835	29	586
Barley to naked wheat ratio (by rachis fragment count)	0.74	0.89	nc	0.70	0.37	0.40	0.34	1.15
Bitter vetch seed weight (g)	1.02	1.10	<0.01	0.07	1.42	0.82	0.05	0.24
Lentil seed weight (g)	0.06	0.06		0.11	0.18	0.70	0.03	0.08
Bitter vetch to lentil ratio (by seed weight)	16.74	18.33	nc	0.64	7.83	1.18	1.61	2.96
Seed >2 mm weight	18.55	33.07	0.47	5.68	58.23	32.16	1.79	5.33
Wood charcoal >2 mm weight	157.28	396.98	5.47	302.27	556.91	569.68	78.22	74.32
Seed to wood charcoal ratio (by weight)	0.118	0.083	0.086	0.019	0.105	0.056	0.023	0.072
Median seed to wood charcoal ratio (by weight)	0.135	0.107	0.075	0.032	0.053	0.043	0.040	0.051
Identified wild seed count	4485	6673	59	1005	14850	10238	366	3574
Total cereal weight	21.31	39.12	0.44	6.70	62.80	36.28	1.30	4.61
Identified wild seed (count) to cereal (weight) ratio	210	171	133	150	236	282	280	775
Including burned crop stores								
Number of flotation samples	32	78	21	43	108	117	27	19
Cyperaceae count	282	560	32	150	2384	1498	117	992
Total identified wild seed count	4485	7169	554	1005	14850	10238	366	3574
Cyperaceae as % of total wild seeds	6.3%	7.8%	5.8%	14.9%	16.1%	14.6%	32.0%	27.8%
Healthy steppe taxa count	1684	1639	11	86	2122	2081	29	671
Overgrazed steppe taxa count	23	32	0	31	789	1544	24	22
Healthy to overgrazed steppe ratio (by count)	73.22	51.22	nc	2.77	2.69	1.35	1.21	30.50
Median healthy to overgrazed steppe ratio (by count)	28.23	20.33	nc	0.57	2.00	1.00	0.25	8.38
Shannon-Weaver index of healthy steppe taxa	0.44	1.43	0.10	0.51	1.88	0.62	0.20	1.42

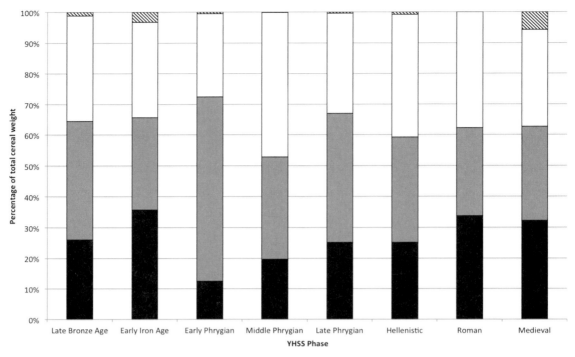

Fig. 5.1. Relative proportions by weight of all cereal taxa from Gordion by period, including contexts from burned buildings (data from Table 5.3). Indeterminate cereal remains are fragments that cannot be confidently identified but the vast majority of these are most likely barley and naked wheat.

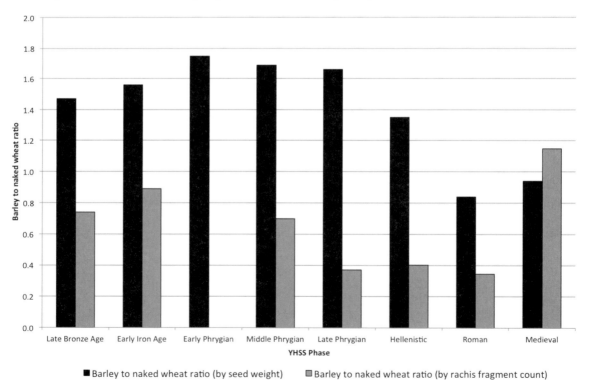

Fig. 5.2. Ratio of barley to naked wheat by seed weight (black) and by rachis fragment count (grey), by period, excluding burned crop stores from the Early Iron Age and Early Phrygian periods. No rachis fragments were identified from Early Phrygian contexts (data from Table 5.2).

Table 5.3. Seed weights for major cultivated cereals and pulses, and counts for millets, by period (data from Online Appendix 2; Marston and Miller 2014: electronic supplementary material 1; Miller 2010: app. F).

YHSS Phase	Late Bronze Age	Early Iron Age	Early Phrygian	Middle Phrygian	Late Phrygian	Hellenistic	Roman	Medieval
Number of flotation samples	32	78	21	43	108	118	27	19
Naked wheat seed weight (g)	5.36	30.67	10.47	1.28	15.23	8.85	0.44	1.44
Barley seed weight (g)	7.90	25.72	50.21	2.17	25.32	11.95	0.37	1.36
Indeterminate cereal seed weight (g)	7.06	26.72	22.69	3.07	19.74	14.07	0.49	1.41
Minor cereal total seed weight (g)	0.23	2.74	0.38	0.01	0.25	0.25	< 0.01	0.26
Broomcorn millet seed count				2	12	2		5
Foxtail millet seed count	1.02	24	3	3	172	93	3	142
Bitter vetch seed weight (g)	1.02	19.07	0.33	0.07	1.42	0.82	0.05	0.24
Lentil seed weight (g)	0.06	0.07	35.46	0.11	0.18	0.70	0.03	0.08
Chickpea seed weight (g)			0.34			0.04		
Indeterminate pulse seed weight (g)	0.35	0.27	0.17	0.18	0.48	0.50	0.05	0.11

Table 5.4. Estimated proportions of 2-row and 6-row barley in samples excavated in 1988–89 and analyzed by Miller; figures calculated from Miller 2010: table 5.6. Note that Roman period samples were not present in contexts excavated in 1988 and 1989.

	Late Bronze Age	Early Iron Age	Early Phrygian	Middle Phrygian	Late Phrygian	Hellenistic	Medieval
Number of whole grains	581	872	9	78	1275	474	59
Number of rachis fragments	264	486	0	8	815	349	87
% 6-row from grain (high)	88	91	0	54	94	78	77
% 6-row from grain (low)	53	48	0	15	51	45	30
% 6-row from rachis	11	7	nc	28	17	19	15

*High estimate removes all indeterminate grains, low estimate assumes all indeterminate grains are straight

in favor of imported wine consumption beginning in the Late Phrygian period, as Gordion became increasingly connected to Mediterranean economic networks and culinary trends (Stewart 2010:48).

Although it is not possible to distinguish the seeds of hard wheat from bread wheat, the ratio of bread to hard wheat can be reconstructed using rachis fragments identifiable to one of the two taxa, which can be distinguished when complete internodes are preserved (Fig. 5.3; see Jacomet [2006a] for identification criteria). These results are limited to a small proportion of the total assemblage and are not equally representative for each period, although only the Early and Middle Phrygian and Roman periods have fewer than 60 observations. Based on these data, hard wheat was only more common than bread wheat during the Late Bronze Age, with bread wheat increasing in frequency after the Late Phrygian period to a dominant role during the Hellenistic, Roman, and Medieval periods.

Other cereals that appear frequently enough to indicate cultivation include einkorn (*Triticum monococcum*), foxtail (*Setaria italica*) and broomcorn millet (*Panicum miliaceum*), and rice (*Oryza sativa*; Table 5.1). Miller (2010:43) argues that einkorn was a crop into the Iron Age and that the rise in the absolute amount of einkorn present during the Early Iron Age may be a marker of Phrygian cultural affiliation and immigration into Central Anatolia, an inference based on the relative frequency of einkorn in Southeastern European contexts, from where the Phrygians are likely to have emigrated during that period (Kroll 1991; Mallory 1989; Vassileva 2005; Voigt and Henrickson 2000a, 2000b). Its ubiquity drops after the Early Iron Age, however, and it is unlikely to have been a primary crop after that time (Table 5.1). In contrast, millets appear infrequently in early phases of occupation but foxtail millet may have been farmed during the Late Phrygian and Hellenistic periods, based on its frequency and ubiquity, and was certainly a crop during the Medieval period, when millets are significantly more ubiquitous than in other periods (Tables 5.1 and 5.3). Rice only appears in the Medieval period, evidently an introduction as part of the "early Islamic agricultural complex" (Watson 1983), and was likely farmed as a summer irrigated crop together with millet (Samuel 2001:402–4), as argued later in this chapter. Rice was still grown at Gordion into the 20th century as an irrigated crop on the Sakarya floodplain (Miller 2010:16).

Other cereals, including emmer (*Triticum dicoccum*), rye (*Secale cereale*), and oat (*Avena sativa*), are unlikely to have been farmed at Gordion in substantial quantities, although rye was farmed in the region as a fodder crop during the 1980s (Miller 2010:67). Emmer is a possible exception, somewhat ubiquitous through the Early Phrygian period (Miller 2010: fig. 5.3), and may have been farmed at that time. Rye and oats appear rarely and in small quantities throughout the sequence (Table 5.1). Rye is a common field weed in wheat fields of Central Anatolia today, possibly a legacy of 1980s farming practices.

Pulses

Pulses are much less common than cereals in the Gordion archaeobotanical assemblage, accounting for only 7% of total economic seeds by count and 5% by weight. The most common pulse, bitter vetch (*Vicia ervilia*), appears in only 39% of samples and lentil (*Lens culinaris*), the other common pulse, appears in less than 20% (Table 5.1). These are, however, the only major identified pulse taxa and comprise nearly all the identified pulses by weight (Table 5.3; Fig. 5.4). Bitter vetch is an early domesticate in Anatolia, appearing at least by the 9th millennium BCE at sites including Çayönü (van Zeist and de Roller 2003) and Gritille (Miller 2002), but also across the ancient Near East and Mediterranean, into southern Europe (Miller 1991b; Miller and Enneking 2014; Zohary and Hopf 2000). Although grown into the 20th century across the eastern Mediterranean, primarily for animal fodder but also eaten on rare occasions, bitter vetch is not cultivated today in the Gordion region (Halstead and Jones 1989:51; Jones and Halstead 1995:111; Miller 2010:13). The bitterness of the pulse makes it unpalatable and it contains the neurotoxin L-canavanine, which affects monogastric animals (i.e., humans and pigs); both limitations can be addressed by leaching the seeds with hot water, yielding an edible and nutritious protein source (Enneking 1995). It is an excellent fodder crop for ruminants, so it is possible that bitter vetch was grown for both human and animal consumption at Gordion. Concentrations of bitter vetch were found in the Early Phrygian Destruction Level (Nesbitt 1989) and the Early Iron Age Burnt Reed House (Miller 2010; reflected in Fig. 5.4) along with other caches of food crops, suggesting some de-

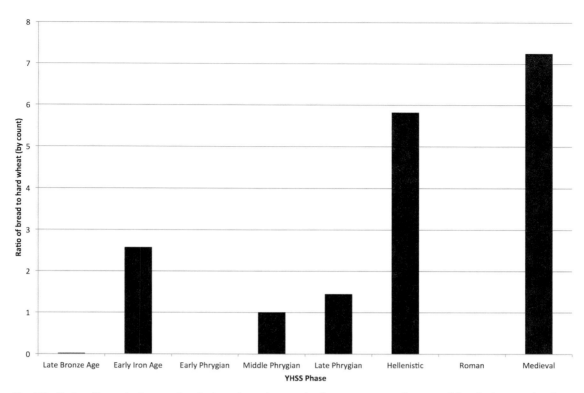

Fig. 5.3. Ratio of bread wheat to hard wheat based on rachis fragment count by period; hard wheat rachis fragments were not identified in Early Phrygian or Roman contexts (data from Online Appendix 2; Marston and Miller 2014: electronic supplementary material 1; Miller 2010: app. F).

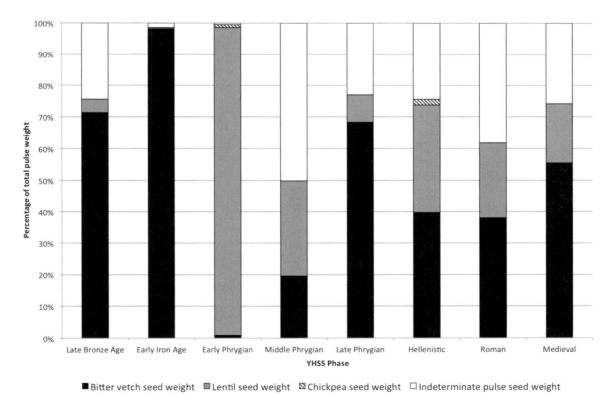

Fig. 5.4. Relative proportions by weight of all pulse taxa from Gordion by period, including contexts from burned buildings (data from Table 5.3). Indeterminate pulse remains are fragments that cannot be confidently identified.

gree of human consumption of bitter vetch at least during earlier periods of occupation.

Lentils require less processing than bitter vetch for human consumption (no leaching) and concentrations from the Early Phrygian Destruction Level (Miller 2010; Nesbitt 1989) confirm the identification of lentil as a food crop. A combination of bitter vetch and lentil seems to have provided the vast majority of pulse cultivation in the region in all periods (Fig. 5.4). Other pulses identified include chickpea (*Cicer arietinum*), a few seeds of which were identified in flotation light and heavy fractions as well as seed concentrations from burned buildings (Miller 2010; Nesbitt 1989), and possible pea (*Pisum sativum*) and grass pea (*Lathyrus sativus*), which seem to represent incidental field weeds or occasional small-scale cultivation rather than regular crops (Tables 5.1 and 5.3). Seeds of fenugreek (*Trigonella foenum-graecum*) are present in a single Medieval sample; this condiment species was domesticated in the Near East prior to the Early Bronze Age, but is only occasionally found in later contexts (Zohary and Hopf 2000:122). The ubiquities of both bitter vetch and lentil are consistent across all periods at the site but lentil becomes increasingly common relative to bitter vetch beginning in the Middle Phrygian period (Fig. 5.5). The economic and environmental implications of this shift are discussed further below.

Bovids and Suids

The primary animal domesticates present in archaeological deposits at Gordion during every period are bovids and suids: cattle (*Bos taurus*), sheep (*Ovis aries*), goats (*Capra hircus*), and pigs (*Sus scrofa*). Together, these species comprise at least 89% of the total identified animal bone assemblage in every period of occupation (Fig. 5.6). The proportion of sheep and goats to cattle and pigs, however, does fluctuate substantially throughout the sequence. The implications of this variability for interpreting changing land use strategies is outlined by Miller and colleagues (2009), who suggest that higher ratios of pigs and cattle to sheep and goats indicate a more intensive agricultural system, while lower ratios are the result of a more extensive system. Following this logic, the Middle and Late Phrygian periods were the most intensive agropastoral periods during the Gordion occupation sequence (Fig. 5.7; Miller et al. 2009). Choices in

animal husbandry practices have implications for agropastoral risk management strategies, as discussed in Chapter 2 and further below.

An additional component of animal management strategies is the ratio of sheep to goats, which has potentially important economic and environmental implications. Although many archaeological skeletal fragments cannot be identified definitely to genus, morphometric analysis can distinguish between the closely related genera for specific bones in the postcranial skeleton with high reliability (Zeder and Lapham 2010), in contrast with generally low reliability for mandibles and teeth (Zeder and Pilaar 2010). Despite the similarity of goat and sheep skeletons, the genera occupy distinct ecological niches. Sheep require better forage and more water than goats but produce more meat and higher quality wool (Zeder and Arter 1994:112). Sheep and goats have distinct environmental impacts as well: goats are browsers and willing to eat a wider variety of vegetation, so can rapidly denude vegetated areas if left unchecked (as was the case in early Iceland [McGovern et al. 2007]). For this reason, goats are currently banned from summer grazing in the Black Sea forest range north of Gordion and only sheep may be herded there (interview with shepherd near Bolu, July 2008). The ratio of sheep to goat bones is relatively stable over time at Gordion, with an average value close to 1.3 (Fig. 5.8). This ratio is higher, however, during two periods: the Late Bronze Age and the Early Phrygian period. Environmental or economic conditions may have favored sheep herding during those periods. Ongoing analyses of both the Roman period faunal assemblage and age profiles, which indicate herd structure management practices and allow the differentiation of dairy, meat, and wool production, will provide further context for understanding changes in animal herding practices at Gordion over time.

Fruits and Nuts

Several types of fruit and nuts were identified in small numbers in the Gordion flotation samples. The most common of these was grape (*Vitis vinifera*), which was present in about 6% of samples examined and is commonly grown in domestic courtyards today throughout the region (Table 5.1). Wine was certainly consumed in substantial quantities at Gordion, as evidenced by imported wine amphorae from the

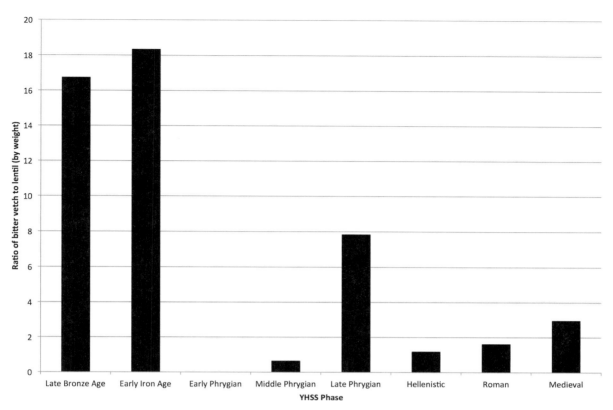

Fig. 5.5. Ratio of bitter vetch to lentil based on total seed weight by period, excluding burned crop stores from the Early Iron Age and Early Phrygian periods; otherwise, no measurable quantities of either species were present in Early Phrygian samples (data from Table 5.2).

Greek world (Lawall 1997, 2012) and traces of tartaric acid in the alcoholic cocktail served at the funeral feast of the king buried in Tumulus MM (McGovern et al. 1999), but there is no indication that wine was produced locally. Although the area around Ankara, 80 km east of Gordion, was famous for wine production into the last century, the immediate environs of Gordion are hotter and drier, making it less suitable for grape production outside of watered household gardens. One cherry pit (*Prunus avium* or *cerasus*) was found in a Late Phrygian sample, possibly a local garden product or import from northern Anatolia, where sweet cherry (*Prunus avium*) grows natively; a greater number of cherry stones were found in the Early Phrygian Destruction Layer (Nesbitt 1989). A few fig (*Ficus carica*) seeds, occurring in about 1% of samples, are present throughout the occupation sequence. Fig was domesticated by the Early Bronze Age (Zohary and Hopf 2000) and fig remains are widely distributed in the Near East (Miller 1991b), making it possible that the fruit tree was planted at Gordion.

Possible pistachio nutshell was present in about 4% of samples (Table 5.1); this is likely the shell of wild pistachio (*Pistacia terebinthus*), which grows in the region today and provides small, but edible, nuts. Other nuts present in the flotation samples include oak (*Quercus* sp.) and almond (*Prunus* sp.) nutshell, both in about 1% of samples. The acorn is unlikely to have been eaten by people but may have been collected for fuel or animal fodder, while the almond shell includes both domesticated almond (*Prunus dulcis*) and possible wild almond shell (*Prunus argentea*); both grow in the area today (Miller 2010). Hazelnuts (*Corylus avellana*) were identified among seed caches from the Early Phrygian Destruction Level (Nesbitt 1989), but the tree is native to northern Anatolia and the nuts were almost certainly imported (Miller 2010:62; Zohary and Hopf 2000:190), as has been documented at Middle Bronze Age Kültepe (Fairbairn et al. 2014). None of these tree crops are likely to have been produced in meaningful quantities at Gordion; likely all of these fruit and nut remains represent the products

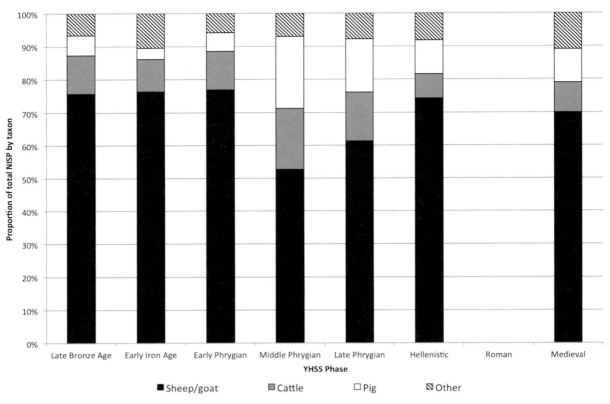

Fig. 5.6. Animal remains from 1988–89 excavation seasons by count of number of identified specimens (NISP). The "other" category includes both hunted food animals and domesticates unlikely to have been eaten regularly (canines, horses, and donkeys). Data for most periods from Miller et al. (2009: table 3), but Medieval from Zeder and Arter (1994: table 3); no faunal data have been published from the Roman period.

of scattered garden trees and possible occasionally gathered wild foods, as well as trade for imported nuts such as hazelnut.

Hunted and Gathered Foods

A small proportion of the caloric intake of the inhabitants of Gordion came from wild foods. Game animals, including deer (red deer, *Cervus elaphus*, fallow deer, *Dama* cf. *dama*, and roe deer, *Capreolus capreolus*), hare (*Lepus* cf. *capensis*), birds (including both wild and domestic species in the Anatidae, the ducks and geese, and Phasianidae, which includes chicken, quails, and pheasants), reptiles (mainly tortoise, *Testudo* sp.), and fish (various species), comprise only about 5% of total identified bones at the site (Fig. 5.6; Miller et al. 2009; Zeder and Arter 1994). The meat portion provided by these animals would be an even smaller percentage of the total meat because, with the excep-

tion of deer, these wild animals provide much less meat per bone than their larger domestic equivalents. The only exception is during the Early Iron Age, when deer comprise nearly 6% of the total animal bone assemblage. This represents a significant meat contribution to the diet, especially in comparison to all other phases where deer is less than 1% of the assemblage by count. It is also likely that fish are underrepresented in these analyses because the screen size used during the excavations was roughly 1 cm, which results in significant loss of smaller remains, such as fish and rodents, and some birds (Reitz and Wing 2008). Rodent and fish bones are found regularly in both heavy and light fractions of flotation samples. Future analysis of these microfaunal assemblages will provide better estimates for the role of fish in particular in the Gordion diet.

In terms of wild plant food, as mentioned above, possibly wild almonds and pistachios were eaten and discarded at Gordion. In addition, one seed of possi-

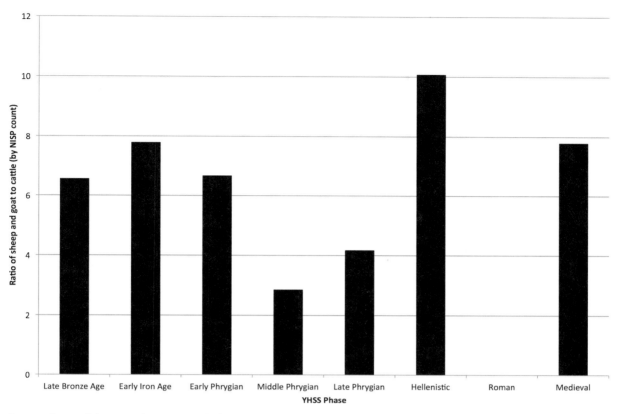

Fig. 5.7. Ratio of sheep and goat to cattle by NISP count (data for most periods from Miller et al. 2009: table 3, but Medieval from Zeder and Arter 1994: table 3).

ble coriander (*Coriandrum* cf. *sativum*) may represent human use of the cultivated or wild plant (*Coriandrum tordylium*)—both of which are possible in Central Anatolia during this time (Davis 1972:330–31; Zohary and Hopf 2000:205–6)—although it may also be the incidental presence of a native plant seed. Other potential condiments do appear in the flotation samples, including seeds of thyme (*Thymus* sp.) and poppy (*Papaver* sp.), as well as possible asparagus (*Asparagus officinalis*), celery (*Apium graveolens*), and carrot (*Daucus carota*). All of the latter taxa may represent wild or domestic plants, since their domestication history is unclear and the wild ancestors of modern domesticates are native to Anatolia (Davis 1965, 1972, 1982; Zohary and Hopf 2000). While any of these plants may have been used as condiments or vegetables, their infrequent presence in the archaeobotanical assemblage is more likely the result of animal grazing. Poppy from one deposit (YH 43843) is a potential exception; a cache of 934 seeds in a Late Phrygian oven suggests possible economic use of the plant. One

poppy capsule can hold thousands of seeds, however, so economic attribution of this find remains uncertain. Unfortunately, it is rarely possible to determine domestic status of poppy from the seeds, as some wild poppies have large-seeded forms; the absence of pores below the stygmatic lobes of the capsule is the only reliable marker of domestication (Zohary and Hopf 2000:136–37). Both wild and cultivated poppies are possible in Central Anatolia, where several dozen species of wild poppy are native and domestic poppies are commonly grown (Davis 1965:219–36).

Plants and Animals in Non-Food Economies

Plant and animal remains in archaeological contexts do not represent simply food discard, but also other economic activities that require the products of agriculture and domestic animals. Paleoethnobotanical and zooarchaeological data from Gordion provide evidence for two regional industries: oil and textile production. In addition, animals may be

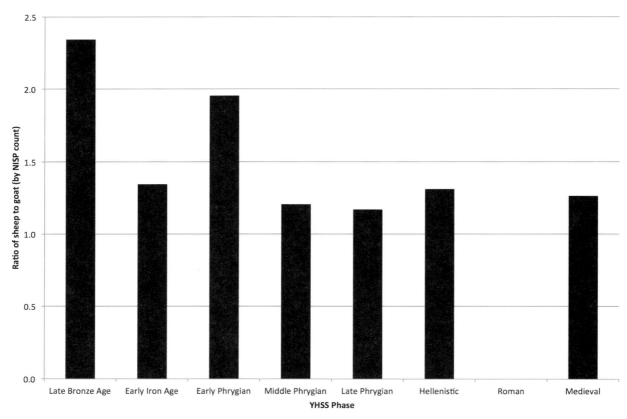

Fig. 5.8. Ratio of sheep to goat by NISP count (data for most periods from Miller et al. 2009: table 3, but Medieval from Zeder and Arter 1994: table 3).

used to assist in agropastoral activities: for traction, transport, or protection. It is this supportive role provided by horses, donkeys, and canines represented in the zooarchaeological assemblage at Gordion. Both donkeys and dogs are particularly important in sheep and goat herding, as they can be trained to keep the animals together and, in the case of dogs, to protect the flock against human or animal threats. This arrangement is common today in Central Anatolia, with many shepherds using dogs, donkeys, or both to tend their flocks.

Oil Production

A variety of plants found frequently at Gordion produce oily seeds that can be pressed to yield vegetable oils. These include wild plants, especially those in the mustard (Brassicaceae) and poppy (Papaveraceae) families, but these are small-seeded plants that would have been potentially difficult to collect in sufficient quantities to make meaningful quantities of oil. Do-

mesticated plants in the Gordion assemblage that can be used for oil production include flax (*Linum usitatissimum*), poppy (*Papaver somniferum*), safflower (*Carthamus tinctorius*), gold of pleasure (*Camelina sativa*), and, in the Medieval period, cotton (*Gossypium arboreum* or *herbaceum*). Entirely absent from these contexts is olive (*Olea europea*), the primary oil crop of the Mediterranean world, which is native to the southern and western coasts of Anatolia and was widely grown as an oil crop in those areas in antiquity, but not in Central Anatolia. Olive oil was imported to Gordion by the Middle Phrygian period; its characteristic fatty acids have been identified in a cooking vessel from the site (McGovern et al. 1999). Large-scale importation of olive oil, however, is not evidenced until the Late Phrygian period, when Greek transport amphorae first appear at Gordion in substantial numbers (Lawall 1997), and it may have been restricted to use by elites or for special occasions, as suggested by Greek literature and its increased cost as an import to Central Anatolia (Foxhall 2007:87–91).

Pending a larger-scale study of transport amphorae and chemical analysis of cooking vessels and other oil-use vessels (i.e., lamps), it is not possible to assess the degree to which olive oil was used at Gordion during different phases of occupation, but even as an import it is likely to have been the vegetable oil most commonly used in the region as little evidence exists for local oil production. Flax, more than 11,000 seeds of which appear in a jar from the Early Phrygian Destruction Level, was almost certainly grown at the site, although it may have been primarily intended as a fiber crop (Miller 2010:46). Flax was grown for oil near Gordion during the 20th century (Miller 2010:13).

Poppy is the other potential oil crop found in a significant quantity at Gordion, although the seeds of domestic and wild poppies are generally indistinguishable (Zohary and Hopf 2000:136–37). One cache of nearly 1,000 seeds provides possible evidence for poppy cultivation at Gordion during the Middle Phrygian period. Poppy seeds can be used as a condiment or crushed for oil and the pods can be scored for opium; any or all of these uses may have been the intention of putative poppy farmers at Gordion. It is important to consider, however, that mustards, flax, and poppy are all thin-walled oily seeds, which are among the least likely to be preserved after heating, so these taxa are very likely significantly underrepresented in the archaeobotanical assemblage in comparison to starchy cereals and pulses (Wilson 1984).

Animal fats were likely the primary oil source at Gordion. Rendered sheep fat, lard, and butter appear in Hittite texts as commodities used in cooking that were cheaper than olive oil (Hoffner 1995:109; Yakar 2000:281). The Hittites used both rendered sheep fat, or tallow, and liquid oils in lamps (Hoffner 1995:112); later Roman lamps from Sagalassos in southwestern Anatolia burned primarily olive oil but also animal fat (Kimpe et al. 2001). Sheep fat appears in cooking vessels used in the funerary feast of the king buried in Tumulus MM (McGovern et al. 1999); this may have represented sheep meat or have been a cooking fat used to prepare the stew.

Beeswax is another animal lipid that may have been produced in the Gordion region. Apiculture is practiced in the region today, intended for honey but also producing beeswax (METU 1965:94–95). Two large ceramic vessels from Hellenistic Gordion are likely to be beehives, with a flight hole in the shoulder

and a crudely scored interior (Stewart 2010:191–92). Related vessels are common in Greece, especially Attica, and throughout the Classical world (Anderson-Stojanović and Jones 2002; Crane 1983; Jones et al. 1973; Rotroff 2001, 2006:124–131). Chemical evidence from some of these Greek vessels confirms the presence of beeswax (Evershed et al. 2003), which is also present in Roman cooking vessels at Sagalassos, likely because it was melted down in those vessels (Kimpe et al. 2002). Extensive linguistic and visual evidence exists for apiculture in the Bronze Age Aegean (Harissis and Harissis 2009) and beeswax residues also have been found in lamps on Neopalatial Crete (Evershed et al. 1997). Future chemical residue analysis on ceramic lamps and cooking vessels from Gordion may help to determine the roles of animal and plant oils in both lighting and cooking across different periods of occupation at the site.

The Textile Industry at Gordion

Textile production was a major industry at Gordion beginning at least by the Early Phrygian period, as evidenced by weaving equipment present in several rooms of the Destruction Level (Burke 2005, 2010; Voigt 2002). More than 135 spindle whorls and a heap of loom weights were found in Terrace Building 7 together with six iron needles in a small vase and three more needles with an iron knife, apparently sewing kits (DeVries 1990:385). A line of loom weights, some with traces of plant fiber attached, indicate that a loom was probably set up in this room at the time of the destruction (DeVries 1990:385). In the adjacent Clay Cut Building 3 nearly 200 spindle whorls were found, including five in a wooden box together with clumps of braided animal hair thread (DeVries 1990:387). This pattern continues throughout that area of the city (Burke 2005: fig. 6-2).

Some textile fragments survive in excavated contexts from the Citadel Mound and more complete pieces have been preserved in burial tumuli, although detailed analyses of these remains have not been well published in recent years (Barber 1991; Bellinger 1962; Ellis 1981; but see also Ballard 2012; Ballard et al. 2010). Various textile techniques and materials are present among these fragments, including bronze-studded leather (not a true textile), wool felt blankets, bags or cloth wrappings of linen and hemp, goat-wool cloth, and combination fabrics incorporating both linen and wool or both sheep and goat wool (Bellinger 1962; Ellis 1981:304). These textiles came in a variety of colors and some individual objects incorporated multiple colors of yarn (Bellinger 1962); local plants may well have provided these dyes, as described further below. Continuing work using modern techniques provides additional information on the original colors of these textiles and their composition (Ballard 2012). It has also been argued that the intricate patterns present on the inlaid furniture of Tumuli MM, P, and W, and the rock-cut reliefs of Midas City may be representations of common patterns on Phrygian textiles (Simpson 2007; Simpson and Spirydowicz 1999). This seems a reasonable suggestion, although the directionality of these artistic influences remains impossible to trace with current archaeological data.

The large numbers of sheep and goats kept at Gordion represent both an efficient adaptation to the semi-arid steppe environment and a diverse source of economic production. Aside from the meat and milk produced from these flocks (Gürsan-Salzmann 2005; Zeder and Arter 1994), wool and goat hair were widely used in textile production, an industry that continues today (Gürsan-Salzmann 1997). Angora goats, bred in the region around Ankara, from which they take their name, produce fine wool that yields high-quality textiles, such as some recovered from excavated contexts at Gordion during the 1957 season (Bellinger 1962:14). It is likely that trade in such textiles was one source of wealth at Gordion during the Phrygian periods, as it was well situated on trade routes between the Black Sea, Aegean, and Mediterranean coasts and later a stop on the Persian royal road connecting Sardis to the east (Graf 1994; Lawall 1997; Young 1963).

Linen appears to have been the main plant fiber used in the Gordion textile industry and so must have been grown locally. Additional evidence for this conclusion is the jar of flax seeds found in the Early Phrygian Destruction Level (Miller 2010:46). The lack of flax in any other flotation sample indicates that this was probably not a food plant or fodder, but instead the seeds were kept only for the purpose of planting flax fields for linen (or, potentially linseed) production, thus existing outside of the cycle of burning and disposal characteristic of kitchen and household waste and preserved only by a chance fire. Heavy, symmetrical biconical spindle whorls, such as those found in the

Early Phrygian Destruction Level, are well suited to spinning the straight fibers of flax, providing additional evidence for its production at Gordion (Burke 2005; Miller et al. 2006). Although Bellinger (1962:13) identified hemp among the textile fragments from the Phrygian burial tumuli, no *Cannabis* seeds have been identified archaeologically at the site, so it is uncertain whether hemp was imported or grown locally for fiber production, food, or fodder.

Cotton (*Gossypium arboreum* or *herbaceum*) was not present as a textile fiber at Phrygian Gordion, according to currently published analyses of Gordion textiles (Bellinger 1962; Ellis 1981), but was grown at Gordion during the Medieval period, based on numerous archaeobotanical remains from the site (Table 5.1; Miller 2010:45). Textile analyses conducted at Gordion have focused on the Phrygian textile fragments preserved in tumuli (Ballard 2012; Ballard et al. 2010; Bellinger 1962), but no comparable burials or preserved textiles exist from the few Medieval contexts excavated at Gordion. Cotton seeds appear in 16% of Medieval samples but no samples prior to that period (Table 5.1). They are also present in one sample from the 1961 excavations (1961-BOT-02) identified by Mark Nesbitt (1989:4); the dating of this latter sample is unknown but of great interest. The evidence that cotton was not introduced into Central Anatolia until the Medieval period is consistent with historical, genetic, and archaeological evidence that cotton was a South Asian domesticate, grown in Ethiopia and the Gulf with cloth imported to the Near East and Mediterranean by the Roman period, but cultivation in the Mediterranean did not begin until the Islamic period (Brite and Marston 2013; Watson 1983:32–34). Cotton became a major textile, with a trading house for cotton alone set up in Baghdad, and travelers' accounts describe it growing in Asia Minor after the 10th century CE (Watson 1983:40). The relative ubiquity of cotton at Gordion during the Medieval period suggests that it was planted in the region, either for export as a raw fiber or for the production of cotton textiles woven on site. Further archaeological investigation of Medieval Gordion is needed to evaluate these alternatives.

One important element of the textile industry is the use of mineral- and plant-based dyes, and although spectrographic and chemical testing of the Gordion textiles has not yet been published, it is clear that multiple dyes were used to color yarn based on the faded patterns visible on the textile fragments, including a red or purple dye (Ellis 1981:296). Safflower (*Carthamus tinctorius*) produces a yellow-red dye from its flowers, and achenes of this plant are present in several flotation samples from the Late Phrygian and early Hellenistic periods, as well as one Medieval sample (Table 5.1). Other plant dyes that may have been used at Gordion include madder (*Rubia tinctorum*), which produces a red dye, woad (*Isatis tinctoria*), which produces a blue dye, and Dyer's rocket or weld (*Reseda luteola*), which produces a yellow dye (Zohary and Hopf 2000:208–10). Seeds identified as *Reseda* were found in several Hellenistic and Medieval samples at Gordion (Online Appendix 2; Miller 2010: app. F2); *Reseda lutea* and *R. microcarpa* have been identified growing wild in the area, but *Reseda luteola* is also native to Central Anatolia (Davis 1965:505–6). It is possible that the roots of any of these species were gathered wild and used for dying yarn. Neither *Isatis* nor *Rubia* have been identified among the flotation samples but both dye species are native to the area (Davis 1965:302, 1982:860–61) and a number of unidentified Rubiaceae seeds might potentially belong to *Rubia*. Further chemical analysis of the surviving textile fragments is needed to gain concrete information on the dye economy at Gordion.

Conclusions on Diet and Economy

The Gordion diet changed little over the millennia of occupation at the site, but this consistency masks evident changes in agricultural practices. The subsistence diet focused on cereals, pulses, bovids, and suids, with minimal evidence for arboriculture, viticulture, and the gathering of wild nuts. Unlike the Mediterranean coasts of Turkey, olive oil and wine were never locally produced, although they were imported to the site in increasing quantities following the Persian conquest of 540 BCE. Animal fats were likely the primary source of oil in the city, much like the Hittite economy, with butter, tallow, lard, and beeswax available. The elite did have access to wine and olive oil as early as the 8th century BCE, as shown by chemical analyses of the funerary feast held for the king buried in Tumulus MM, but this was clearly not the norm for most inhabitants of Gordion. A thriving textile industry may have provided much of the early

wealth at the site and both wool (of sheep and goat) and plant fibers (linen, perhaps hemp, and later cotton) were produced and woven in the region. A variety of wild plants may have been gathered for use as dyes in this industry and safflower was likely grown locally for use as a pigment and/or oil.

Agricultural Risk Management Strategies

Following prior work on risk management at Gordion (Marston 2011), I distinguish strategies that manage risk through diversification from those that manage risk through intensification. Among diversification strategies, I consider crop and herd animal diversification to manage rainfall risk, evidence for spatial diversification through the presence of seeds of wild plants from distinct ecological niches, and temporal diversification through multicropping. Intensification strategies include irrigation and foddering of animals with agricultural products.

Diversification Strategies

Crop Diversification

The primary agricultural products of the Gordion region were the cereals barley and a free-threshing, or naked, wheat (evidently primarily bread wheat), as these far exceed the seeds of other crops in frequency and ubiquity (Tables 5.1, 5.2; Fig. 5.1). That cereals should be the primary products of the Gordion region is not surprising; prior to the introduction of mechanized irrigation in the 1980s they were planted on the vast majority of agricultural lands in the area around Yassıhöyük (Gürsan-Salzmann 2005; METU 1965). Both are viable rainfed crops, although bread wheat does require more water and is susceptible to crop failure if rainfall totals fewer than 250 mm, which during the period 1929–2009 happened in 6% of years at Polatlı, which lies 200 m higher in elevation than Gordion and receives more rainfall (Fig. 3.4). Thus, we can surmise that perhaps 10% or more of years during the 20th century were unfavorable for rainfed wheat agriculture in the Gordion region, making diversification of cereals a valuable risk-minimization strategy. A ratio of barley to naked wheat, whether assessed by weight (or count) of seeds or by count of rachis fragments, indicates the degree of vulnerability to drought conditions and the riskiness of cereal agriculture. This ratio (Fig. 5.2) remained relatively consistent from the Late Bronze Age through at least the Middle Phrygian period. During the Late Phrygian and Hellenistic periods this ratio declines substantially to its lowest value by both measures during the Roman period. These data suggest that Roman cereal agriculture was significantly riskier than that practiced during the periods when Gordion was a large city and capital of the Phrygian kingdom (i.e., the Middle Phrygian). This may be an effect of changing site size and economy and indicates that traditional ecological knowledge in the Gordion region favored risk mitigation but Roman period agriculture was historically anomalous for the area.

Similar rainfall sensitivity metrics can be developed for pulses and for herd animals. The pulses primarily grown in the Gordion region, both archaeologically and into the 20th century, were bitter vetch and lentil (Tables 5.1 and 5.2). There is a clear chronological trend in the use of bitter vetch at Gordion, which decreased in ubiquity, absolute frequency, and frequency relative to lentils after the Early Iron Age (Table 5.1; Figs. 5.4 and 5.5). Nonetheless, bitter vetch remains a frequent agricultural product throughout the occupation sequence. Vetch is considered only a famine crop in the modern world (e.g., Halstead and Jones 1989; M.L. Smith 2006) and lentils were certainly more easily palatable with less preparation required, but vetch has the advantage of producing good yields even under drought conditions, when lentils produce very low yields (Enneking 1995; Siddique et al. 1993). The lowest values of the bitter vetch to lentil ratio occurred during the Middle Phrygian, Hellenistic, and Roman periods, although both the Middle Phrygian and Roman values are derived from less than 0.2 g of identified pulses, rendering those values more speculative.

Among herd animals, sheep and goats can thrive with less water than cattle, rendering the ratio of sheep and goats to cattle a measure of drought sensitivity and risk management (Mace 1993; Miller et al. 2009). The value of this ratio was >6 for all periods until the Middle Phrygian, when the ratio dropped below 3 (Fig. 5.7). The substantial increase in cattle relative to sheep and goats, and the concomitant rise in pig remains at Gordion (Fig. 5.6), indicate that the Middle

Phrygian (and, to a lesser extent, the Late Phrygian) herding system was considerably more vulnerable to drought conditions than that employed during other periods of occupation.

Spatial Diversification

In contrast to strong indications of varying levels of crop and animal diversification, evidence for spatial diversification is circumstantial at present. One promising avenue to clarify spatial patterning of animal pasturing is to use strontium isotope analyses of sheep and goat teeth to detect spatial patterning of pasture locations (Balasse et al. 2002; Bentley 2006). Preliminary research at Gordion indicates that at least some sheep and goats whose mandibles were deposited on site lived in widely different geological areas over the first year of their lives (DiBattista 2014). This might indicate seasonal transhumance, in which sheep and goats are herded into the mountains for summer forage and return to the warmer lowlands for the winter, as practiced traditionally in the region (Gürsan-Salzmann 2005). Alternatively, looking at herd animal diet provides another proxy measure of animal mobility. Animals penned at the city would have been foddered, eating a high proportion of agricultural produce relative to seeds of wild plants. Animals grazing across different ecological communities carried seeds of plants from those areas in their dung. Through an analysis of botanical assemblages resulting from burned dung, the wild seed to cereal ratio, a measure of foddering intensity, and the relative frequency or ubiquity of seeds of plants characteristic of specific areas on the landscape can be calculated, giving some suggestion of relative mobility of herd animals during the history of Gordion.

The ratio of wild seeds to cereal in contexts derived from dung fuel serves as a proxy measure of the degree to which animals grazed on wild plants or ate agricultural products (Miller 1997, 1999b; Miller and Marston 2012). Median values of this ratio are relatively consistent from the Early Iron Age to the Hellenistic, with the Middle Phrygian value demonstrating the greatest emphasis on foddering. The Late Bronze Age, Roman, and especially Medieval periods have higher values, suggesting greater pastoral mobility and perhaps greater spatial separation between herders and farmers in these periods (Table 5.2; Fig. 5.9).

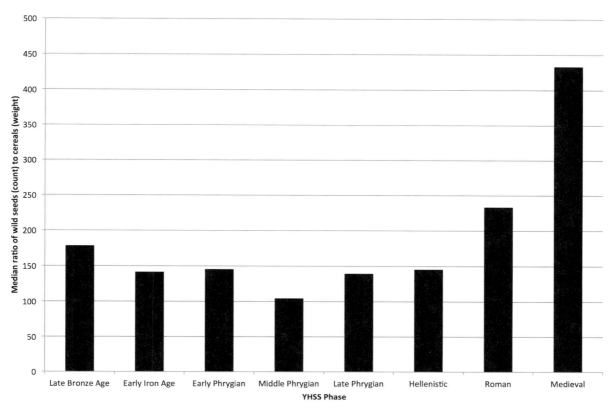

Fig. 5.9. Median ratio of identified wild seeds (count) to cereal (weight) by period, excluding burned crop stores from the Early Iron Age and Early Phrygian periods (data from Table 5.2).

Most of these wild seeds are characteristic of steppe grassland, the predominant vegetation community in the region (see Ch. 3), but some seeds found frequently in archaeological samples are indicative of other habitats. Notably the Malvaceae (including *Malva* and *Lavatera*) are found primarily as members of the riparian plant community, while *Gypsophila* is characteristic of plants growing in the gypsum steppe, as well as other steppe communities, depending on species (Miller 2010).

The ubiquity of *Gypsophila* (Fig. 5.10) suggests increased use of gypsum steppe environments during the Late Bronze Age, Early Iron Age, and Medieval period, with relatively low values during the Middle Phrygian period, the peak of population and wealth at the site. Although gypsum steppe was grazed during all periods save the Roman (and the lack of *Gypsophila* in Roman samples may be an effect of sample size), grazing in that ecological zone appears to have been most prevalent during the earliest and latest occupation periods. These data suggest that the ubiquity of

Gypsophila, and by proxy gypsum steppe grazing, may track regional population to some extent, with higher values correlated with lower population densities, as discussed in the next chapter. The Malvaceae provide a good proxy for access of riparian environments, as some genera (e.g., *Lavatera*) are found exclusively along the Sakarya River today (Miller 2010:129). Although their ubiquity fluctuates throughout the Gordion occupation sequence, it is consistently low and there is no apparent pattern in this measure, suggesting that access to riparian grazing environments was relatively consistent across periods (Fig. 5.10).

Temporal Diversification

Evidence for temporal diversification at Gordion comes primarily from datasets beyond botanical remains. Storage structures, including bins, pits, and vessels, are consistently present in domestic structures over the occupation sequence at Gordion. Structures from the Late Bronze Age, Early Iron Age, and Early

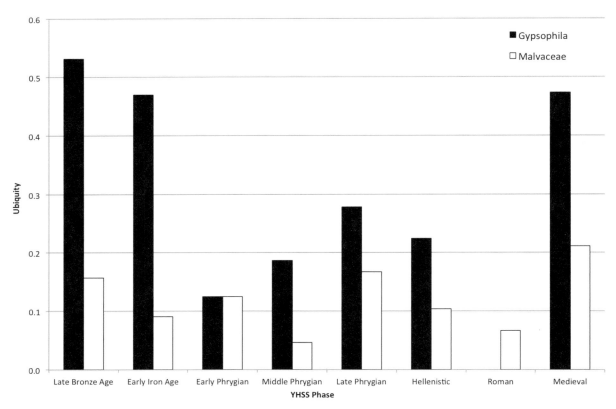

Fig. 5.10. Ubiquity of *Gypsophila* and Malvaceae seeds as proxies for gypsum steppe and riparian habitats, respectively (data from Online Appendix 2; Marston and Miller 2014: electronic supplementary material 1; Miller 2010: app. F).

Phrygian period, which were burned in situ, contain carbonized masses of grain (Miller 2010; Nesbitt 1989; Voigt 2011; Young 1958a, 1962a). Herd animals provide another source of temporal diversification, as they can be fed on fodder during harsh winter months when forage is unavailable; these animals provide milk and meat during the winter. Although kill pattern data from osteological analysis of animal bones—which would suggest the season in which young animals died—are not yet published, ethnographic analogy suggests that animals would have been kept in the lowlands near Gordion and foddered during winter months (Anderson and Ertuğ-Yaras 1998; Forbes 1998; Gürsan-Salzmann 2005). The presence of considerable quantities of cereal remains in contexts containing burned ruminant dung (Fig. 5.9) provides support for this inference.

Botanical remains can also provide evidence for temporal diversification of farmed crops. While year-round crop scheduling is impossible in the continental climate of Central Anatolia (cf. Baksh and Johnson

1990; Clawman 1985; Flowers et al. 1982; McCorriston 2006), multicropping of autumn- or spring-sown, summer-harvested, cereals and summer-sown crops with short germination times (e.g., millets, cotton) for autumn harvest is possible. This agricultural system is only evident during the Medieval period, when rice and cotton are found for the first time and millet increases significantly in ubiquity (Table 5.1). These data suggest the addition of a second harvest to the traditional summer harvest of cereals and pulses at Gordion.

Intensification Strategies

Although agricultural intensification is a complex concept, the basic definition is increasing production from a unit of land (Brookfield 1972). This may result in an increase (e.g., Allen 2004; Kirch 2006) or a decrease (e.g., Boserup 1965, 1981; Johnston 2003) in labor efficiency, depending on the specific agricultural strategy chosen and the interactions between specific ecological and economic systems (Brookfield 2001;

Morrison 1994, 1996; Stone and Downum 1999). Overproduction of agricultural products, while "inefficient" in terms of caloric returns per unit of labor, is a productive form of risk management when that surplus can be stored for later use. This may occur through conventional storage, e.g., of grains (Kuijt 2009), through temporally spaced reciprocal exchange (Kaplan and Hill 1985; Winterhalder 1990, 1997), or through foddering of animals with agricultural products (Halstead 1989). These mechanisms effectively time-shift agricultural surpluses to periods of scarcity, converting intensification into a form of temporal (and perhaps, in the case of sharing and mobile herds, spatial) diversification. The wild seed to cereal ratio of archaeobotanical remains resulting from burned animal dung provides a relative measure of foddering intensity (Miller 1997; Miller et al. 2009). This value fluctuates significantly over time at Gordion, indicating differences in foddering between periods, with foddering slightly more prevalent during the Middle Phrygian period than preceding and succeeding periods (Fig. 5.9).

Another mechanism for agricultural intensification is irrigation. Complex systems of irrigation existed in many parts of the semi-arid and arid Near East (e.g., Harrower 2008; Watson 1983; Wilkinson 2003, 2006), which provide opportunities for both risk minimization, reducing the risk of crop failure due to drought, and intensification, by increasing crop yield per unit of land (see Fig. 2.3). Aside from identifying irrigation features (e.g., channels, canals, raised fields) in the landscape, and dating them, botanical remains also offer evidence for irrigation. Two distinct archaeobotanical proxies for irrigation are seeds of crops that require irrigation to grow in a given environment and weeds that preferentially colonize wet areas, such as the banks of irrigation canals. As discussed above, the presence of cotton and rice in Medieval contexts indicates a summer-sown, irrigated crop planted after traditional summer-harvested, dry-farmed cereal and pulse crops (Table 5.1; Watson 1983). In earlier periods, however, prior to the introduction of the obligate irrigated tropical cultivars rice and cotton, the only proxy for irrigation is the presence of weeds of wet areas.

Sedges (the family Cyperaceae) are commonly identified in Near Eastern archaeobotanical assemblages and serve as a proxy for the frequency of moist areas that form their ecological niche; irrigation provides a common mechanism for the expansion of this ecological zone (Miller 2010; Miller and Marston 2012). The frequency of Cyperaceae as a percentage of the total number of wild seeds varies over time (Table 5.2, Fig. 5.11), with substantial, apparently permanent, increases in sedges during the Middle Phrygian and Roman periods, suggesting expansion of irrigation networks during those periods. While Medieval irrigation may have been targeted specifically for the cultivation of summer crops (cotton, rice, and perhaps millet), both the Roman and Middle Phrygian are periods of high regional population, which suggests that irrigation investments during those periods may have been an effort to increase food availability to feed a growing population (Marston 2015; Marston and Miller 2014). Roman period irrigation may have had a second goal as well: minimizing the risk associated with bread wheat production. As noted above, the ratio of barley to naked wheat declines significantly during the Roman period, suggesting the prioritization of naked wheat agriculture (likely bread wheat, as no hard wheat rachis fragments have been identified from Roman contexts). As bread wheat is a risky crop in Central Anatolia, due to its higher water requirements than hulled wheats and barley, Roman farmers may have attempted to reduce the risk of crop failure through large-scale irrigation. Increasing wheat crop yields may have been an added benefit of this practice, despite the additional labor required to expand and maintain the irrigation canal network.

Impacts of Agriculture on Local Environments

Agricultural practices disturb natural vegetation communities, displacing native plants and creating new ecological niches that selectively favor certain species over others. Additionally, grazing animals disturb vegetation communities but their impact differs across ecological zones. Indicators of these ecological changes appear in the archaeobotanical record in several ways. We can identify the presence of weedy colonizers that take advantage of new ecological niches created by agriculture to expand; of field weeds that may be harvested with, and unintentionally propagated with, agricultural crops; and of grazing-vulnerable and grazing-resistant plants that change in frequency in response to animal predation.

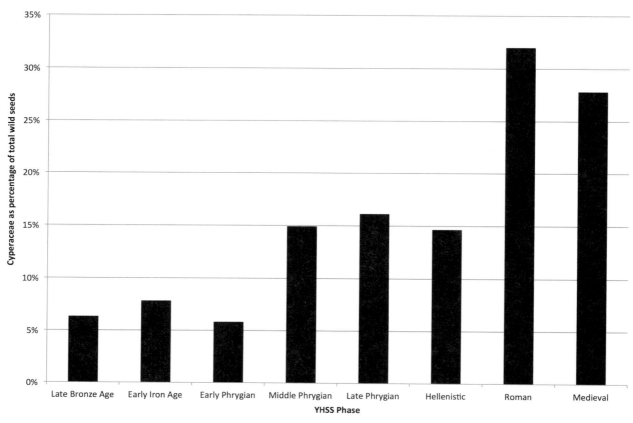

Fig. 5.11. Frequency of sedges (Cyperaceae) as a percentage of total identified wild seeds by period, as a proxy for irrigation intensity (data from Table 5.2).

Field Weeds and Colonizers of Disturbance

Wild seeds from the paleoethnobotanical assemblage include those of ruderal plants and those that are field weeds. The seeds of field weeds may be harvested together with a crop and unintentionally propagated through future batches of seed reserved for planting. Harvesting activities place significant evolutionary pressure on the weed species and can create a coevolutionary relationship between the crop and weed species resulting in a weed that is uniquely suited to colonize fields of that domesticated crop (Hillman 1984; Hillman and Davies 1990; Riehl 2014; Rindos 1984). Modern use of herbicides and relatively pure seed has minimized the presence of such field weeds in agricultural fields around Gordion today, but rye (*Secale cereale*) and wild oat (*Avena sterilis* and *A. barbata*) remain common minor components of wheat fields. Rye is a particularly good example of this phenomenon: closely related to wheat, it has a similar

height (or slightly taller), rachis structure, and seed size, so remains a part of the cleaned wheat seed when the field is harvested and threshed by combines. This adaptation ensures its continuing propagation in future wheat stocks and plantings.

Miller (2010:7, 49–50) only distinguishes two field weed taxa in her analysis of modern vegetation communities around Gordion, as there are so few modern field weeds identifiable for comparative study due to herbicide use. She notes *Alhagi* (Fabaceae) and *Eremopyrum* (Poaceae) as segetal field weeds (weeds of crop fields). Although rye is a common field weed in the region today, it is present in only sixteen samples from Gordion, with low ubiquity in all periods where present, thus, providing little insight into past agricultural practices (Table 5.1). Other authors base their evidence for field weeds on the co-occurrence of wild taxa with grain samples. Fuller (2007:907) lists "arable weed" taxa that are typical components of cereal fields in the Neolithic Levant, including small legumes (such as *Trifolium*, *Melilotus*, and *Trigonella*), Boraginaceae, small-seeded grasses (such as *Hordeum murinum*, *Bromus*, and *Lolium*), Polygonaceae, *Silene*, *Galium*, and *Centaurea*. Van Zeist and Roller (1993) list *Lolium*, *Phalaris*, *Vicia*, and *Rumex* as common field weeds in cereal fields of Lower Egypt. Chernoff and Harnischfeger (1996:171) consider *Silene*, *Chenopodium* or *Amaranthus*, *Malva*, *Galium*, *Polygonum* or *Rumex*, and various small grasses and legumes to have been field weeds in Central Anatolia.

By these criteria, I identify *Galium* (Rubiaceae) as another potential field weed. It appears in high densities in three burned crop caches, which are the only burned caches of pulses (YH 33243 and 33575 of lentil, and 33335 of bitter vetch [Miller 2010:256]). Among these samples, *Galium* comprises more than 80% of the total wild seed component in each (including seeds identified by Miller as possible *Asperula*, which may actually be another species of *Galium*). Based on these data, I suggest that *Galium* may have been a preferential weed of pulse fields. The seed is large (1–3 mm in diameter) and nearly spherical, similar in size and shape to both bitter vetch and lentil, and may be reserved in the process of sieving agricultural products from pulse fields. Assessing the co-occurrence of pulses and *Galium* in the 484 Gordion samples, we find that these taxa are more likely to occur together in samples than predicted by random chance (Table 5.5; $\chi^2 = 15.03$, $p < 0.001$).

Miller, in contrast, notes that *Galium* has a broad ecological range as a ruderal plant and suggests that it can be used as a general proxy for ecological disturbance (Miller 2010:49, 56). Assessing the frequency of *Galium* over time, however, Miller's (2010:61) initially noted trend (higher *Galium* frequency during the Middle Phrygian period) is not evident in the larger, combined dataset presented here, as measured by either frequency or ubiquity (Fig. 5.12). If *Galium* is both a field weed, as identified through burned crop caches and co-occurrence analysis above, and a ruderal plant of disturbed areas, then *Galium* by frequency or ubiquity is unlikely to be an effective marker of landscape disturbance. Instead, we should turn to other ruderal taxa to find plants that serve solely as markers of disturbance.

Ruderal taxa in the Gordion region include *Adonis*, *Aegilops*, *Alhagi*, *Eryngium*, *Glaucium*, *Heliotropium*, *Hordeum murinum*, *Medicago*, *Onopordum*, *Rumex*, and *Taeniatherum*, based on Miller's years of ecological survey (Miller 2010: table 5.13). Most of these taxa are also natural components of the Central Anatolian steppe vegetation. In addition to Miller's list above, I would include *Galium* and many grasses, including *Alopecurus*, *Lolium*, *Setaria*, and *Avena*; these grasses colonize cereal fields throughout the world as noxious weeds (USDA 2010). *Rumex*, though ruderal, is also a plant of moist soils and irrigated fields, so like *Galium* serves as a mixed proxy for disturbance, as it is also a marker of irrigation. Since all of these taxa characterize more than one ecological zone and/or landscape process, none serve as a suitable proxy for landscape disturbance alone. A more focused and illustrative marker instead focuses solely on the impact of ruminant grazing on grassland ecology.

Markers of Overgrazing

The effects of overgrazing on steppe vegetation are substantial. In its most extreme form, overgrazing replaces a healthy, diverse steppe with scattered tufts of a few antipastoral species and large areas of bare soil. Characteristic plants of overgrazed steppe include antipastoral species that are not eaten by livestock due to their chemical content (*Artemisia*, thyme, and *Peganum harmala*), thorniness (*Alhagi pseudalhagi* and *Onopordum*), or low-lying habit (e.g., *Poa bulbosa*). Botanical survey in the Gordion region (Miller 1999a, 2010) and controlled grazing studies elsewhere in Central Anatolia (Fırıncıoğlu et al. 2008; Fırıncıoğlu et al. 2007; Fırıncıoğlu et al. 2009)

Table 5.5. Chi-square analysis of co-occurrence of pulses and *Galium* across all samples. Numbers reflect count of samples in each category. The chi-square value is highly significant, indicating that the positive correlation between the distribution of these taxa is highly unlikely to be due to chance.

	Galium	No *Galium*	Total
Pulse	117	40	157
No Pulse	184	143	327
Total	301	183	484

Chi-square = 15.03, $p < 0.001$

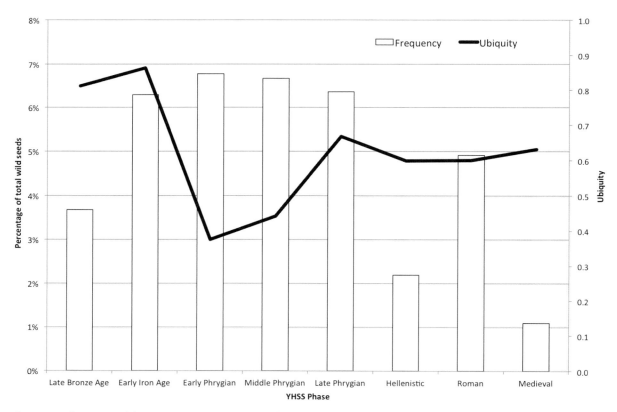

Fig. 5.12. Ubiquity and frequency as a percentage of total identified wild seeds of *Galium*, by period (data from On-line Appendix 2; Marston and Miller 2014: electronic supplementary material 1; Miller 2010: app. F).

document the quantitative effects of grazing on the native vegetation.

I have identified three proxy measures for assessing the degree of overgrazing in the Gordion paleoethnobotanical assemblage over time: the ubiquity of antipastoral species, the ratio of healthy to overgrazed steppe indicator taxa, and diversity indices of wild seeds in dung-derived deposits. In prior work (Marston 2010:396, 450), I utilized ubiquity of three marker species for overgrazing—camelthorn (*Alhagi*), wild rue (*Peganum harmala*), and wild barley (*Hordeum murinum*)—as a proxy measure of grassland degradation. This measure, however, proved only a coarse measure of grazing pressure, due to the insensitivity of ubiquity as a marker for changes in absolute abundance between periods. Additionally, wild barley is a ruderal plant and colonizer of disturbed areas, so is inconclusive as a direct marker of overgrazing. Only camelthorn shows significant changes in ubiquity over time, but a more effective measure

of grazing pressure proved to be a ratio of seeds of healthy taxa to seeds of overgrazed steppe taxa. Other paleoethnobotanical assemblages, however, might prove more responsive to simple changes in ubiquity of marker species, and this method should still be considered for other sites.

The second proxy measure of steppe health is the ratio of taxa indicative of healthy steppe to those indicating overgrazing, which records simultaneous decline of grazing-vulnerable species and increase of antipastoral species in response to grazing pressure (Table 5.2; Fig. 5.13). Healthy steppe taxa include *Ajuga, Alyssum, Androsace, Anthemis, Ceratocephalus, Eremopyrum, Euphorbia, Glaucium, Kochia, Matricaria, Matthiola, Medicago, Phleum, Salsola, Senecio, Stipa, Thymelaea, Trigonella,* and *Ziziphora*; overgrazed steppe taxa are *Alhagi* and *Peganum*. The diachronic trend in this measure indicates that steppe is healthiest during the earliest occupation phases (the Late Bronze and Early Iron Ages) and declines

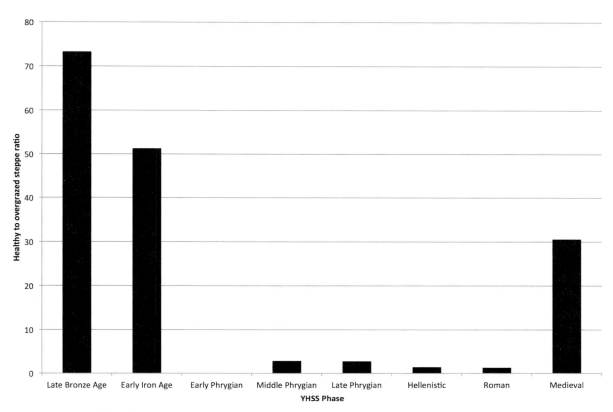

Fig. 5.13. Ratio of healthy to overgrazed steppe taxa as a proxy for degree of overgrazing; high values indicate low grazing pressure and low values high grazing pressure. Healthy steppe taxa include *Ajuga, Alyssum, Androsace, Anthemis, Ceratocephalus, Eremopyrum, Euphorbia, Glaucium, Kochia, Matricaria, Matthiola, Medicago, Phleum, Salsola, Senecio, Stipa, Thymelaea, Trigonella,* and *Ziziphora*; overgrazed steppe taxa are *Alhagi* and *Peganum* (data from Table 5.2).

substantially during Phrygian periods through the Roman; there is a dramatic increase in steppe health during the Medieval period, likely an effect of the abandonment of Gordion and general depopulation of the region for approximately 800 years following the Roman period. I argue that this measure closely tracks regional population but with a lag representing a legacy effect, as explained in further detail in the following chapter.

The final proxy measure of steppe health is diversity: botanical survey demonstrates that a healthy steppe is substantially more diverse than an overgrazed steppe. In earlier work (Marston 2010:397), I applied both the Simpson's and Shannon-Weaver diversity indices to quantify taxonomic diversity in wild seed assemblages, but concluded that the Shannon-Weaver index is a better marker for this assemblage, as it is more sensitive to differing presence of infrequent taxa. The results of this analysis for

healthy steppe taxa from Gordion are shown in Figure 5.14. Higher values correspond to greater diversity in the sample (Popper 1988), which we expect for healthier steppe, based on ecological survey that demonstrates the botanical diversity of ungrazed Anatolian steppe (Miller 1999a, 2010). The trend in these data is generally similar to that shown by the healthy to overgrazed steppe ratio (Fig. 5.13): a high value in the Early Iron Age, declining into the Middle Phrygian and again declining through the Hellenistic and Roman and rebounding in the Medieval. Two substantial differences between the results of these two analyses lie in the Late Bronze Age, where the healthy to overgrazed ratio is high but species diversity is lower, and the Late Phrygian, which shows the opposite pattern. The Late Bronze Age pattern is explained by the high number of *Trigonella* seeds in that period (comprising more than 75% of all healthy steppe seeds), which suppresses diversity values for

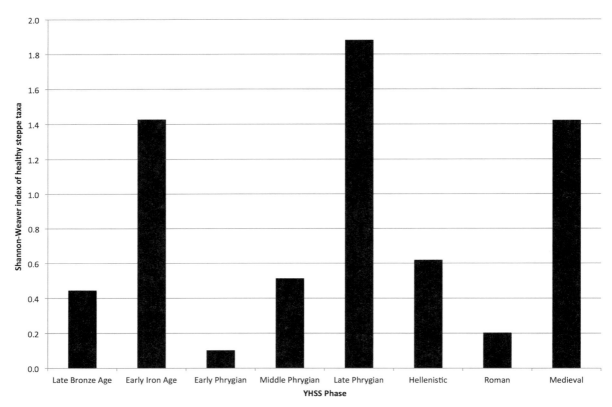

Fig. 5.14. Shannon-Weaver diversity index of healthy steppe taxa; high values indicate greater diversity and/or higher evenness. Healthy steppe taxa include *Ajuga, Alyssum, Androsace, Anthemis, Ceratocephalus, Eremopyrum, Euphorbia, Glaucium, Kochia, Matricaria, Matthiola, Medicago, Phleum, Salsola, Senecio, Stipa, Thymelaea, Trigonella,* and *Ziziphora* (data from Table 5.2).

that period by making every other taxon proportionally less of the total. This illustrates a limitation of diversity values, especially with large and diverse datasets (Marston 2014; Popper 1988). In contrast, the Late Phrygian increase in diversity is supported by a large and diverse seed assemblage from many samples; there are, however, also many overgrazed steppe taxa seeds from this period. One potential explanation for this pattern is that certain areas subject to frequent grazing during the Late Phrygian were overgrazed, but others (perhaps new grazing lands) were healthy, producing high indicators of steppe diversity at the same time as a relatively low healthy to overgrazed ratio. An expansion of grazing lands in this period is plausible, given the continued large size of the city (Voigt 2002) and evidence for ongoing erosion of substantial areas upstream of Gordion (Marsh and Kealhofer 2014).

Overall, the results of these analyses are ambiguous during several periods. The diversity measure should be considered the less reliable of these analyses due to its sensitivity to large sample sizes and to identification level, two factors that do not affect other analyses. The healthy to overgrazed ratio shows diachronic variation, suggesting that it may be most sensitive in tracking changes in steppe quality over time. In general, however, both analyses show similar diachronic trends: healthy steppe in the earliest phases of occupation and increasing evidence for overgrazing through the Middle Phrygian period. At that point, they diverge with increased steppe diversity during the Late Phrygian before a decline in the Hellenistic, while the healthy to overgrazed ratio shows a continuous decline through the Roman. All measures show substantially decreased steppe quality during the Roman period, likely a result of the substantial increase in regional settlement density, and possibly a change in pastoral strategy, during that period (Kealhofer 2005b; Marston 2015; Marston and Miller 2014). Steppe quality appears to have improved by the Medieval period, likely an effect of the abandonment of Gordion between the 6th and 12th centuries (Goldman 2005).

Anthropogenic Landscape Change

Bringing together several independent datasets, it is possible to reconstruct an overall pattern of anthropogenic landscape change that culminated in the large-scale disturbance evident during the Roman period. The earliest evidence for human impact on the Gordion region comes from sediment records from hillslopes above the Sakarya Valley (Marsh 2005, 2012; Marsh and Kealhofer 2014). Marsh and Kealhofer (2014) reconstructed alluviation speeds from stream sediment cores and identified significant instability in local uplands beginning in the Chalcolithic and continuing through the Early Bronze Age, prior to the archaeological record of substantial habitation at the site of Gordion. Upland landscape change appears to precede population growth, with landscape disturbance peaking around 2000 BCE while population increases through the Late Phrygian period, some 1500 years later. As a result, Marsh and Kealhofer (2014:697) conclude that landscape vulnerability, rather than just population and agricultural expansion, must be considered as factors in erosion rates; upland landscape instability can thus be understood best as a threshold event achievable by even relatively small-scale agricultural populations, with a fairly low ecological and geomorphological hurdle to overcome.

A subsequent episode of landscape transformation is documented in the Early Iron Age with evidence for dramatic changes in wood fuel use at Gordion: juniper sharply declined as a fuel wood following the Late Bronze Age (Fig. 4.7) while wood from both open steppe woodland and riparian communities began to increase beginning in the subsequent Early and Middle Phrygian periods (Fig. 4.8). These changes suggest significant anthropogenic disturbance of local woodland composition and may reflect different wood acquisition strategies. Such changes to forest structure affect regional hydrology and may be an important contributing factor to sedimentation and alluviation visible in the geomorphological record.

The Middle and Late Phrygian periods were the peak of site size at Gordion (Voigt 2002), as well as regional population (Marsh and Kealhofer

2014:697). This era also witnessed a substantial increase in irrigation (Fig. 5.11), in cattle herding relative to sheep and goats (Fig. 5.7), and in lentil cultivation (Fig. 5.5, during the Middle Phrygian). These changes in agricultural strategies suggest efforts to feed a larger urban population and appear to have had profound impacts on local landscapes. Steppe health declined precipitously from the Early Iron Age (Fig. 5.13). At the same time, alluviation of the Sakarya River increased 100-fold, from 2 mm per century to 2 mm per year, during the middle of the 1st millennium BCE, reflecting erosion throughout a large regional catchment upstream of Gordion (Marsh 1999; Marsh and Kealhofer 2014:690). Marsh and Kealhofer (2014:698) suggest that environmental degradation moved up the Sakarya Valley during this period, explaining the renewed cycle of erosion 2,000 years after the vulnerable and easily transported soils of the uplands immediately proximate to Gordion had eroded. The comprehensive nature of economic and agricultural transformation of Gordion during the Phrygian era cannot be understated: even though the agricultural system practiced during this era shows significant continuity with earlier periods in terms of crops and animal species, new agricultural strategies for intensification were introduced, suggesting that the social organization of agriculture may have changed dramatically on par with craft production and funerary practice (Henrickson and Blackman 1996; Marston 2012a; Voigt 2007, 2011).

The Roman and Medieval periods represent two distinct agricultural systems with widely differing environmental implications. The Roman period was characterized by intensive bread wheat cultivation (Figs. 5.2 and 5.3), a substantially increased level of irrigation (Fig. 5.11), and increased grazing, rather than foddering, of herd animals (Fig. 5.9). These practices resulted in severe overgrazing (Figs. 5.13 and 5.14) and increased rates of alluviation in the Sakarya River drainage (Marsh 2005). The Medieval agricultural system, in contrast, introduced new cultivars (Table 5.1) and a summer cropping system supported by large-scale irrigation (Fig. 5.11), with sheep and goats pastured extensively (Figs. 5.7 and 5.9) but, whether due to low population levels across the region (Marsh and Kealhofer 2014; Marston 2015) or to specific agricultural strategies employed by Medieval farmers, regional steppe appears healthy (Fig. 5.13) and alluviation rates of the Sakarya declined (Marsh 2005).

Conclusions

Through the analysis of archaeological plant and animal remains, we now have an excellent understanding of diet and pastoral economy at Gordion. Agricultural strategies, which are not directly visible, can also be reconstructed through the analysis of plant and animal remains in the context of well-developed, testable formulations of theoretical expectations. In the case of Gordion, the use of marker species indicative of specific practices, such as weeds of irrigation or overgrazed steppe, allowed the identification of changes in agricultural practices over the occupation history of the site. Cumulative pressure on ecological communities, such as steppe vegetation, led to significant ecological and geomorphological change in the Gordion region. While behavioral ecology models, especially those related to risk management, help to explain decision making behind some changes in farming practices, resilience thinking provides a more holistic framework for understanding the complex feedbacks between environmental change and agricultural economies over both the short and long terms. In the final chapter, I integrate the environmental and agricultural histories of Gordion from the perspective of resilience thinking. Identifying effects of scale, thresholds, and legacy effects not only give new insight into landscape change at Gordion but also make Gordion a suitable case study for considering environmental change more broadly, both past and present.

6

Risk, Resilience, and Sustainability in Agricultural Systems

Agricultural practices leave material traces in the archaeological record that can be reconstructed through environmental archaeology. Behavioral ecology provides a framework for assessing decision making by placing these strategies within their environmental and economic settings. In this chapter, I revisit the results of this study from the perspective of resilience thinking, considering how agricultural risk and sustainability serve to elucidate patterns of decision making within coupled human and natural systems. I return to the original questions laid out in Chapter 1 to consider in detail how agricultural practitioners chose farming and herding strategies; what social, economic, and environmental factors influenced those decisions; how, when, and where agriculture affected local landscapes; and how and on what scales environmental change influenced the development of agricultural economies at Gordion. I conclude that environmental change significantly affected land-use strategies over time, but that close attention to the speed and scale of human and environmental responses is critical to understanding the environmental impact of agriculture and its sustainability within the Gordion region.

Risk and Resilience in Agricultural Systems

Risk and resilience provide two compatible analytical frameworks for assessing agricultural decision making. Minimizing risk in agriculture is a goal of most farmers (Cashdan 1990b; Halstead and O'Shea 1989a; Marston 2011), as is ensuring the continued success of their agricultural systems in the face of environmental and social change. Although risk can be modeled using behavioral ecology, allowing the identification of specific risk-management practices and

their change over time, a resilience perspective helps to frame agricultural decision making within ongoing, reciprocal interactions between human societies and local environments. Resilience thinking highlights mismatches, threshold events, and legacy effects as key interactions and points of change that deserve additional analysis. It also draws attention to spatial, temporal, and organizational scale as structuring principles of human-environmental interactions.

Risk Management and Intensification

I drew on behavioral ecology models for risk management to identify metrics for risk management that can be measured using available archaeological data from Gordion, as described in Chapter 5 and in more detail in an earlier publication (Marston 2011). The first is a measure of crop diversification: the hulled barley to naked wheat ratio, which is based on historical (Gallant 1989, 1991; Garnsey 1999; Garnsey and Morris 1989) and ethnographic (Halstead 1990; Halstead and Jones 1989; Jones and Halstead 1995) evidence that farmers in the Mediterranean and Near East can plant a mixture of cereal crops to maximize returns in good rainfall years but minimize the risk of crop failure in dry years. This ratio remains consistent across occupation periods at Gordion until the Hellenistic, indicating a relatively conservative risk-management system for cereal agriculture through the Late Phrygian period but a riskier system in the Roman period (Fig. 5.2).

Diversification can be identified using other metrics as well and over multiple scales. On an organizational scale, a balance between farming and herding can reduce the risk associated with both sedentary agricultural and mobile pastoral economies. Pastoral systems appear to have been relatively consistent over time at Gordion. The high ratio of sheep and goat bones to those of cattle, which require more water and fodder, indicates a relatively low-risk pastoral system appropriate to the semi-arid plateau of Central Anatolia, albeit with a notable deviation towards higher risk during the Middle and Late Phrygian periods (Fig. 5.7). Spatial diversification is suggested by sheep and goat mobility (DiBattista 2014), frequent seeds of both gypsum steppe and riparian plant communities in animal dung burned as fuel (Fig. 5.10), and increased reliance on grazing versus foddering for herd animals in the Roman and Medieval periods (Fig. 5.9). Temporal diversification is evident during the Medieval period when a multicropping system of both winter- or spring-sown cereals and pulses with summer-sown millet and cotton was practiced; the availability of animal resources throughout the year in every period provides another form of temporal diversification (Ch. 5).

Agricultural intensification is an alternative strategy for risk reduction. Intensification can reduce agricultural risk through both increasing mean production and reducing the chance of crop failure due to drought (Fig. 2.3). One effective mechanism for such intensification in semi-arid regions is irrigation (Marston 2011; Miller and Marston 2012). The ratio of weeds of irrigation, here Cyperaceae, to total wild seeds serves as an effective proxy for the expansion of irrigation infrastructure (Fig. 5.11). The presence of obligately irrigated crops including cotton and rice during the Medieval period help to confirm this identification. These measures demonstrate the importance of irrigation to the agricultural systems of the Middle and Late Phrygian, Roman, and especially Medieval periods at Gordion (Fig. 5.11).

These paleoethnobotanical and zooarchaeological measures can be used to assess risk tolerance on a broader regional and hemispherical scale (Marston 2011, 2015). Analogous metrics can be developed for New World sites, where a diversity of indigenous cultigens were adopted and farming strategies may have utilized that diversity, together with other strategies, to reduce risk of crop failure (Scarry 2008; Stone and Downum 1999). Identifying diachronic change in risk management can help to illuminate the decision-making processes of ancient farmers, allowing hypotheses regarding climate change or political and economic change to be tested using archaeological

remains, further clarifying the role of agriculture in shaping coupled human and natural systems.

Scalar Mismatches

A resilience thinking approach illuminates interactions between variables that operate on the same scale, or across scales, within coupled human and natural systems (Anderies and Hegmon 2011; Marston 2015; Walker et al. 2012). Mismatches occur when variables interact in a manner that disrupts functions of a social or ecological system (Cumming et al. 2006:3; see further explanation in Chapter 2). Two such mismatches are evident at Gordion: the first a temporal and spatial mismatch within grazing systems during the Middle Phrygian period and the second, a spatial and organizational mismatch in agriculture during the Roman period.

Temporal and Spatial Mismatch: The Middle Phrygian Period

Temporal mismatches occur when variables interact at different speeds and intensities such that systemic changes result. In the case of Middle Phrygian Gordion, the interacting variables are the rates of population growth, urban growth, agricultural expansion, woodland growth, grassland growth, and erosion. While population and urban expansion lie squarely in the human system, and woodland and grassland growth, and erosion, in the natural system, agriculture straddles these two systems as a co-option of ecosystem function by deliberate human intervention. As such, agricultural systems comprise the most direct tool for exploring interactions in the coupled natural and human system at Phrygian Gordion. The catalyzing factor for change in this coupled system is population growth at Gordion, which expanded dramatically during the Middle Phrygian period (Voigt 2002, 2011, 2013), as did evidence for rural activity attested by survey data (Kealhofer 2005b; Marsh and Kealhofer 2014). As noted by Marsh and Kealhofer (2014:695, 697) and my earlier work (Marston 2015:594), it is the rapid acceleration of population growth and landscape use that sets the Middle Phrygian period apart from earlier population growth.

The agricultural system of the Middle Phrygian responded quickly to the needs of the larger population. Agricultural intensification is evident in multiple metrics: increased irrigation of crops (Fig. 5.11); more lentil for human consumption instead of bitter vetch for animal fodder (Fig. 5.5); greater numbers of pigs and cows, kept close to the city, compared to sheep and goat, grazed further afield (Figs. 5.6 and 5.7); and increased evidence for foddering of those animals with agricultural products (Fig. 5.9). Seed to charcoal ratios decline significantly (Miller 1999b, 2010:59; Miller and Marston 2012:100–101), indicating less use of dung for fuel, perhaps because sheep and goat herds were pushed further afield as a result of urban and agricultural expansion, making their dung unavailable for daily use for much of the year. The use of oak (Fig. 4.7) and riparian thicket (Fig. 4.8) plant communities for fuel increases, suggesting a focus on wood sources more easily managed for sustainable fuel production via coppicing and pollarding. An increase in the use of open steppe woodland trees (Fig. 4.8), however, also suggests land clearance for agriculture.

These interventions had direct environmental effects, in addition to landscape clearance and woodland structure modification. Animal grazing may have been constrained due to the increase in lands devoted to farming and animal herds may have also increased significantly in size, but whichever combination of these events occurred the result was increasingly severe overgrazing of local grasslands (Fig. 5.13). Correlated with this decline in steppe health is a substantial increase in sediment loads deposited in the Sakarya River, leading to aggradation of the river from the mid-first millennium BCE (Marsh 1999, 2005:169; Marston 2015:591).

In terms of how these variables interact, the increasing rate of population growth (a component of the human system) led to spatial expansion but also intensification and additional pressure on local resources, including a faster rate of consumption (of wood fuel, of grasses) in the ecological system. The result was that those ecological resources became increasingly impoverished, ultimately feeding back to the human system and resulting in the reduced availability of resources locally, necessitating further spatial expansion of the agricultural system, which then compounded these environmental effects across the landscape. Within this pattern of environmental change, human agency played a significant role in adaptation via agricultural decision making (Cancian 1980; Mar-

ston 2011, 2012a; Riehl 2009; M.L. Smith 2006; VanDerwarker et al. 2013). How farmers and herders responded to both changing economic demands for resources and different opportunities and constraints imposed by environmental change was conditioned on the social, political, and economic structures of the Middle Phrygian period—in short, the political economy (Marston 2012a). The speed of this response was rapid: farmers can change their crop cycle or expand irrigation ditches within a single year, and herders can shift grazing lands from day to day, while changes in herd structure (increase in animal numbers, change in management strategies) take only a few years to implement. Thus, here we find a "fast" cycle of human adaptation to economic and environmental change.

The mismatch ultimately lies in the differences in speed between the "fast" feedback cycles of human response and the slower processes of ecological and soil systems. While grassland communities in Central Anatolia are incredibly quick to respond to rainfall with growth and flowering cycles, and to temporary relief from animal predation (Miller 1999a, 2012), the possibility for complete denuding of vegetation on certain soil substrates is a real risk in the area (Fig. 3.11a). Such exposed soils are highly vulnerable to rapid transport via wind or water, resulting in rapid erosion, potentially on a regional scale. Destabilization of hillslope sediments occurred early in the occupation of the Gordion region despite minimal evidence for human presence (Marsh 2012; Marsh and Kealhofer 2014), while larger-scale erosion far upstream in the Sakarya watershed does not appear until the rapid population and agricultural expansion of the Middle Phrygian, presumably a result of a significant expansion in both scale and intensity of land use during that period (Marsh 1999, 2005). While both the decline in grasses due to grazing and their regrowth under suitable rainfall are "fast" processes and erosion can also be "fast," soil regeneration is a "slow" process that requires conditions of environmental stability. As plants do not regrow without an appropriate soil substrate, erosion limits available grazing areas, forcing grazing to become more intensive in areas that remain, or extensive across a broader spatial scale, and leads to changes in plant community ecology. As such, the relatively lower prevalence of *Gypsophila* during the Phrygian periods compared to the earliest phases of human settlement at Gordion (Fig. 5.10) may reflect reduced plant growth on gypsum substrates due to grazing pressure.

This mismatch is temporal in origin but became also a spatial mismatch as the effects of environmental change close to Gordion rippled out across the region. We have multiple suggestions that the agricultural system dramatically increased in spatial extent during this period: Middle Phrygian pottery is found more broadly across the landscape compared to earlier types (Kealhofer 2005b), the decreased seed to charcoal ratio suggests less animal dung was available for fuel at Gordion as sheep and goats were kept further afield (Table 5.2), the increase in open steppe woodland trees in the Gordion fuel assemblage indicates landscape clearance for agriculture (Fig. 4.8), and erosion of the Sakarya river catchment begins in earnest, suggesting expansion of grazing upstream (i.e., on upland plateaus and hillslopes to the south of Gordion; Fig. 3.5). Future research holds significant potential to explore spatial patterns of agriculture in greater detail: carbon and nitrogen stable isotope measurements on crop seeds can indicate the presence of crops from different fields (Fiorentino et al. 2012), while strontium analysis of animal teeth indicates the degree of pastoral movement across geological zones, including seasonal transhumance (Balasse et al. 2002; Chase et al. 2014; DiBattista 2014; Meiggs 2007). These data, together with further analysis of animal herd structure that address herd management strategies (e.g., Arbuckle 2012; Chase et al. 2014), will indicate spatial measures of adaptation and expansion of agricultural practices across the Gordion region, providing additional detail on how the temporal mismatch was either mitigated or exacerbated by spatial expansion. Ultimately, the interaction between these variables led to a "fast" feedback system that was successful in terms of meeting local subsistence needs but brought with it long-term implications, namely changes to plant communities and significant regional erosion.

Spatial and Organizational Mismatch: The Roman Period

The second scalar mismatch evident at Gordion occurred during the Roman period. In this case, the interacting variables are rates of extraurban population expansion, agricultural specialization and intensification, pastoral strategies, taxation systems, grassland growth, and erosion. Unlike the previous example, most of these variables lie in the human system, with grassland growth and erosion the only responding elements of the ecological system. During the Roman period, Gordion had a low urban population (Goldman 2000, 2005; Voigt 2002) but high regional population density (Kealhofer 2005b; Marsh and Kealhofer 2014). In this case, the ultimate cause for regional agricultural and environmental change appears to have been the Roman economic system, particularly taxation practices (Marston 2012a; Marston and Miller 2014). The agricultural economy departed from the diversified system of earlier periods with an emphasis on bread wheat production (Figs. 5.2 and 5.3) accompanied by a significant new investment in irrigation (Fig. 5.11). Although we lack local primary historical documents attesting to landholding and agricultural practices, as present for Roman Egypt (Arlt and Stadler 2013; Blouin 2014; Garnsey 1988; Monson 2012), historical studies of Roman Anatolia suggest that bread wheat was the preferred taxable agricultural product for this region and that it was produced on large rural landholdings (Garnsey 1988; Mitchell 1993:149–58, 232–34). A similar pattern is hypothesized to exist for animal husbandry, with a focus on sheep resulting from demand for wool or woolen goods as a taxation product (Mitchell 1993:146); unfortunately, no faunal evidence from Roman Gordion has yet been published so it is not possible to identify whether sheep herding increases during this period. Certainly, there is no reason to suspect a focus on cattle as suggested by Bennett (2013). Increased grazing is, however, apparent during this period: there is little evidence of dung fuel use at Gordion, based on reduced seed to charcoal ratios (Table 5.2). This suggests that animals were kept further from the site (perhaps in rural enclosures), while steppe grasslands became severely overgrazed (Fig. 5.13). This overgrazing is indicative of larger herds or increased competition between irrigated farmland and grazing ranges forcing intensified grazing on smaller areas. Additionally, animal diet contained less fodder during this period, suggesting an increase in wild forage tied to extensive grazing (Fig. 5.9).

The mismatch here lies within the organizational scale of agricultural decision making. Taxation demands were imposed on this region from without, either Ancyra (the provincial capital) or Rome itself, which severely constrained farmers' choices regarding crops to plant. With considerable energy and land devoted to bread wheat production for export, traditional risk-management strategies including crop and spatial diversification were eliminated in favor of intensive irrigation agriculture of bread wheat in the areas suitable for such production. Other agricultural activities were pushed into different ecological zones, including growing crops on different soils or in areas with poor drainage, and grazing herds either further afield or on lands unsuitable for farming, such as hillslopes and dry plateaus, with less productive and stable soils. With the loss of local autonomy in decision making came two attendant challenges. The first is the inability of local farmers and herders to make use of traditional ecological knowledge gathered through generations of experience in the Gordion region (Berkes and Turner 2006; Lepofsky and Kahn 2011; Turner and Berkes 2006). The second is the elimination of feedback loops between environmental change and agricultural practice at the local level, which allows rapid modification of agricultural strategies in response to environmental or economic change, as we saw earlier in the rapid transformation of the Middle Phrygian agricultural system. With these key losses, the Roman agricultural system was more vulnerable than that of earlier periods: both drought and flood posed dangers to farmlands in low-lying irrigable areas, and herding on less stable soils over a large area upstream may have been the cause of increased erosion rates in the Sakarya basin during this period (Marsh 2005). Herds were constrained from bottomlands (evidenced by lower prevalence of Malvaceae, Fig. 5.10) and overgrazing of gypsum soils may have already led to a loss of vegetation cover, as evidenced by the absence of *Gypsophila* seeds (Fig. 5.10) but high prevalence of *Alhagi* and *Peganum* (Fig. 5.13), indicating a severely degraded steppe with lower taxonomic diversity (Fig. 5.14).

Questions regarding the hypothesized spatial patterning of farming and grazing during the Roman period, however, remain unanswered with direct evidence. Additional light could be shed on this topic using the same isotopic approaches described above, namely carbon and nitrogen on crop seeds and strontium on animal teeth, and reconstruction of the Roman herding system through osteological analysis is much awaited. In addition, landscape reconstruction through remote sensing and spatial and hydrological modeling can clarify the relative value of specific ar-

eas for various agricultural practices (Casana 2014; DiBattista 2014; Harrower et al. 2012). Further historical research may also illuminate landholding structures and taxation systems in the region in further detail (cf. Haldon et al. 2014).

Despite the limited nature of some of the evidence regarding spatial patterning of land use, there is ample indication for the presence of organizational and spatial mismatches between human and natural systems during the Roman period. Resilience frameworks predict these outcomes given the spatial divorce in agricultural decision making between Gordion and those setting taxes for the region. Holling and Gunderson (2002a:27–28) argue that fixed rules for resource management, in place of adaptive feedbacks that allow rapid responses to change in coupled systems, tend to emphasize the efficiency of function (i.e., high crop yields) rather than long-term existence of function (i.e., diversified, low-risk, sustainable agriculture) and predictably lead to rapid change within coupled human and natural systems. Nonlocal, extractive decision making often imposes such rules on communities spatially and organizationally separate from those making such decisions (Ostrom 1990; Ramankutty 2001). In such an instance, coupled systems are characterized by overconnectedness (Holling and Gunderson 2002a:34–35; Redman and Kinzig 2003), in which stability is sought through constant maintenance of specific connections between systems (e.g., the need to maintain irrigation systems for wheat agriculture) that ultimately leave the system fragile and vulnerable to sudden and dramatic change if a single key variable is changed or connection is broken. Such an overconnected state is evident at Roman Gordion, leaving the agricultural and economic systems of the region vulnerable to external shocks, whether economic or environmental, and increasing the risk of sudden, dramatic change in the system. Thresholds of instability are effectively lowered to a point where they can be breached more easily, due to the loss of traditional mechanisms for maintaining system function, such as agricultural diversification.

Thresholds of Environmental and Economic Stability

Exploring the mismatches described above allows us to identify thresholds of environmental and economic stability. Two thresholds of environmental stability are evident: the first of upland soils and small watercourses, which aggraded rapidly during early human occupation of the Gordion region (Chalcolithic through Middle Bronze Age), and the second of soils in the Sakarya watershed upstream of Gordion, which eroded rapidly from the mid-first millennium BCE. Although geological in character, both erosion and alluviation events represent the results of an ecological threshold being crossed. The earlier erosional event represents a very low threshold boundary, crossed as soon as significant human (and, more importantly, herd animal) populations enter the northwestern Central Anatolian Plateau during the Chalcolithic (Marsh and Kealhofer 2014). The landscape encountered by early settlers contained plant and soil communities of varying stability, and those most vulnerable, generally thin soils with limited plant growth on hillslopes, were the first to be transported (Marsh and Kealhofer 2014:697–98). Contrary to initial expectation, increasing population densities and agricultural intensification in later periods, beginning in the Late Bronze Age, had a reduced impact on soil erosion in these drainages. As the most vulnerable soils had previously eroded, increasing land use in more stable areas of the Gordion landscape (such as the Sakarya floodplain) in later periods may have had little effect on these drainages.

The second threshold, represented by the alluviation and aggradation of the Sakarya River, is also ecological in origin. Its timing, coincident with the first major expansion of agricultural intensification during the Middle Phrygian period, clarifies the set of related processes by which this threshold was breached. As described above, spatial expansion of farming and especially herding, which most directly affects grassland communities and soil stability, is likely to have moved upstream during the Middle Phrygian. Overgrazing on thin gypsum soils such as those found in the plateaus south of Gordion (Fig. 3.5, evident as whiter soils) can lead to rapid and catastrophic soil loss if a sufficient transport agent (wind or water) is present. While areas upstream of Gordion were likely grazed from early occupation of the region, during the Middle Phrygian a critical grazing threshold was crossed that moved soils from conditions of stability to instability, resulting in rapid erosion and aggradation of the Sakarya valley.

Thresholds in the economic system are evident as well, although perhaps more challenging to identify than the ecological ones outlined above due to the complexity and frequent equifinality of economic and political processes. One is evident in the Roman period example described earlier: a threshold of stability and, perhaps, sustainability in the agricultural economy. The rigidity of the Roman agricultural system, constrained by taxation decreed by extralocal actors, left it vulnerable to collapse given sufficient economic or environmental shock. Unfortunately for this study, such shocks are not clearly evident at Late Roman Gordion, so it remains unclear why the site and region were evidently depopulated during the 4th century CE. One possibility is that changing geopolitics and increased conflict in the Roman world that resulted in the division of the eastern and western empires beginning in the early 4th century also led to changing economic incentives in Anatolia, resulting in the collapse of the taxation-driven agricultural system (Mitchell 1993:247–48) and migration from the Gordion region to other areas (e.g., cities of the Aegean coast or Constantinople). Alternately, environmental shocks may have caused the threshold event leading to regional depopulation, perhaps a series of droughts or floods that broke the fragile economic system of Roman Gordion. Several significant drought and famine events, as well as floods, are recorded in 4th century records, e.g., drought in 368–369 in Phrygia followed by floods across northwestern Anatolia in 370 (Haldon et al. 2014:154). Both the precise timing and cause for the abandonment of Roman Gordion and the surrounding countryside remain unclear, but a threshold is evident: the economic network that governed the intensive agricultural system of the Roman countryside in Central Anatolia clearly failed during the 4th century, resulting in the abandonment of Gordion for some 800 years.

Legacy Effects

Legacy effects are the long-term results of interactions between social and environmental processes that result in changed conditions for subsequent populations. Archaeology is a tremendously potent analytical tool for exploring legacy effects, because through diachronic archaeological data we can consider interactions between past dynamics that changed environmental conditions and how they affected subsequent societies and ecosystems. As archaeology is a discipline of materiality, focused on study of the tangible products of human-environmental interaction (such as agricultural remains), archaeologists simultaneously study both environmental and cultural change over time, highlighting both the spatial and temporal pace of expansion and contraction in those relationships. Legacy effects are the results of these scalar processes and because archaeology allows us to explore environmental and social end states, such as societal collapse, we can begin to place collapse (or other societal events) in long-term environmental context and explore the role of legacy effects in those outcomes.

The question of how legacy effects within the cultural sphere affect subsequent societies is an open one. While environmental legacy effects and their implications for both subsequent ecosystems and human systems are well documented (e.g., Cuddington 2011; Foster et al. 2003; Messner and Stinchcomb 2014; Waylen et al. 2015), cultural parallels are few. From one perspective, practices such as traditional ecological knowledge employed in the service of agriculture and land use might be considered a legacy of prior populations, although the visibility of those practices archaeologically is limited (for an example see Lepofsky and Kahn 2011). Similarly, cultural continuity in shared material culture, such as ceramic traditions linked to culinary practices (Hegmon et al. 2008; Stewart 2010) or patterns of architectural construction and the use of space (Atalay and Hastorf 2006; Bogaard et al. 2009; Love 2013a, 2013b), could be understood in this light. While this question deserves more attention from social scientists (including archaeologists), for the purposes of this study I consider only environmental legacy effects and their impact on both ecosystems and societies over time.

Two environmental legacy effects are evident at Gordion. The first is landscape transformation tied to deforestation, forest succession, and landscape clearance, beginning in the Late Bronze Age and continuing through the Late Phrygian period. Juniper use for fuel declines substantially following the Late Bronze Age (Fig. 4.7), while juniper is still used in monumental civic and elite construction through Middle Phrygian period. This transition suggests that juniper became a rarer resource, likely obtained further from the city at increased cost, and so was restricted to

higher-status construction contexts and rarely available for fuel (Marston 2009; Miller 1999b, 2010). A different ecological transition in woodland community can be documented through increased use of open steppe woodland trees for fuel from the Early to Late Phrygian period (Fig. 4.8), a product of landscape clearance that I conclude is likely to represent the spatial expansion of agriculture. Finally, the increase in the use of both riparian trees and oak from the Early Phrygian to the Hellenistic period suggests that inhabitants of Gordion focused fuelwood collection on species that represent the most renewable sources of wood in the region, capable of sustained harvest via techniques such as coppicing and pollarding (Smith 2014; Thiébault 2006). The legacy effect of these processes is the thinning of large, old-growth trees suitable for construction from the immediate landscape of Gordion, while maximizing fuel resources locally. Such changes privilege fuel over other potential uses of plants, such as fruit production and construction. Instead we see increased travel time required for the acquisition of timbers suitable for large structure construction (Ch. 4; Marston 2009). Inhabitants of Gordion even in the Phrygian periods faced a woodland landscape significantly different than that present during the earliest periods of settlement in the Gordion region. This had implications for several economies, including construction and both domestic and industrial fuel acquisition and use.

The second legacy effect at Gordion is soil erosion, which occurred in two main phases as described earlier. The legacy effect of early upland erosion moved agricultural production to flatter, lower-lying sediments, constraining the spatial extent of agriculture in the immediate vicinity of Gordion. The second phase of erosion, resulting in the aggradation of the Sakarya, may have again resulted in the movement of agricultural lands, as the floodplain of the Sakarya became increasingly vulnerable to flooding as the river began to aggrade and meander. This further constrained the availability of local agricultural lands or else necessitated additional investment in water diversion infrastructure. As a result, land-use decisions made by earlier populations in the region significantly altered the landscape available to later farmers and herders, imposing costs and constraints on subsequent agropastoral systems due to past sequences of environmental change.

Assessing Agricultural Sustainability

Defining an Archaeology of Sustainability

A definition for sustainability and a discussion of its specific relationship to resilience is lacking in many recent archaeological publications (e.g., Marston 2015; Middleton 2012; Smith 2009b), which instead use sustainability as a commonplace term referring to the long-term implications of a practice: those that endure are sustainable, those that do not are unsustainable. Such a teleological approach is likely unsatisfying to many. In contrast, Holling and colleagues define sustainability in a resilience context as "the capacity to create, test, and maintain adaptive capability" (Holling et al. 2002b:76), but such a definition is unsuitable for archaeology as it defines sustainability in the context of present dynamics rather than enduring outcomes that might become visible in a material culture assemblage. In this context, a basic definition of sustainability as "the capacity to endure" seems most appropriate as a concept easily operationalized for archaeological investigation. Of course, the duration of what constitutes an "enduring" practice remains relative: while some scholars might consider persistence throughout several generations to be enduring, in other regions scholars may hold only millennial-scale continuity as endurance. I make no claims about the value of these definitions and the challenges in their use here, but rather argue that operationalization of such a definition is key to archaeological research. My focus is on agricultural sustainability as one example of a productive way to identify sustainable versus unsustainable practices, allowing more productive dialog between scholars of the past and the present (cf. Chase and Scarborough 2014; Cooper and Sheets 2012; Costanza et al. 2007).

The challenge for archaeology lies in finding clearly identifiable material evidence for the endurance or ephemerality of cultural traditions and decision making that can be classified as sustainable or not. Environmental sustainability, the primary concern of scholars of the present, is a vague notion, whether defined as process (as by Holling et al. 2002b:76) or as outcome. One challenge in applying environmental sustainability to the past is that the term is encoded with presentist notions of how people "should" interact with environments responsibly, which both discriminates against certain traditional practices (e.g., swidden or slash-and-burn agriculture [Erickson 2006; Netting 1993]) and undermines the ability and agency of indigenous groups in shaping their landscape in a deliberate and premeditated fashion (e.g., through deliberate burning [Bird et al. 2005; Gremillion 2015; Roos et al. 2014; Sullivan and Forste 2014]). Archaeology has an advantage here: if the definition of sustainability focuses instead on ultimate outcomes, then archaeologists can reconstruct past dynamics and actions leading to final outcomes (such as societal collapse) already preserved in the archaeological record. While scholars may dispute the nature and causality of past practices in reaching those outcomes, notably in such high-profile case studies as Easter Island (Diamond 2005; Hunt 2007; Hunt and Lipo 2010; Mieth and Bork 2010), archaeology provides real data to explore the pathways and chained events by which sustainable practices endure or unsustainable practices are adopted.

Agriculture is perhaps the most tangible field for the archaeological exploration of sustainability. Sustainable agriculture is relatively clear to define: sustainable practices result in the ongoing supply of food and other resources (e.g., fibers) for a population, while unsustainable practices result in food insecurity and/or significant environmental change that renders an agricultural practice untenable for the future (e.g., Marston and Miller 2014; Mieth and Bork 2010). A sustainable agricultural system is likely to be resilient to various shocks, from drought to warfare (Langlie and Arkush 2016; M.L. Smith 2006; VanDerwarker et al. 2013). Through the reconstruction of agricultural practices and the decision-making processes behind them, coupled with evidence for paleoenvironmental change, as detailed throughout this book, archaeologists can identify the agricultural practices that led to specific environmental and economic outcomes and, thus, identify which were supportive of long-term agricultural stability and which led to environmental degradation over the long term. From this standpoint, resilience thinking informs models of agricultural sustainability by highlighting processes (mismatches) and moments (thresholds) that led to unsustainable outcomes for future generations (legacy effects). Such processes may be observed in the past, but can also inform consideration of the likely future outcomes of similar processes acting in the present.

Bridging the Past and the Present

Archaeological contributions to the study of environmental sustainability, especially agricultural sustainability, in the present day are meaningful. This is due, in no small part, to the fact that archaeologists can identify outcomes of past agricultural practices within their social and environmental settings, which we can use to predict outcomes of present practices. Legacy effects resulting from past (historical or archaeological) land use have implications for present and future agriculture, and so are important to understand for contemporary planning and require archaeology to do so. Similarly, thresholds are more evident in retrospect, after they have been crossed, than in advance (Barnosky et al. 2012; Scheffer and Carpenter 2003; Walker and Meyers 2004). If archaeology can be used to identify the dynamics contributing to past threshold events, we become more informed about local processes leading to dramatic environmental and social change and become more able to predict thresholds in the future as well, a practice adopted in many high-profile attempts to predict future tipping points for the planet (e.g., Barnosky et al. 2012; Foley et al. 2005; Liu et al. 2007a; Steffen et al. 2015). Archaeology is a valuable tool in establishing the time depth necessary to understand the role of humans in reaching thresholds in the past.

What is the future of agricultural sustainability at Gordion? The modern village of Yassıhöyük which lies at the ancient site still bases its economy on agriculture, although the crops farmed and methods for farming have been transformed in the last century through mechanization and connections with the global economy. Cash crops for export, including irrigated onions and sugar beets, dominate portions of the landscape. The input of nitrogen fertilizers on these fields has the potential for causing significant (and perhaps irreversible) local ecological change through freshwater eutrophication. A decline in herding and increase in farming of cash crops, together with fewer, larger landholdings and the use of migrant labor from Southeastern Turkey and Syria for fieldwork have also transformed the local economy (Gürsan-Salzmann 2005; Miller 2011). Successes in local ecological preservation include newfound interest in organic farming among some of the new generation of family farmers, alongside continued efforts to link the preservation of steppe ecosystems and archaeological landscapes in the region (Erder et al. 2013; Miller 2012). The future of this region looks to be hotter and drier (Fujihara et al. 2008; Kimura et al. 2007; Kovats et al. 2014:1278), two trends unlikely to be favorable for current agricultural regimes. Borrowing past practices, including crop diversification, rainfed farming of drought-tolerant barley and hulled wheats, and extensive grazing of sheep and goats is likely to be a more successful strategy than intensive, irrigated cultivation of bread wheat, onions, and sugar beets. How local farmers respond to climate change is itself a worthy study, as deep ethnographic perspectives on their ties to the land and the ways in which they employ traditional ecological knowledge in the context of a globalized liberal economy is likely to provide insights into past systems of agricultural decision making in that landscape. While the present can benefit from deeper knowledge of the past, the study of the past is made more robust through study of decision making today.

Future Directions

With this book, building on Miller's (2010) volume on agriculture and plant economy at Gordion, we conclude the publication of botanical remains from the 1988–2005 excavation campaigns at Gordion. With that accomplished, however, several more specific questions about Gordion remain to be answered. How did changing agricultural strategies involve animal herd management in concert with farming? Ongoing analysis of herd mortality profiles through age-at-death and sex estimation will inform many of the hypotheses regarding intensification and extensification of agriculture posed above. What was the spatial structure of land use during each period of occupation? While patterns of landscape change and animal diet, as preserved in burned dung, give suggestions about human and animal impacts on specific landscape components, future isotopic work can help to source fields of production, as well as fertilization and irrigation, for agricultural crops (Bogaard et al. 2007; Fiorentino et al. 2012; Fiorentino et al. 2015), and distinguish geological locations of animal grazing (Bogaard et al. 2014; DiBattista 2014; Madgwick et al. 2012). Finally, several periods of great interest remain poorly sampled, with the Early and

Middle Bronze Age completely unstudied (with the exception of two Middle Bronze Age samples [Miller 2010]) and the Roman and Medieval periods under-represented. Additional work on new samples from these eras would be of great interest, especially the Medieval, which saw the introduction of a new agricultural system under Islamic rule. The study of early Islamic Anatolia is a nascent but growing field (e.g., Miller 1998; Ramsay and Eger 2015) and additional data from Gordion would make a meaningful contribution to this conversation.

The approach I adopt in this study, which integrates botanical remains with both behavioral ecology and resilience frameworks, is a successful model for future research on human-environmental engagement, especially that centered on agricultural systems. Behavioral ecology modeling allows specific economic and environmental relationships to be modeled in a flexible framework that highlights critical variables informing human decision making and illuminates why certain practices, but not others, might be adopted at a certain time and place. Resilience thinking provides an overarching perspective on processes inherent to human-environmental relationships that drive reciprocal change in both economic and ecological systems. More importantly, it helps us understand why dramatic change occurs under certain conditions but not others. This approach can be applied productively to other archaeological case studies on both local (site-based) and regional scales; I have suggested that this analytical approach is particularly amenable to regional scales of inquiry (Marston 2015:599) in both the Old and New Worlds.

This study highlights the need for broader scales of interpretation. To understand changes in agricultural practices during imperial rule, such as that of the Romans, requires understanding economic and political networks on a scale beyond that of the site and its immediate landscape. Regional political economy plays a critical role in structuring local decision making and environmental responses (Marston 2012a; Marston and Miller 2014). Broader scales of inquiry are needed for imperial periods of the eastern Mediterranean world, from the Hittite to the Islamic. Fortunately, as the database of site-specific archaeobotanical studies has grown, regional syntheses are becoming more feasible for the region (e.g., Fall et al. 2015; Klinge and Fall 2010; Miller and Marston 2012; Riehl 2009). Historical periods of the eastern Mediterranean remain understudied in comparison with earlier sites, however, so the study of empire and environment remains a nascent investigation in this part of the world. Historians, however, are already beginning to take up this challenge, applying techniques of global history (cf. Morris 2010; Scheidel 2009) to detailed regional studies of imperial engagements with environment (e.g., Blouin 2014; Gallant 1991; Garnsey 1988; Monson 2012). Archaeological synthesis of both site-based and regional historical and archaeobotanical datasets is now both possible and necessary to understand the dynamics of human-environmental interactions across the diverse landscapes, economies, and cultures of the imperial eastern Mediterranean.

Finally, such place-based and regional studies of past cycles of human-environmental engagement offer potential lessons and warnings for the present and future. Agricultural strategies practiced in the past had environmental implications, and studying the successes and failures of specific practices and integrated agricultural systems within their cultural and economic context provides insights for land use today. Since archaeology exposes the ultimate sustainability of land-use strategies, it is possible to use insight from the past to inform the likelihood of sustainability for analogous practices today. Archaeology can be a partner with contemporary environmental studies in the goal of sustainable development of agricultural systems to meet the real challenges of population growth, urbanization, environmental degradation, and climate change that face populations worldwide today. Studies such as this begin to offer insights from specific corners of the world that can inform local, and potentially regional, decision making in the future.

Appendix A

Phytogeographic Communities at Gordion

Note: Data in this appendix comes from Miller (2010: app. C1). "Riverine" combines Miller's "Sakarya Bank" and "Sakarya Plain" categories. "Healthy Steppe" combines Miller's "Tumulus MM" and "Grassy Steppe" categories. "Overgrazed Steppe" is the same as Miller's "Grazed" category. "Gypsum Steppe" combines Miller's "S. Ridge" and "N. Ridge/Kızlarkaya" categories. "Field Edge" is the same as Miller's "Irrigated Field Edge." All identifications in this appendix are those of Miller.

Family	Species	Riverine	Healthy Steppe	Overgrazed Steppe	Gypsum Steppe	Field Edge
Amaranthaceae	*Amaranthus albus/graecizans*					X
Amaranthaceae	*Amaranthus* – misc.					X
Apiaceae	*Bifora radians* Bieb.					X
Apiaceae	*Bupleurum* cf. *flavum* Forssk		X		X	
Apiaceae	*Bupleurum* cf. *turcicum* Snogerup		X		X	
Apiaceae	*Caucalis platycarpos* L.					X
Apiaceae	*Conium maculatum* L.		X			X
Apiaceae	*Daucus carota* L.	X				X
Apiaceae	cf. *Echinophora* sp.			X	X	X
Apiaceae	*Eryngium campestre* L.		X	X	X	
Apiaceae	*Eryngium creticum* Lam.		X	X	X	
Apiaceae	*Falcaria vulgaris* Bernh.		X			X
Apiaceae	*Malabaila carvifolia* Boiss. & Mal./*Peucedanum palimboides* Boiss.				X	
Apiaceae	*Malabaila* cf. *secacul* Banks & Sol.		X			
Apiaceae	*Malabaila* sp.		X			
Apiaceae	*Scandix* sp.		X			
Apiaceae	*Torilis leptophylla* (L.) Reichb.		X		X	
Apiaceae	*Turgenia latifolia* (L.) Hoffm.		X			X
Asclepiadaceae	*Cynanchum acutum* L. subsp. *acutum*	X				X
Asteraceae	*Achillea* cf. *biebersteinii* Afan.	X				X
Asteraceae	*Achillea wilhelmsii* C. Koch	X	X	X		

Family	Species	Riverine	Healthy Steppe	Overgrazed Steppe	Gypsum Steppe	Field Edge
Asteraceae	*Achillea* perennial (like *A. teretifolia* Willd.)		X	X		
Asteraceae	*Achillea tenuifolia* Lam.		X			
Asteraceae	*Achillea* – various		X			
Asteraceae	*Acroptilon repens* (L.) DC.					X
Asteraceae	*Anthemis* (nm 1232)	X	X		X	
Asteraceae	*Artemisia* sp.		X	X	X	
Asteraceae	*Carduus nutans* L.	X	X	X		
Asteraceae	*Carduus pycnocephalus* L.	X	X			
Asteraceae	*Carduus* (nm 2495)	X	X	X		
Asteraceae	*Centaurea calcitrapa* L.	X				
Asteraceae	*Centaurea patula*		X		X	
Asteraceae	*Centaurea pseudoreflexa*		X		X	
Asteraceae	*Centaurea pulchella* Ledeb.		X		X	
Asteraceae	*Centaurea* cf. *rigida* Banks & Sol.	X	X			
Asteraceae	*Centaurea solstitialis* L. subsp. *solstitialis*	X	X	X		
Asteraceae	*Centaurea virgata* Lam.		X	X	X	
Asteraceae	*Centaurea* (nm 1227; blue)		X			X
Asteraceae	*Cichorium intybus*		X			X
Asteraceae	*Cirsium* sp.	X				X
Asteraceae	*Cnicus benedictus* L.		X			
Asteraceae	*Cousinia halysensis* Hub.-Mor.		X	X	X	
Asteraceae	*Crepis* sp.		X			X
Asteraceae	Crepis/Scorzonera		X			
Asteraceae	*Cymbolaena griffithii* (A. Gray) Wagenitz		X	X		
Asteraceae	*Echinops* sp.		X			
Asteraceae	*Gundelia tournefortii* L.		X	X	X	
Asteraceae	*Jurinea pontica* Hausskn. & Freyn ex Hausskn.				X	
Asteraceae	*Koelpinea linearis* Pallas		X			
Asteraceae	*Lactuca serriola* L.		X			X
Asteraceae	cf. *Leucocyclus* sp.				X	
Asteraceae	*Matricaria aurea* (L.) Schultz Bip.	X				X
Asteraceae	*Matricaria* cf. *macrotis* Rech. fil.		X	X	X	
Asteraceae	*Onopordum anatolicum* (Boiss.) Eig	X	X			
Asteraceae	*Pulicaria vulgaris* Gaertner	X				
Asteraceae	*Scorzonera laciniata* L.		X			
Asteraceae	*Scorzonera/Crepis*		X			

Family	Species	Riverine	Healthy Steppe	Overgrazed Steppe	Gypsum Steppe	Field Edge
Asteraceae	*Senecio* cf. *vernalis* Waldst. & Kit.		X			
Asteraceae	*Taraxacum* sp.		X			
Asteraceae	*Tragopogon dubius* Scop.		X			X
Asteraceae	*Xanthium spinosum* L.	X				X
Asteraceae	*Xanthium strumarium* ssp. *cavanillesii* (Schouw) D. Löve & P. Dansereau	X				X
Asteraceae	*Xeranthemum inapertum* (L.) Miller		X		X	
Boraginaceae	*Anchusa arvensis* (L.) Bieb.					X
Boraginaceae	*Echium italicum* L.		X			
Boraginaceae	*Heliotropium* spp.		X			X
Boraginaceae	*Lappula patula* (Lehm.) Aschers. ex Gürke		X			
Boraginaceae	*Moltkia* cf. *coerulea* (Willd.) Lehm.		X	X	X	
Boraginaceae	*Nonea caspica* (Willd.) G. Don		X	X		
Boraginaceae	*Rochelia disperma* (L. fil.) C. Koch		X	X		
Brassicaceae	*Alyssum* cf. *linifolium* Steph. ex Willd.		X			X
Brassicaceae	*Alyssum sibiricum* Willd.			X	X	
Brassicaceae	*Alyssum* (nm 1648)	X	X	X		
Brassicaceae	*Alyssum* (nm 2393)		X			
Brassicaceae	*Alyssum* – various	X	X		X	
Brassicaceae	*Boreava orientalis* Jaub. & Spach					X
Brassicaceae	*Camelina rumelica* Vel.			X		
Brassicaceae	*Cardaria draba* (L.) Desv.		X			X
Brassicaceae	*Conringia orientalis* (L.) Andrz.					X
Brassicaceae	*Descurainia sophia* (L.) Webb & Berth. ex Prantl		X	X		X
Brassicaceae	*Erysimum repandum* L.	X				
Brassicaceae	*Isatis* cf. *cappadocica* Desv.		X			X
Brassicaceae	cf. *Lepidium* (nm 2682)					X
Brassicaceae	*Matthiola longipetala* (Vent.) DC. subsp. *bicornis*		X			
Brassicaceae	*Rapistrum rugosum* (L.) All.					X
Brassicaceae	*Sinapis arvensis* L.					X
Brassicaceae	*Sisymbrium altissimum* L.		X			X
Brassicaceae	*Sisymbrium* sp. (nm 2482)		X			
Brassicaceae	Brassicaceae (nm 2526)		X			
Capparidaceae	*Capparis spinosa* L. var. *spinosa*			X	X	
Caryophyllaceae	*Bufonia* cf. *virgata* Boiss.		X			
Caryophyllaceae	*Cerastium dichotomum* L.					X
Caryophyllaceae	*Dianthus zonatus* Fenzl				X	

Family	Species	Riverine	Healthy Steppe	Overgrazed Steppe	Gypsum Steppe	Field Edge
Caryophyllaceae	*Dianthus zonatus* Fenzl/*floribundus* Boiss.		X	X	X	
Caryophyllaceae	*Gypsophila eriocalyx* Boiss.		X		X	
Caryophyllaceae	*Gypsophila* cf. *lepidioides* Boiss.				X	
Caryophyllaceae	*Gypsophila pilosa* Hudson					X
Caryophyllaceae	*Gypsophila viscosa* Murray		X			
Caryophyllaceae	*Gypsophila* (nm 2193)	X				
Caryophyllaceae	*Lychnis* sp.					X
Caryophyllaceae	*Minuartia anatolica* (Boiss.) Woron		X	X	X	
Caryophyllaceae	*Minuartia hamata* (Hausskn.) Mattf.		X		X	
Caryophyllaceae	*Silene conoidea* L.					X
Caryophyllaceae	*Silene* cf. *subconica* Friv.		X	X		
Caryophyllaceae	*Silene supina* Bieb.		X			
Caryophyllaceae	*Spergularia media* (L.) C. Presl	X				
Caryophyllaceae	*Stellaria media* (L.) Vill.		X			
Caryophyllaceae	Caryophyllaceae (nm 2530)		X			
Chenopodiaceae	*Atriplex laevis* C.A. Meyer					X
Chenopodiaceae	*Atriplex* cf. *leucoclada* Boiss.			X		
Chenopodiaceae	*Atriplex* – various	X	X			X
Chenopodiaceae	*Beta vulgaris* L.					X
Chenopodiaceae	*Camphorosma* sp.	X		X		
Chenopodiaceae	*Chenopodium album* L.					X
Chenopodiaceae	*Chenopodium* cf. *vulvaria* L.	X				
Chenopodiaceae	*Chenopodium* – various					X
Chenopodiaceae	*Kochia prostrata* (L.) Schrad.		X			
Chenopodiaceae	*Krascheninnikovia ceratoides* (L.) Güldenst.		X			
Chenopodiaceae	*Noaea mucronata* (Forssk.) Aschers. & Schewinf.		X		X	
Chenopodiaceae	*Petrosimonia* sp.	X				
Chenopodiaceae	*Salsola laricina* Pallas		X	X	X	
Chenopodiaceae	*Suaeda altissima* (L.) Pall.	X				X
Chenopodiaceae	Chenopodiaceae – various	X				X
Cistaceae	*Fumana paphlagonica* Bornm. & Janchen				X	
Cistaceae	*Helianthemum salicifolium* (L.) P. Mill.		X	X	X	
Convolvulaceae	*Convolvulus arvensis* L.	X		X		X
Convolvulaceae	*Convolvulus galaticus* Rostan ex Choisy	X				
Convolvulaceae	*Convolvulus scammonia* L.	X				X

Family	Species	Riverine	Healthy Steppe	Overgrazed Steppe	Gypsum Steppe	Field Edge
Convolvulaceae	*Convolvulus* (nm 2395)		X	X	X	
Convolvulaceae	*Convolvulus* – various perennial				X	
Cuscutaceae	*Cuscuta* sp.	X	X			
Cyperaceae	*Cyperus* (nm 1408)	X				
Cyperaceae	*Eleocharis mitrocarpa* Steudel/ *palustris* (L.) Roemer & Schultes	X				
Cyperaceae	*Scirpoides holoschoenus* (L.) Sojak	X				
Dipsacaceae	cf. *Dipsacus cephalarioides* Matthews & Kupicha		X			
Dipsacaceae	*Dipsacus* cf. *laciniatus* L.	X				
Dipsacaceae	*Scabiosa* cf. *rotata* Bieb.		X		X	
Dipsacaceae	*Scabiosa* cf. *argentea* L.		X		X	
Dipsacaceae	*Scabiosa* – various		X			
Elaeagnaceae	*Elaeagnus angustifolia* L.					X
Equisetaceae	*Equisetum fluviatile* L.	X				
Euphorbiaceae	*Euphorbia macroclada* Boiss.		X		X	
Euphorbiaceae	*Euphorbia* (nm 1681)		X			
Euphorbiaceae	*Euphorbia* (nm 1772)		X			
Fabaceae	*Alhagi pseudalhagi* (Bieb.) Desv.	X	X	X	X	X
Fabaceae	*Astracantha* sp.		X		X	
Fabaceae	*Astragalus asterias* Stev. ex Ledeb.		X		X	
Fabaceae	*Astragalus hamosus* L.			X		
Fabaceae	*Astragalus lycius* Boiss.		X	X	X	
Fabaceae	*Astragalus macrocephalus* Willd.		X			
Fabaceae	*Astragalus odoratus* Lam.					X
Fabaceae	*Astragalus triradiatus* Bunge		X			
Fabaceae	*Astragalus* (nm 2347)		X			
Fabaceae	*Galega officinalis* L.	X				X
Fabaceae	*Genista sessilifolia* DC.				X	
Fabaceae	*Glycyrrhiza echinata* L.	X				
Fabaceae	*Hedysarum cappadocicum* Boiss.		X		X	
Fabaceae	*Hedysarum varium* Willd.		X	X	X	
Fabaceae	*Lotus corniculatus* L.	X				
Fabaceae	*Medicago constricta* Dur.		X	X		
Fabaceae	*Medicago sativa* L.					X
Fabaceae	*Melilotus alba* Desr.	X				
Fabaceae	*Onobrychis armena* Boiss. & Huet		X	X	X	
Fabaceae	*Onobrychis tournefortii* (Willd.) Desv.		X		X	
Fabaceae	*Ononis spinosa* L.	X				
Fabaceae	*Trifolium resupinatum* L.	X				

Family	Species	Riverine	Healthy Steppe	Overgrazed Steppe	Gypsum Steppe	Field Edge
Fabaceae	*Trifolium* sp.	x				
Fabaceae	*Trigonella astroites* Fisch. & Mey.		x			
Fabaceae	*Trigonella capitata* Boiss.		x			x
Fabaceae	*Trigonella coerulescens* (Bieb.) Hal.		x			
Fabaceae	*Trigonella crassipes* Boiss.	x	x			
Fabaceae	*Trigonella* cf. *fischeriana* Ser.		x			
Fabaceae	*Trigonella monantha* C.A. Meyer		x			
Fabaceae	*Trigonella* cf. *orthoceras* Kar. & Kir.		x			
Frankeniaceae	*Frankenia hirsuta* L.				x	
Geraniaceae	*Erodium cicutarium* (L.) L'Hérit.	x	x			x
Globulariaceae	*Globularia orientalis* L.				x	
Hypericaceae	cf. *Hypericum scabroides* Robson et Poulter				x	
Illecebraceae	*Scleranthus* cf. *annuus* L.		x	x		
Illecebraceae	*Herniaria incana* Lam.		x			
Illecebraceae	*Paronychia* (nm 1699)				x	
Juncaceae	*Juncus* cf. *gerardi* Loisel.	x				
Juncaceae	*Juncus* cf. *inflexus* L.	x				
Lamiaceae	*Acinos rotundifolius* Pers.		x			
Lamiaceae	*Ajuga chamaepitys* L. Schreber		x	x		
Lamiaceae	*Marrubium parviflorum* Fisch. et Mey.		x	x	x	
Lamiaceae	*Marrumbium vulgare* L.	x	x			x
Lamiaceae	*Mentha aquatica* L.	x				
Lamiaceae	*Molucella laevis* L.	x				
Lamiaceae	*Nepeta congesta* Fisch. & Mey.		x			
Lamiaceae	*Nepeta stricta* (Banks & Sol.) Hedge & Lamond		x			
Lamiaceae	*Phlomis pungens* Willd.	x		x		x
Lamiaceae	*Salvia ceratophylla* L.		x			
Lamiaceae	*Salvia cryptantha* Montbret & Aucher ex Bentham				x	
Lamiaceae	*Scutellaria orientalis* L. subsp. *pinnatifida*		x	x	x	
Lamiaceae	*Sideritis montana* L.				x	
Lamiaceae	*Stachys cretica* L.	x				
Lamiaceae	*Teucrium polium* L.		x	x	x	
Lamiaceae	*Thymus* – various		x		x	
Lamiaceae	*Wiedemannia orientalis* Fisch. & Mey.		x			x
Lamiaceae	*Ziziphora taurica* Bieb.		x	x	x	
Liliaceae	*Allium rotundum* L.		x			

Family	Species	Riverine	Healthy Steppe	Overgrazed Steppe	Gypsum Steppe	Field Edge
Liliaceae	*Allium* (nm 2380)		X		X	
Liliaceae	cf. *Asparagus tenuifolius* Lam.	X				
Liliaceae	cf. *Bellevalia* sp.		X			
Linaceae	*Linum bienne* Mill.		X		X	
Lythraceae	*Lythrum salicaria* L.	X				
Malvaceae	cf. *Lavatera bryonifolia* Miller	X				
Orobanchaceae	*Orobanche* sp.		X			
Papaveraceae	*Glaucium corniculatum* (L.) Rud.		X			X
Papaveraceae	*Glaucium haussknechtii* Bornm. & Fedde				X	
Papaveraceae	*Hypecoum imberbe* Sibth. & Sm.		X			
Papaveraceae	*Hypecoum pendulum* L.		X			
Papaveraceae	*Papaver* cf. *dubium* L.		X			
Papaveraceae	*Papaver hybridum* L.		X			X
Papaveraceae	*Papaver rhoeas* L.					X
Papaveraceae	*Papaver* sp.		X			
Plantaginaceae	*Plantago lanceolata* L.	X				
Plantaginaceae	*Plantago major* L.	X				
Plantaginaceae	*Plantago media* L.	X				
Plumbaginaceae	*Acantholimon* cf. *acerosum* (Willd.) Boiss.		X		X	
Plumbaginaceae	*Acantholimon* sp.				X	
Plumbaginaceae	*Limonium lilacinum* (Boiss. & Bal.) Wagenitz					X
Poaceae	*Aegilops cylindrica* Host	X	X			X
Poaceae	*Aegilops umbellulata* Zhuk.		X			
Poaceae	*Aegilops triuncalis* L./*geniculatum* Roth		X			
Poaceae	*Aeluropus littoralis* (Gouan) Parl.	X				
Poaceae	*Aegilops* (nm 2349)		X			
Poaceae	*Agropyron cristatum* (L.) Gaertner subsp. *pectinatum*		X	X	X	X
Poaceae	cf. *Alopecurus myosuroides* Hudson	X				
Poaceae	*Amblyopyrum muticum* (Boiss.) Eig/ *Elymus* sp.	X	X	X		
Poaceae	*Avena sativa* L.					X
Poaceae	*Briza humilis* M. Bieb.		X		X	
Poaceae	*Bromus cappadocicus* Boiss.		X	X	X	
Poaceae	*Bromus japonicus* Thunb. ssp. *anatolicus*	X	X	X	X	

Family	Species	Riverine	Healthy Steppe	Overgrazed Steppe	Gypsum Steppe	Field Edge
Poaceae	*Bromus tectorum* L.	x	x	x		
Poaceae	*Bromus tomentellus* Boiss.		x	x	x	
Poaceae	*Calamagrostis epigejos* (L.) Roth	x				
Poaceae	*Cynodon dactylon* (L.) Pers.	x	x	x		
Poaceae	*Echinaria capitata* (L.) Desf.		x		x	
Poaceae	*Echinochloa crus-galli* (L.) P. Beauv.	x				
Poaceae	*Elymus* cf. *hispidus* (Opiz) Melderis		x		x	
Poaceae	cf. *Eragrostis* sp.					x
Poaceae	*Eremopyrum bonaepartis* (Sprengel) Nevski	x	x	x		x
Poaceae	*Festuca ovina* L.		x	x	x	
Poaceae	*Hordeum distichum* L.					x
Poaceae	*Hordeum geniculatum* All.	x				
Poaceae	*Hordeum murinum* L.	x	x	x		x
Poaceae	*Leymus* sp.	x				
Poaceae	*Lolium* sp.	x				
Poaceae	*Phalaris arundinacea* L.	x				
Poaceae	*Phleum* cf. *boissieri* Bornm.		x			
Poaceae	*Phleum pratense* L.	x	x	x		x
Poaceae	*Phragmites australis* (Cav.) Trin. ex Stendel	x				
Poaceae	*Poa bulbosa* L.		x			
Poaceae	*Poa bulbosa* L. – proliferous		x	x	x	
Poaceae	*Polypogon monspeliensis* (L.) Desf.		x			
Poaceae	*Sclerochloa* sp.			x		x
Poaceae	*Secale cereale* L.					x
Poaceae	*Setaria verticillata* (L.) P. Beauv.	x				
Poaceae	*Stipa arabica* Trin. et Rupr.		x	x	x	
Poaceae	*Stipa holoseriaca* Trin.		x	x	x	
Poaceae	*Stipa lessingiana* Trin. & Rupr.			x	x	
Poaceae	*Taeniatherum caput-medusae* (L.) Nevski		x			
Poaceae	*Triticum aestivum* L.					x
Poaceae	*Triticum durum* Desf.					x
Poaceae	*Triticum turgidum* L.					x
Poaceae	Poaceae – various	x	x			
Polygonaceae	*Polygonum arenarium* Waldst. & Kit.	x				
Polygonaceae	*Polygonum pulchellum* Lois.					x
Polygonaceae	*Rumex gracilescens* Rech.	x				x
Polygonaceae	*Rumex pulcher* L.	x				x

Family	Species	Riverine	Healthy Steppe	Overgrazed Steppe	Gypsum Steppe	Field Edge
Primulaceae	*Androsace maxima* L.		X	X		
Ranunculaceae	*Aconitum nasutum* Fisch. Ex Reichb.		X			
Ranunculaceae	*Adonis* – misc.	X	X			
Ranunculaceae	Ceratocephalus		X		X	
Ranunculaceae	*Nigella arvensis* L.		X			
Ranunculaceae	*Consolida raveyi* (Boiss.) Schröd .		X		X	
Ranunculaceae	*Consolida orientalis* (Gay) Schröd.		X			X
Ranunculaceae	*Consolida* cf. *saccata* (Huth) Davis	X	X			X
Ranunculaceae	*Delphinium* cf. *cinereum* Boiss.			X		
Ranunculaceae	*Nigella arvensis* L.	X	X	X	X	X
Ranunculaceae	*Nigella nigellastrum* (L.) Willd.				X	
Ranunculaceae	*Nigella segetalis* Bieb.					X
Resedaceae	*Reseda microcarpa* Müller	X				
Resedaceae	*Reseda lutea* L.	X	X		X	X
Rosaceae	*Crataegus* sp.					X
Rosaceae	*Potentilla reptans* L.	X				
Rubiaceae	*Asperula* – various		X		X	
Rubiaceae	*Galium verum* L.	X	X			
Rubiaceae	*Galium* (nm 1242)		X			X
Rubiaceae	*Galium* (nm 2657)		X			
Salicaceae	*Salix* sp.	X				X
Scrophulariaceae	*Bungea trifida* (Vahl) C.A. Meyer				X	
Scrophulariaceae	*Linaria kurdica* Boiss. & Hohen.	X				
Scrophulariaceae	*Linaria simplex* (Willd.) DC.		X			
Scrophulariaceae	*Verbascum* sp.					X
Scrophulariaceae	*Veronica multifida* L.		X			
Solanaceae	*Datura stramonium* L.					X
Solanaceae	*Solanum dulcamara* L.	X				
Tamaricaceae	*Tamarix* sp.	X				X
Ulmaceae	*Ulmus glabra* Huds.					X
Thymeleaceae	*Thymelaea passerina* (L.) Cosson & Germ.		X	X	X	
Zygophyllaceae	*Tribulus terristris* L.					X
Zygophyllaceae	*Peganum harmala* L.		X	X		

Appendix B

Wood Charcoal Identification Guide

Gordion wood identifications were made using comparative collections, both my own (currently housed at Boston University [Boston University 2015]) and that of Naomi F. Miller (Penn Museum), and published descriptions and photomicrographs of wood sections in several wood anatomy texts (Fahn et al. 1986; Panshin and de Zeeuw 1970; Schoch et al. 2004; Schweingruber 1990; Schweingruber et al. 2006). Most identifications were possible to the genus level. It is rarely possible to distinguish between the woods of different species within a genus, although this is occasionally an option if particular species have distinctive characters. In two cases, species identifications were made based on phytogeographic grounds, as only one species of the genus grows in the Gordion region.

The format of this guide and some of the characters noted come from the previously published work of Miller (1991a; 2010) on Gordion charcoal excavated in 1988 and 1989. I consulted with Miller about the identification of some unknown types, but all final determinations (and any possible errors) are my own. Asterisks mark distinctive characters that are necessary for secure identification of the type.

Coniferous Woods

Pinus cf. *nigra*

Low magnification

x-section	*resin canals present in latewood, *early- to latewood transition distinct

High magnification

x-section	thin-walled epithelial cells surround resin canals
r-section	*cross-field pits uniseriate, fenestriform, marginal ray tracheids present
t-section	not examined

Juniperus sp.

Low magnification

x-section	*resin canals absent

High magnification

x-section	*intracellular spaces present
r-section	*cross-field pits cupressoid, no marginal ray tracheids
t-section	ray height usually less than 6 rows

Notes: other conifers without resin ducts that might be present at Gordion include fir (*Abies* sp.), yew (*Taxus* sp.), and cedar (*Cedrus libani*). Cedar often has traumatic resin ducts, as well as marginal ray tracheids, which were not observed here. Yew has spiral thickenings, and fir lacks intracellular spaces and has taxodioid/piceoid cross-field pits, which differ from the type observed here. In the modern surroundings of Gordion, *Pinus nigra*, *Juniperus oxycedrus*, and *Juniperus excelsa* are the only conifers present. Cedar has been identified from the wooden bier within Tumulus MM but not from any non-elite burial contexts at Gordion; the wood previously identified as yew is now identified as pine (Blanchette and Simpson 1992; Miller 2010:18; Simpson and Spirydowicz 1999).

Dicotyledenous Woods

Quercus sp.

Low magnification

x-section   *ring porous, *dendritic pore bands, growth ring distinct *narrow and wide multiseriate rays

High magnification

x-section   not observed

r-section   not observed

t-section   not observed

Salicaceae (*Salix* and *Populus*)

Low magnification

x-section   *diffuse porous, *pores small, in radial multiples or small groups, growth ring indistinct, *uniseriate rays

High magnification

x-section   as above

r-section   *homocellular rays (*Populus*) or *rays with 1–2 rows of square (upright) marginal ray cells (*Salix*); large, simple ray-vessel pits and intervessel pits

t-section   *rays uniseriate, 10–15 cells high

Notes: The distinction between *Salix* and *Populus* wood can only be made through differentiation of homo- and heterogeneous rays in the radial section. This is not always possible to determine for a given charcoal fragment, and some pieces of *Populus* may have a row of square marginal cells (Schweingruber 1990:673). Although I identified some pieces as either willow or poplar, the majority of fragments could not be securely identified and I report all Salicaceae wood at the family level for consistency.

Maloideae (most likely *Crataegus*, *Pyrus*, or *Malus*)

Low magnification

x-section   *diffuse porous, *pores small, solitary, thin rays

High magnification

x-section   *pores often angular in section

r-section   *thin spiral thickenings in vessels, rays homocellular or with one row of square marginal cells

t-section   *rays usually bi- to 3-seriate, ~5–15 cells high

Notes: It is not possible to distinguish between the woods of hawthorn, wild or cultivated apple, or pear (Schweingruber 1990:617).

Prunus sp. (includes *Amygdalus* spp.)

Low magnification

x-section   *ring to semi-ring porous, *latewood pores numerous, mostly solitary, *uni- and multiseriate rays

High magnification

x-section   as above

r-section   *thin spiral thickenings in vessels, rays generally slightly heterocellular

t-section   *rays uni- and ~5-seriate, *30+ cells high

Notes: It is possible to distinguish between groups of *Prunus* species based on the pore distribution and multiseriate ray width. The fragments observed here fall into the *Prunus armeniaca/dulcis/persica* group, suggesting they may be from trees with edible fruit (Schweingruber 1990:643).

Tamarix sp.

Low magnification

x-section   *ring porous, *pores sparse, solitary or in groups, growth ring distinct, *wide multiseriate rays

High magnification

x-section   as above

r-section   rays heterocellular, *numerous small intervessel and ray-vessel pits

t-section   *rays 5- to 20-seriate, *up to ~100 cells high

Fraxinus sp.

  Low magnification

x-section	*ring porous, *earlywood pores large, solitary or in radial pairs, *latewood pores small and sparse, thin rays

  High magnification

x-section	multiseriate rays
r-section	small, numerous ray-vessel pits, no spiral thickenings
t-section	*rays bi- to 3-seriate, ~5–15 cells high

Ulmus sp.

  Low magnification

x-section	*ring porous, growth ring fairly distinct, *earlywood pores arranged in groups of 2 or 3, occasionally solitary, latewood pores in groups of ~10; *pores and parenchyma in oblique bands that often continue across rays

  High magnification

x-section	as above
r-section	*fine spiral thickenings, *rays homo- to heterocellular, with a single row of squarish marginal cells, exclusively libriform fibers
t-section	*ray height 10–15 rows, *rays 2- to 4-seriate, rays frequent

Rhamnus sp.

  Low magnification

x-section	*diffuse porous, *dendritic pore bands, growth ring distinct, *biseriate rays

  High magnification

x-section	as above
r-section	*spiral thickenings in vessels
t-section	*rays biseriate

Notes: Miller (2010:78) remarks that her *Rhamnus* fragments look like *Rhamnus catharticus* depicted by Schweingruber (1990:615). Samples from this study look similar, but other indistinguishable species of *Rhamnus* exist in Central Anatolia (Davis 1965–2000).

cf. *Celtis* sp.

  Low magnification

x-section	*ring porous, growth ring fairly distinct, *earlywood pores arranged in groups of 2 or 3, occasionally solitary, *pores and parenchyma in oblique bands that often continue across rays

  High magnification

x-section	as above
r-section	*fine spiral thickenings, *rays heterocellular, multiple rows of upright and square marginal cells
t-section	*ray height 10–15 rows, *rays 2- to 4-seriate

Notes: *Celtis* and *Ulmus* can be differentiated primarily based on the presence of multiple rows of marginal ray cells in *Celtis*. I observed that in one sample, which otherwise resembles other fragments of elm. It is possible that this is an unusual piece of elm—more observations are required to confirm the presence of hackberry at Gordion.

Carpinus cf. *betulus*

  Low magnification

x-section	*diffuse porous, *pores sparse, in radial multiples, *aggregate rays, diffuse apotracheal parenchyma

  High magnification

x-section	as above
r-section	*spiral thickenings, simple perforation plates, rays homo- to slightly heterocellular
t-section	*rays uniseriate to aggregate, *5+ cells high

Notes: The lack of tangential parenchyma bands suggest that this is *Carpinus betulus* rather than *C. orientalis* (Schweingruber 1990).

Chenopodiaceae

  Low magnification

x-section   *wood with included phloem, *pores
                in small groups inside phloem bands,
                *growth rings indistinct

  High magnification

x-section   not observed

r-section   not observed

t-section   not observed

Monocotyledonous Woods

Poaceae (probably *Phragmites* sp.)

  Low magnification

x-section   *vascular bundles surrounded by
                asymmetric sclerenchyma sheath,
                *two large metaxylem vessels, *one to
                two small protoxylem vessels

  High magnification

x-section   not observed

r-section   not observed

t-section   not observed

Notes: This may be one of a number of grass stems based on wood anatomy, but the diameter of the stems suggests *Phragmites*, which grows today in irrigation ditches and along riverbanks at Gordion.

Note About the Online Appendices

Electronic supplemental datasets are available for this publication through The Digital Archaeological Record (tDAR) at: doi:10.6067/XCV8X63Q02 (tDAR ID: 427491). Two online appendices are included, each of which presents one of the two archaeobotanical datasets presented in this volume: hand-picked wood charcoal, and seeds and plant parts from flotation samples.

Online Appendix 1: wood charcoal
Online Appendix 1 (doi:10.6067/XCV8NP26HW; tDAR ID: 427493) is the results of the analysis of 445 hand-picked wood charcoal samples from Gordion excavated between 1993 and 2002; this includes all wood charcoal encountered during excavation during that decade. These results are discussed primarily in Chapter 4.

Online Appendix 2: seeds and plant parts
Online Appendix 2 (doi:10.6067/XCV8SF2Z8H; tDAR ID: 427494) is the sample-by-sample analysis of 220 flotation samples from Gordion excavated between 1993 and 2002 selected for analysis in this project. These results are discussed primarily in Chapter 5.

In addition, results from Miller (2010) are incorporated into analyses throughout this volume. Data associated with that publication are also available online through tDAR at doi:10.6067/XCV87H1M0V (tDAR ID: 376588).

Bibliography

Adger, W.N. 2000. Social and Ecological Resilience: Are They Related? *Progress in Human Geography* 24(3):347–64.

———. 2006. Vulnerability. *Global Environmental Change* 16(3):268–81.

Alcock, J. 2006. A Textbook History of Animal Behaviour. In *Essays in Animal Behaviour: Celebrating 50 Years of Animal Behaviour*, ed. J.R. Lucas and L.W. Simmons, pp. 5–21. Amsterdam: Elsevier.

Alcock, S.E. 2002. *Archaeologies of the Greek Past: Landscape, Monuments, and Memories*. Cambridge: Cambridge University Press.

Alexander, R.D. 1974. The Evolution of Social Behavior. *Annual Review of Ecology and Systematics* 5:325–83.

Allcock, S.L. 2013. Living with a Changing Landscape: Holocene Climate Variability and Socio-Evolutionary Trajectories, Central Turkey. Ph.D. diss., Plymouth University, School of Geography, Earth and Environmental Sciences, Plymouth, UK.

Allen, M.S. 2004. Bet-Hedging Strategies, Agricultural Change, and Unpredictable Environments: Historical Development of Dryland Agriculture in Kona, Hawaii. *Journal of Anthropological Archaeology* 23:196–224.

Anderies, J.M., and M. Hegmon. 2011. Robustness and Resilience across Scales: Migration and Resource Degradation in the Prehistoric U.S. Southwest. *Ecology and Society* 16(2):22.

Anderson, M.K. 2005. *Tending the Wild*. Berkeley: University of California Press.

Anderson, S., and F. Ertuğ-Yaras. 1998. Fuel, Fodder and Faeces: An Ethnographic and Botanical Study of Dung Fuel Use in Central Anatolia. *Environmental Archaeology* 1:99–109.

Anderson-Stojanović, V.R., and J.E. Jones. 2002. Ancient Beehives from Isthmia. *Hesperia* 71(4):345–76.

Arbuckle, B.S. 2012. Pastoralism, Provisioning, and Power at Bronze Age Acemhöyük, Turkey. *American Anthropologist* 114(3):462–76.

Arlt, C., and M.A. Stadler, eds. 2013. *Das Fayûm in Hellenismus und Kaiserzeit: Fallstudien zu multikulturellem Leben in der Antike*. Wiesbaden: Harrassowitz Verlag.

Asensi Amorós, M.V. 2002. Wood from Ancient Egypt: The Ramesseum and the Valley of the Queens (18th Dynasty–Roman Period): A Preliminary Report. In *Charcoal Analysis: Methodological Approaches, Palaeoecological Results and Wood Uses: Proceedings of the Second International Meeting of Anthracology, Paris, September 2000*, ed. S. Thiébault, pp. 273–77. Oxford: Archaeopress.

Asouti, E. 2005. Woodland Vegetation and the Exploitation of Fuel and Timber at Neolithic Çatalhöyük: Report on the Wood Charcoal Macro-Remains. In *Inhabiting Çatalhöyük: Reports from the 1995–99 Seasons*, ed. I. Hodder, pp. 213–58. Cambridge & London: McDonald Institute for Archaeological Research and British Institute at Ankara.

———. 2012. Rethinking Human Impact on Prehistoric Vegetation in Southwest Asia: Long-Term Fuel/Timber Acquisition Strategies at Neolithic Çatalhöyük. In *Wood and Charcoal: Evidence for Human and Natural History*, ed. Y.C.E. Badal, M. Macías, and M. Ntinou, pp. 33–41. Saguntum: Papeles del Laboratorio de Arqueología de Valencia. Valencia: University of Valencia.

———. 2013. Evolution, History and the Origin of Agriculture: Rethinking the Neolithic

(Plant) Economies of South-West Asia. *Levant* 45(2):210–18.

Asouti, E., and P. Austin. 2005. Reconstructing Woodland Vegetation and Its Exploitation by Past Societies, Based on the Analysis and Interpretation of Archaeological Wood Charcoal Macro-Fossils. *Environmental Archaeology* 10:1–18.

Asouti, E., and A. Fairbairn. 2002. Subsistence Economy in Central Anatolia During the Neolithic: The Archaeobotanical Evidence. In *The Neolithic of Central Anatolia: Internal Developments and External Relations During the 9th–6th Millennia cal BC: Proceedings of the International Canew Table Ronde, Istanbul, 23–24 November 2001*, ed. F. Gérard and L. Thissen, pp. 181–92. Istanbul: Ege Yarınları.

Asouti, E., and J. Hather. 2001. Charcoal Analysis and the Reconstruction of Ancient Woodland Vegetation in the Konya Basin, South-Central Anatolia, Turkey: Results from the Neolithic Site of Çatalhöyük East. *Vegetation History and Archaeobotany* 10(1):23–32.

Atalay, I. 2001. The Ecology of Forests in Turkey. *Silva Balcanica* 1:25–34.

Atalay, S., and C.A. Hastorf. 2006. Food, Meals, and Daily Activities: Food Habitus at Neolithic Çatalhöyük. *American Antiquity* 71(2):283–319.

Baird, D. 2001. Settlement and Landscape in the Konya Plain, South Central Turkey, from the Epipalaeolithic to the Medieval Period. In *"Çatalhöyük'ten Günümüze Çumra" Kongresi*, ed. A. Basim, pp. 269–76. Çumra: Çumra Belediyesi Kültür Hizmeti.

Bakker, J., D. Kaniewski, G. Verstraeten, V. De Laet, and M. Waelkens. 2012a. Numerically Derived Evidence for Late-Holocene Climate Change and Its Impact on Human Presence in the Southwest Taurus Mountains, Turkey. *The Holocene* 22(4):425–38.

Bakker, J., E. Paulissen, D. Kaniewski, V. De. Laet, G. Verstraeten, and M. Waelkens. 2012b. Man, Vegetation and Climate During the Holocene in the Territory of Sagalassos, Western Taurus Mountains, SW Turkey. *Vegetation History and Archaeobotany* 21(4–5):249–66.

Baksh, M., and A. Johnson. 1990. Insurance Policies among the Machiguenga: An Ethnographic Analysis of Risk Management in a Non-Western Society. In *Risk and Uncertainty in Tribal and Peasant Economies*, ed. E.A. Cashdan, pp. 193–227. Boulder, CO: Westview Press.

Balasse, M., S.H. Ambrose, A.B. Smith, and T.D. Price. 2002. The Seasonal Mobility Model for Prehistoric Herders in the South-Western Cape of South Africa Assessed by Isotopic Analysis of Sheep Tooth Enamel. *Journal of Archaeological Science* 29(9):917–32.

Ballard, M.W. 2012. King Midas' Textiles and His Golden Touch. In *The Archaeology of Phrygian Gordion: Royal City of Midas*, ed. C.B. Rose, pp. 165–70. Philadelphia: University of Pennsylvania Museum of Archaeology and Anthropology.

Ballard, M.W., H. Alden, R.H. Cunningham, W. Hopwood, J. Koles, and L. Dussubieux. 2010. Appendix Eight. Preliminary Analyses of Textiles Associated with the Wooden Furniture from Tumulus MM. In *The Gordion Wooden Objects, Volume 1: The Furniture from Tumulus MM*, ed. E. Simpson, pp. 203–23. Leiden: Brill.

Bamforth, D.B. 2002. Evidence and Metaphor in Evolutionary Archaeology. *American Antiquity* 67(3):435–52.

Bar-Matthews, M., and A. Ayalon. 2004. Speleothems as Palaeoclimate Indicators: A Case Study from Soreq Cave Located in the Eastern Mediterranean Region, Israel. In *Past Climate Variability through Europe and Africa*, ed. R. Battarbee, F. Gasse, and C.E. Stickley, pp. 363–91. Dordrecht, Netherlands: Springer.

———. 2011. Mid-Holocene Climate Variations Revealed by High-Resolution Speleothem Records from Soreq Cave, Israel and Their Correlation with Cultural Changes. *The Holocene* 21(1):163–71.

Bar-Matthews, M., A. Ayalon, and A. Kaufman. 1997. Late Quaternary Paleoclimate in the Eastern Mediterranean Region from Stable Isotope Analysis of Speleothems at Soreq Cave, Israel. *Quaternary Research* 47:155–68.

Bar-Matthews, M., A. Ayalon, A. Kaufman, and G.J. Wasserburg. 1999. The Eastern Mediterranean Paleoclimate as a Reflection of Regional Events: Soreq Cave, Israel. *Earth and Planetary Science Letters* 166(1–2):85–95.

Barber, E.J.W. 1991. *Prehistoric Textiles: The Development of Cloth in the Neolithic and Bronze Ages with Special Reference to the Aegean*. Princeton: Princeton University Press.

Barefoot, A.C., and F.W. Hankins. 1982. *Identification of Modern and Tertiary Woods*. Oxford: Clarendon Press.

Barlow, K.R. 2002. Predicting Maize Agriculture among the Fremont: An Economic Comparison of Farming and Foraging in the American Southwest. *American Antiquity* 67(1):65–88.

———. 2006. A Formal Model for Predicting Agriculture among the Fremont. In *Behavioral Ecology and the Transition to Agriculture*, ed. D.J. Kennett and B. Winterhalder, pp. 87–102. Berkeley: University of California Press.

Barnosky, A.D., E.A. Hadly, J. Bascompte, E.L. Berlow, J.H. Brown, M. Fortelius, W.M. Getz, J. Harte, A. Hastings, P.A. Marquet, N.D. Martinez, A. Mooers, P. Roopnarine, G. Vermeij, J.W. Williams, R. Gillespie, J. Kitzes, C. Marshall, N. Matzke, D.P. Mindell, E. Revilla, and A.B. Smith. 2012. Approaching a State Shift in Earth's Biosphere. *Nature* 486(7401):52–58.

Baruch, U. 1986. The Late Holocene Vegetational History of Lake Kinneret (Sea of Galilee), Israel. *Paléorient* 26(2):37–48.

———. 1990. Palynological Evidence of Human Impact on the Vegetation as Recorded in Late Holocene Lake Sediments in Israel. In *Man's Role in the Shaping of the Eastern Mediterranean Landscape*, ed. S. Bottema, G. Entjes-Nieborg, and W. van Zeist, pp. 283–93. Rotterdam: A.A. Balkema.

Bazzaz, F.A. 1996. *Plants in Changing Environments: Linking Physiological, Population, and Community Ecology*. Cambridge: Cambridge University Press.

Beach, T.P., and S. Luzzadder-Beach. 2008. Geoarchaeology and Aggradation around Kinet Höyük, an Archaeological Mound in the Eastern Mediterranean, Turkey. *Geomorphology* 101(3):416–28.

Beaton, J.M. 1973. The Nature of Aboriginal Exploitation of Mollusk Populations in Southern California. Master's thesis, University of California, Los Angeles, Department of Anthropology, Los Angeles.

Beatty, J. 1980. Optimal-Design Models and the Strategy of Model Building in Evolutionary Biology. *Philosophy of Science* 47(4):532–61.

Bellinger, L. 1962. Textiles from Gordion. *Bulletin of the Needle and Bobbin Club* 46:5–34.

Belovsky, G.E. 1978. Diet Optimization in a Generalist Herbivore: The Moose. *Theoretical Population Biology* 14(1):105–34.

Bennett, J. 2013. Agricultural Strategies and the Roman Military in Central Anatolia During the Early Imperial Period. *OLBA* 21:315–43.

Bennett, J., and A.L. Goldman. 2009. A Preliminary Report on the Roman Military Presence at Gordion, Galatia. In *Limes: The 20th International Congress of Roman Frontier Studies*, ed. Á. Morillo, N. Hanel, and E. Martín, pp. 1605–16. Madrid: Consejo Superior de Investigaciones Científicas.

Bentley, R.A. 2006. Strontium Isotopes from the Earth to the Archaeological Skeleton: A Review. *Journal of Archaeological Method and Theory* 13:135–87.

Berkes, F. 2007. Understanding Uncertainty and Reducing Vulnerability: Lessons from Resilience Thinking. *Natural Hazards* 41(2):283–95.

Berkes, F., and N.J. Turner. 2006. Knowledge, Learning and the Evolution of Conservation Practice for Social-Ecological System Resilience. *Human Ecology* 34(4):479–94.

Bernard, V., S. Renaudin, and D. Marguerie. 2006. Evidence of Trimmed Oaks (*Quercus* sp.) in North Western France During the Early Middle Ages (9th–11th Centuries A.D.). In *Charcoal Analysis: New Analytical Tools and Methods for Archaeology: Papers from the Table-Ronde Held in Basel 2004*, ed. A. Dufraisse, pp. 103–8. Oxford: Archaeopress.

Betts, H.S. 1913. Studies in Physical Properties: Wood Fuel Tests. *Review of Forest Service Investigations* 2:39–42.

Bianchi, G.G., and I.N. McCave. 1999. Holocene Periodicity in North Atlantic Climate and Deep-Ocean Flow South of Iceland. *Nature* 397(6719):515–17.

Binford, L.R. 1967. Smudge Pits and Hide Smoking: The Use of Analogy in Archaeological Reasoning. *American Antiquity* 32(1):1–12.

———. 1978. *Nunamiut Ethnoarchaeology*. New York: Academic Press.

Bird, D.W., and R. Bliege Bird. 2002. Children on the Reef: Slow Learning or Strategic Foraging? *Human Nature* 13(2):269–97.

Bird, D.W., R. Bliege Bird, and B.F. Codding. 2009. In Pursuit of Mobile Prey: Martu Hunting Strategies and Archaeofaunal Interpretation. *American Antiquity* 74(1):3–29.

Bird, D.W., R. Bliege Bird, and C.H. Parker. 2005. Aboriginal Burning Regimes and Hunting Strategies in Australia's Western Desert. *Human Ecology* 33(4):443–64.

Bird, D.W., B.F. Codding, R. Bliege Bird, D.W. Zeanah, and C.J. Taylor. 2013. Megafauna in a Continent of Small Game: Archaeological Implications of Martu Camel Hunting in Australia's Western Desert. *Quaternary International* 297:155–66.

Bird, D.W., and J.F. O'Connell. 2006. Behavioral Ecology and Archaeology. *Journal of Archaeological Research* 14(2):143–88.

Bishop, R.R., M.J. Church, and P.A. Rowley-Conwy. 2015. Firewood, Food and Human Niche Construction: The Potential Role of Mesolithic Hunter-Gatherers in Actively Structuring Scotland's Woodlands. *Quaternary Science Reviews* 108:51–75.

Blanchette, R.A., and E. Simpson. 1992. Soft Rot and Wood Pseudomorphs in an Ancient Coffin (700 BC) from Tumulus MM at Gordion, Turkey. *IAWA Bulletin* 13:201–13.

Bliege Bird, R., and D.W. Bird. 2008. Why Women Hunt: Risk and Contemporary Foraging in a Western Desert Aboriginal Community. *Current Anthropology* 49(4):655–93.

Bliege Bird, R., D.W. Bird, E.A. Smith, and G.C. Kushnick. 2002. Risk and Reciprocity in Meriam Food Sharing. *Evolution and Human Behavior* 23(4):297–321.

Bliege Bird, R.L., B. Scelza, D.W. Bird, and E.A. Smith. 2012. The Hierarchy of Virtue: Mutualism, Altruism and Signaling in Martu Women's Cooperative Hunting. *Evolution and Human Behavior* 33:64–78.

Bliege Bird, R., and E.A. Smith. 2005. Signaling Theory, Strategic Interaction, and Symbolic Capital. *Current Anthropology* 46(2):221–48.

Bliege Bird, R., E.A. Smith, and D.W. Bird. 2001. The Hunting Handicap: Costly Signaling in Human Foraging Strategies. *Behavioral Ecology and Sociobiology* 50(1):9–19.

Blouin, K. 2014. *Triangular Landscapes: Environment, Society, and the State in the Nile Delta under Roman Rule*. Oxford: Oxford University Press.

Bogaard, A., E. Henton, J.A. Evans, K.C. Twiss, M.P. Charles, P. Vaiglova, and N. Russell. 2014. Locating Land Use at Neolithic Çatalhöyük, Turkey: The Implications of 87Sr/86Sr Signatures in Plants and Sheep Tooth Sequences. *Archaeometry* 56(5):860–77.

Bogaard, A., M. Charles, K.C. Twiss, A. Fairbairn, N. Yalman, D. Filipović, G.A. Demirergi, F. Ertuğ, N. Russell, and J. Henecke. 2009. Private Pantries and Celebrated Surplus: Storing and Sharing Food at Neolithic Çatalhöyük, Central Anatolia. *Antiquity* 83(321):649–68.

Bogaard, A., T.H.E. Heaton, P. Poulton, and I. Merbach. 2007. The Impact of Manuring on Nitrogen Isotope Ratios in Cereals: Archaeological Implications for Reconstruction of Diet and Crop Management Practices. *Journal of Archaeological Science* 34(3):335–43.

Bogaard, A., C. Palmer, G. Jones, M. Charles, and J.G. Hodgson. 1999. A FIBS Approach to the Use of Weed Ecology for the Archaeobotanical Recognition of Crop Rotation Regimes. *Journal of Archaeological Science* 26(9):1211–24.

Bonhôte, J., B. Davasse, C. Dubois, V. Izard, and J.-P. Métailié. 2002. Charcoal Kilns and Environmental History in the Eastern Pyrenees (France). In *Charcoal Analysis: Methodological Approaches, Palaeoecological Results and Wood Uses: Proceedings of the Second International Meeting of Anthracology, Paris, September 2000*, ed. S. Thiébault, pp. 219–28. Oxford: Archaeopress.

Bookman, R., Y. Enzel, A. Agnon, and M. Stein. 2004. Late Holocene Lake Levels of the Dead Sea. *Geological Society of America Bulletin* 116(5–6):555–71.

Boone, J.L. 1998. The Evolution of Magnanimity: When Is It Better to Give Than to Receive? *Human Nature* 9(1):1–21.

Borgerhoff Mulder, M., and R. Schacht. 2012. Human Behavioural Ecology. In *Encyclopedia of Life Sciences*. Chichester, UK: John Wiley & Sons.

Boserup, E. 1965. *The Conditions of Agricultural Growth: The Economics of Agrarian Change under Population Pressure*. Chicago: Aldine.

———. 1981. *Population and Technological Change: A Study of Long-Term Trends*. Chicago: University of Chicago Press.

Boston University. 2015. Environmental Archaeology Laboratory Collections Database. http://sites.bu.edu/ealab/collections/database/. Accessed November 30, 2015.

Bottema, S. 1995. Holocene Vegetation of the Van Area: Palynological and Chronological Evidence from Söğütlü, Turkey. *Vegetation History and Archaeobotany* 4(3):187–93.

Bottema, S., and H. Woldring. 1984. Late Quaternary Vegetation and Climate of Southwestern Turkey Part II. *Palaeohistoria* 26:123–49.

———. 1990. Anthropogenic Indicators in the Pollen Record of the Eastern Mediterranean. In *Man's Role in the Shaping of the Eastern Mediterranean Landscape*, ed. S. Bottema, G. Entjes-Nieborg, and W. van Zeist, pp. 231–64. Rotterdam: A.A. Balkema.

Bottema, S., H. Woldring, and B. Aytuğ. 1986. Palynological Investigations on the Relations between Prehistoric Man and Vegetation in Turkey: The Beyşehir Occupation Phase. In *Proceedings of the 5th Optima Congress, September 1986*, ed. H. Demiriz and N. Özhatay, pp. 315–28. İstanbul: Istanbul Üniversitesi.

———. 1994. Late Quaternary Vegetation History of Northern Turkey. *Palaeohistoria* 35/36:13–72.

Boyd, R., and P.J. Richerson. 1985. *Culture and the Evolutionary Process*. Chicago: University of Chicago Press.

———. 2005. *The Origin and Evolution of Cultures*. New York: Oxford University Press.

Boyer, P., N. Roberts, and D. Baird. 2006. Holocene Environment and Settlement on the Çarşamba Alluvial Fan, South-Central Turkey: Integrating Geoarchaeology and Archaeological Field Survey. *Geoarchaeology* 21(7):675–98.

Brite, E.B., and J.M. Marston. 2013. Environmental Change, Agricultural Innovation, and the Spread of Cotton Agriculture in the Old World. *Journal of Anthropological Archaeology* 32(1):39–53.

Brookfield, H.C. 1972. Intensification and Disintensification in Pacific Agriculture: A Theoretical Approach. *Pacific Viewpoint* 13:30–48.

———. 2001. Intensification, and Alternative Approaches to Agricultural Change. *Asia and Pacific Viewpoint* 42:181–92.

Broughton, J.M. 1994. Declines in Mammalian Foraging Efficiency During the Late Holocene, San Francisco Bay, California. *Journal of Anthropological Archaeology* 13(4):371–401.

———. 1997. Widening Diet Breadth, Declining Foraging Efficiency, and Prehistoric Harvest Pressure: Ichthyofaunal Evidence from the Emeryville Shellmound, California. *Antiquity* 71(274):845–62.

Brown, K. 2014. Global Environmental Change I: A Social Turn for Resilience? *Progress in Human Geography* 38(1):107–17.

Bruno, M.C. 2014. Beyond Raised Fields: Exploring Farming Practices and Processes of Agricultural Change in the Ancient Lake Titicaca Basin of the Andes. *American Anthropologist* 116(1):130–45.

Burke, B. 2001. Anatolian Origins of the Gordian Knot Legend. *Greek, Roman, and Byzantine Studies* 42:255–61.

———. 2005. Textile Production at Gordion and the Phrygian Economy. In *The Archaeology of Midas and the Phrygians: Recent Work at Gordion*, ed. L. Kealhofer, pp. 69–81. Philadelphia: University of Pennsylvania Museum of Archaeology and Anthropology.

———. 2010. *From Minos to Midas: Ancient Cloth Production in the Aegean and in Anatolia*. Oxford: Oxbow Books.

Butzer, K.W. 1982. *Archaeology as Human Ecology: Methods and Theory for a Contextual Approach*. Cambridge: Cambridge University Press.

———. 1996. Ecology in the Long View: Settlement Histories, Agrosystemic Ecological Performance. *Journal of Field Archaeology* 23(2):141–50.

Cahill, N. 2002. *Household and City Organization at Olynthus*. New Haven: Yale University Press.

Campbell, W.B. 1918. The Fuel Value of Wood. *Canadian Forestry Journal* 14(4):1632–33.

Cancian, F. 1980. Risk and Uncertainty in Agricultural Decision Making. In *Agricultural Decision Making: Anthropological Contributions to Rural Development*, ed. P.F. Bartlett, pp. 161–76. New York: Academic Press, Inc.

Cannon, M.D. 2003. A Model of Central Place Forager Prey Choice and an Application to Faunal Remains from the Mimbres Valley, New Mexico. *Journal of Anthropological Archaeology* 22:1–25.

Cappers, R.T.J. 2006. *Roman Foodprints at Berenike: Archaeobotanical Evidence of Subsistence and Trade in the Eastern Desert of Egypt*. Los Angeles:

Cotsen Institute of Archaeology at UCLA.

Cappers, R.T.J., R.M. Bekker, and J.E.A. Jans. 2006. *Digital Seed Atlas of the Netherlands*. Eelde: Barkhuis Publishing.

Caraco, T. 1981. Energy Budgets, Risk and Foraging Preferences in Dark-Eyed Juncos (*Junco hyemalis*). *Behavioral Ecology and Sociobiology* 8(3):213–17.

———. 1982. Aspects of Risk-Aversion in Foraging White-Crowned Sparrows. *Animal Behaviour* 30:719–27.

———. 1983. White-Crowned Sparrows (*Zonotrichia leucophrys*): Foraging Preferences in a Risky Environment. *Behavioral Ecology and Sociobiology* 12(1):63–69.

Caraco, T., S. Martindale, and T.S. Whittam. 1980. An Empirical Demonstration of Risk-Sensitive Foraging Preferences. *Animal Behaviour* 28:820–30.

Carpenter, S.R., and W.A. Brock. 2006. Rising Variance: A Leading Indicator of Ecological Transition. *Ecology Letters* 9(3):311–18.

Carpenter, S., B. Walker, J.M. Anderies, and N. Abel. 2001. From Metaphor to Measurement: Resilience of What to What? *Ecosystems* 4(8):765–81.

Carrión Marco, Y. 2006. Tres Montes (Navarra, Spain): Dendrology and Wood Uses in an Arid Environment. In *Charcoal Analysis: New Analytical Tools and Methods for Archaeology: Papers from the Table-Ronde Held in Basel 2004*, ed. A. Dufraisse, pp. 83–93. Oxford: Archaeopress.

Casana, J. 2008. Mediterranean Valleys Revisited: Linking Soil Erosion, Land Use and Climate Variability in the Northern Levant. *Geomorphology* 101(3):429–42.

———. 2014. A Landscape Context for Paleoethnobotany: The Contribution of Aerial and Satellite Remote Sensing. In *Method and Theory in Paleoethnobotany*, ed. J.M. Marston, J. d'Alpoim Guedes, and C. Warinner, pp. 315–35. Boulder: University Press of Colorado.

Case, D.O. 2012. *Looking for Information: A Survey of Research on Information Seeking, Needs and Behavior*. 3rd ed. Bingley, UK: Emerald Group Publishing.

Cashdan, E.A. 1990a. Introduction. In *Risk and Uncertainty in Tribal and Peasant Economies*, ed. E.A. Cashdan, pp. 1–16. Boulder, CO: Westview Press.

———, ed. 1990b. *Risk and Uncertainty in Tribal and Peasant Economies*. Boulder, CO: Westview Press.

Casimir, M.J., ed. 2008. *Culture and the Changing Environment: Uncertainty, Cognition and Risk Management in Cross-Cultural Perspective*. New York: Berghahn Books.

Chabal, L. 1988. L'étude paléoécologique de sites protohistoriques à partir des charbons de bois: la question de l'unité de mesure. Dénombrements de fragments ou pesées? In *Wood and Archaeology: Acts of the European Symposium Held at Louvain-la-Neuve, October 1987*, ed. T. Hackens, A.V. Munaut, and C. Till, pp. 189–205. Strasbourg: Conseil de l'Europe.

———. 1992. La représentativité paléo-écologique des charbons de bois archéologiques issus du bois du feu. *Bulletin de la Société Botanique de France* 139:213–36.

Chabal, L., L. Fabre, J.-F. Terral, and I. Théry-Parisot. 1999. L'anthracologie. In *La Botanique*, ed. C. Bourquin-Mignot, J.-E. Brochier, L. Chabal, S. Crozat, L. Fabre, F. Guibal, P. Marinval, H. Richard, J.-F. Terral, and I. Théry, pp. 43–104. Paris: Errance.

Chagnon, N.A., and W. Irons. 1979. *Evolutionary Biology and Human Social Behavior: An Anthropological Perspective*. North Scituate, MA: Duxbury Press.

Charles, M., G. Jones, and J.G. Hodgson. 1997. FIBS in Archaeobotany: Functional Interpretation of Weed Floras in Relation to Husbandry Practices. *Journal of Archaeological Science* 24(12):1151–61.

Charnov, E.L. 1976. Optimal Foraging: The Marginal Value Theorem. *Theoretical Population Biology* 9(2):129–36.

Chase, A.F., and V. Scarborough. 2014. Diversity, Resiliency, and IHOPE-Maya: Using the Past to Inform the Present. *Archeological Papers of the American Anthropological Association* 24(1):1–10.

Chase, B., D. Meiggs, P. Ajithprasad, and P.A. Slater. 2014. Pastoral Land-Use of the Indus Civilization in Gujarat: Faunal Analyses and Biogenic Isotopes at Bagasra. *Journal of Archaeological Science* 50:1–15.

Chernoff, M.C., and T.M. Harnischfeger. 1996. Preliminary Report on Botanical Remains from Çadır Höyük (1994 Season). *Anatolica* 22:159–79.

Chesson, M.S., and N. Goodale. 2014. Population

Aggregation, Residential Storage and Socioeconomic Inequality at Early Bronze Age Numayra, Jordan. *Journal of Anthropological Archaeology* 35:117–34.

Childe, V.G. 1934. *New Light on the Most Ancient East: The Oriental Prelude to European Prehistory*. London: Kegan Paul.

Clark, C.W. 1990. Uncertainty in Economics. In *Risk and Uncertainty in Tribal and Peasant Economies*, ed. E.A. Cashdan, pp. 47–63. Boulder, CO: Westview Press.

Clark, J.S., and P.D. Royall. 1995. Transformation of a Northern Hardwood Forest by Aboriginal (Iroquois) Fire. *The Holocene* 5(1):1–9.

Clawman, D.L. 1985. Harvest Security and Intraspecific Diversity in Traditional Tropical Agriculture. *Economic Botany* 39(1):56–67.

Clements, F.E. 1916. *Plant Succession: An Analysis of the Development of Vegetation*. Washington, DC: Carnegie Institution of Washington.

Codding, B.F., and T.L. Jones. 2007. Man the Showoff? Or the Ascendance of a Just-So-Story: A Comment on Recent Applications of Costly Signaling Theory in American Archaeology. *American Antiquity* 72(2):349–57.

Colledge, S., J. Conolly, and S. Shennan. 2004. Archaeobotanical Evidence for the Spread of Farming in the Eastern Mediterranean. *Current Anthropology* 45:S35–S58.

Cooper, J., and P. Sheets, eds. 2012. *Surviving Sudden Environmental Change*. Boulder: University Press of Colorado.

Costanza, R., L. Graumlich, and W.L. Steffen, eds. 2007. *Sustainability or Collapse? An Integrated History and Future of People on Earth*. Cambridge, MA: MIT Press.

Cowie, R.J. 1977. Optimal Foraging in Great Tits (*Parus major*). *Nature* 268(5616):137–39.

Crane, E. 1983. *The Archaeology of Beekeeping*. London: Duckworth.

Cronin, T.M. 2010. *Paleoclimates: Understanding Climate Change Past and Present*. New York: Columbia University Press.

Cuddington, K. 2011. Legacy Effects: The Persistent Impact of Ecological Interactions. *Biological Theory* 6(3):203–10.

Cullen, H.M., P.B. deMenocal, S. Hemming, G. Hemming, F.H. Brown, T. Guilderson, and F. Sirocko. 2000. Climate Change and the Collapse of the Akkadian Empire: Evidence from the Deep Sea. *Geology* 28(4):379–82.

Cumming, G.S., D.H.M. Cumming, and C.L. Redman. 2006. Scale Mismatches in Social-Ecological Systems: Causes, Consequences, and Solutions. *Ecology and Society* 11(1):14.

d'Alpoim Guedes, J., and R.N. Spengler. 2014. Sampling Strategies in Paleoethnobotanical Analysis. In *Method and Theory in Paleoethnobotany*, ed. J.M. Marston, J. d'Alpoim Guedes, and C. Warinner, pp. 77–94. Boulder: University Press of Colorado.

Dalfes, H.N., G. Kukla, and H. Weiss, eds. 1997. *Third Millennium BC Climate Change and Old World Collapse*. Berlin: Springer.

Dandoy, J.R., P. Selinsky, and M.M. Voigt. 2002. Celtic Sacrifice. *Archaeology* 55(1):44–49.

Darbyshire, G. 2007. The Siege Mound of Küçük Hüyük and the Lower Town of Gordion. Paper presented at the Archaeology of Phrygian Gordion Conference, University of Pennsylvania Museum, Philadelphia.

Davis, P.H. 1965. *Flora of Turkey and the East Aegean Islands. Volume 1*. Edinburgh: Edinburgh University Press.

———. 1965–2000. *Flora of Turkey and the East Aegean Islands*. Edinburgh: Edinburgh University Press.

———. 1967. *Flora of Turkey and the East Aegean Islands. Volume 2*. Edinburgh: Edinburgh University Press.

———. 1972. *Flora of Turkey and the East Aegean Islands. Volume 4*. Edinburgh: Edinburgh University Press.

———. 1982. *Flora of Turkey and the East Aegean Islands. Volume 7*. Edinburgh: Edinburgh University Press.

Dawkins, R. 1976. *The Selfish Gene*. New York: Oxford University Press.

Dean, J.R., M.D. Jones, M.J. Leng, H.J. Sloane, C.N. Roberts, J. Woodbridge, G.E.A. Swann, S.E. Metcalfe, W.J. Eastwood, and H. Yiğitbaşıoğlu. 2013. Palaeo-Seasonality of the Last Two Millennia Reconstructed from the Oxygen Isotope Composition of Carbonates and Diatom Silica from Nar Gölü, Central Turkey. *Quaternary Science Reviews* 66:35–44.

Decker, M. 2009. Plants and Progress: Rethinking the Islamic Agricultural Revolution. *Journal of World History* 20(2):187–206.

Delhon, C. 2006. Palaeo-Ecological Reliability of Pedo-Anthracological Assemblages. In *Charcoal Analysis: New Analytical Tools and Methods for Archaeology: Papers from the Table-Ronde Held in Basel 2004*, ed. A. Dufraisse, pp. 9–24. Oxford: Archaeopress.

Denham, W.W. 1971. Energy Relations and Some Basic Properties of Primate Social Organization. *American Anthropologist* 73(1):77–95.

Dercksen, J.G. 2008a. Observations on Land Use and Agriculture in Kaneš. In *Old Assyrian Studies in Memory of Paul Garelli*, ed. C. Michel, pp. 140–57. Leiden: NINO.

———. 2008b. Subsistence, Surplus and the Market for Grain and Meat at Ancient Kanesh. *Altorientalische Forschungen* 35(1):86–102.

Deur, D., and N. Turner, eds. 2005. *Keeping It Living: Traditions of Plant Use and Cultivation on the Northwest Coast of North America*. Seattle: University of Washington Press.

DeVries, K. 1990. The Gordion Excavation Seasons of 1969–1973 and Subsequent Research. *American Journal of Archaeology* 94(3):371–406.

———. 2005. Greek Pottery and Gordion Chronology. In *The Archaeology of Midas and the Phrygians: Recent Work at Gordion*, ed. L. Kealhofer, pp. 36–55. Philadelphia: University of Pennsylvania Museum of Archaeology and Anthropology.

DeVries, K., P.I. Kuniholm, G.K. Sams, and M.M. Voigt. 2003. New Dates for Iron Age Gordion. *Antiquity* 77(296).

Diamond, J.M. 2005. *Collapse: How Societies Choose to Fail or Succeed*. New York: Viking.

DiBattista, A. 2014. A Strontium Approach to Changes in Pastoral Strategy at Gordion. B.A. thesis, Boston University, Department of Archaeology, Boston.

Diehl, M.W., and J.A. Waters. 2006. Aspects of Optimization and Risk in the Early Agricultural Period in Southeastern Arizona. In *Behavioral Ecology and the Transition to Agriculture*, ed. D.J. Kennett and B. Winterhalder, pp. 63–86. Berkeley: University of California Press.

Doolittle, W.E., J. Neely, P.R. Fish, and K.R. Adams. 2004. *Safford Valley Grids: Prehistoric Cultivation in the Southern Arizona Desert*. Tucson: University of Arizona Press.

Dubowski, Y., J. Erez, and M. Stiller. 2003. Isotopic Paleolimnology of Lake Kinneret. *Limnology and Oceanography* 48(1):68–78.

Dufraisse, A., ed. 2006a. *Charcoal Analysis: New Analytical Tools and Methods for Archaeology: Papers from the Table-Ronde Held in Basel 2004*. Oxford: Archaeopress.

———. 2006b. Charcoal Anatomy Potential, Wood Diameter and Radial Growth. In *Charcoal Analysis: New Analytical Tools and Methods for Archaeology: Papers from the Table-Ronde Held in Basel 2004*, ed. A. Dufraisse, pp. 47–59. Oxford: Archaeopress.

———. 2008. Firewood Management and Woodland Exploitation During the Late Neolithic at Lac de Chalain (Jura, France). *Vegetation History and Archaeobotany* 17(2):199–210.

———. 2012. Firewood and Woodland Management in Their Social, Economic and Ecological Dimensions: New Perspectives. In *Wood and Charcoal: Evidence for Human and Natural History*, ed. Y. Carrión, E. Badal, M. Macías, and M. Ntinou, pp. 65–73. Saguntum: Papeles del Laboratorio de Arqueología de Valencia. Valencia: University of Valencia.

Dunning, N.P, T.P. Beach, and S. Luzzadder-Beach. 2012. Kax and Kol: Collapse and Resilience in Lowland Maya Civilization. *Proceedings of the National Academy of Sciences of the United States of America* 109(10):3652–57.

Dusar, B., G. Verstraeten, K. D'Haen, J. Bakker, E. Kaptijn, and M. Waelkens. 2012. Sensitivity of the Eastern Mediterranean Geomorphic System Towards Environmental Change During the Late Holocene: A Chronological Perspective. *Journal of Quaternary Science* 27(4):371–82.

Dusinberre, E.R.M. 2005. *Gordion Seals and Sealings: Individuals and Society*. Philadelphia: University of Pennsylvania Museum of Archaeology and Anthropology.

Eastwood, W.J., M.J. Leng, N. Roberts, and B. Davis. 2007. Holocene Climate Change in the Eastern Mediterranean Region: A Comparison of Stable Isotope and Pollen Data from Lake Gölhisar, Southwest Turkey. *Journal of Quaternary Science* 22(4):327–41.

Eastwood, W.J., N. Roberts, and H.F. Lamb. 1998. Paleoecological and Archaeological Evidence for Human Occupance in Southwest Turkey: The Beyşehir Occupation Phase. *Anatolian Studies* 48:69–86.

Eastwood, W.J., N. Roberts, H.F. Lamb, and J.C. Tibby. 1999. Holocene Environmental Change in Southwest Turkey: A Palaeoecological Record of Lake and Catchment-Related Changes. *Quaternary Science Reviews* 18(4–5):671–95.

Ecology and Society. 2015. Editorial Team. http://www.ecologyandsociety.org/about/editors.php. Accessed June 30, 2015.

Ellis, R. 1981. Appendix V: The Textile Remains. In *Gordion Excavations Reports, Vol. I: Three Great Early Tumuli [P, MM, W]*, ed. R.S. Young, pp. 294–310. Philadelphia: University of Pennsylvania Museum of Archaeology and Anthropology.

Elston, R.G., and P.J. Brantingham. 2002. Microlithic Technology in Northern Asia: A Risk-Minimizing Strategy of the Late Paleolithic and Early Holocene. In *Thinking Small: Global Perspectives on Microlithization*, ed. R.G. Elston and S.L. Kuhn, pp. 103–16. Arlington, VA: American Anthropological Association.

Emlen, J.M. 1966. Role of Time and Energy in Food Preference. *American Naturalist* 100(916):611–17.

England, A., W.J. Eastwood, C.N. Roberts, R. Turner, and J.F. Haldon. 2008. Historical Landscape Change in Cappadocia (Central Turkey): A Palaeoecological Investigation of Annually Laminated Sediments from Nar Lake. *The Holocene* 18(8):1229–45.

Enneking, D. 1995. *The Toxicity of Vicia Species and Their Utilisation as Grain Legumes*. Centre for Legumes in Mediterranean Agriculture (CLIMA), Occasional Publication No. 6. Perth: University of Western Australia.

Enzel, Y., R. Bookman, D. Sharon, H. Gvirtzman, U. Dayan, B. Ziv, and M. Stein. 2003. Late Holocene Climates of the Near East Deduced from Dead Sea Level Variations and Modern Regional Winter Rainfall. *Quaternary Research* 60(3):263–73.

Erder, E.H., A. Gürsan-Salzmann, and N.F. Miller. 2013. A Conservation Management Plan for Gordion and Its Environs. *Conservation and Management of Archaeological Sites* 15(3–4):329–47.

Erdkamp, P. 2005. *The Grain Market in the Roman Empire: A Social, Political and Economic Study*. Cambridge: Cambridge University Press.

Erentöz, C. 1975. Türkiye Jeoloji Haritası (Geological Map of Turkey), Ankara Map. Ankara: Maden Tetkik ve Arama Enstitüsü (Mine Investigation and Research Institute).

Erickson, C.L. 2006. The Domesticated Landscapes of the Bolivian Amazon. In *Time and Complexity in Historical Ecology: Studies in the Neotropical Lowlands*, ed. W.L. Balée and C.L. Erickson, pp. 235–79. New York: Columbia University Press.

Ethier, D.M., C.J. Kyle, and J.J. Nocera. 2014. Tracking Animal Movement by Comparing Trace Element Signatures in Claws to Spatial Variability of Elements in Soils. *Science of the Total Environment* 468:699–705.

Evershed, R.P., S.N. Dudd, V.R. Anderson-Stojanovic, and E.R. Gebhard. 2003. New Chemical Evidence for the Use of Combed Ware Pottery Vessels as Beehives in Ancient Greece. *Journal of Archaeological Science* 30(1):1–12.

Evershed, R.P., S.J. Vaughan, S.N. Dudd, and J.S. Soles. 1997. Fuel for Thought? Beeswax in Lamps and Conical Cups from Late Minoan Crete. *Antiquity* 71(274):979–85.

Fabre, L., and J.-C. Auffray. 2002. An Anthracological Method for the Study of Charcoal Kilns in Relation to Historical Forestry Management. In *Charcoal Analysis: Methodological Approaches, Palaeoecological Results and Wood Uses: Proceedings of the Second International Meeting of Anthracology, Paris, September 2000*, ed. S. Thiébault, pp. 193–99. Oxford: Archaeopress.

Faegri, K., P.E. Kaland, and K. Krzywinski. 1989. *Textbook of Pollen Analysis*. 4th ed. New York: Wiley.

Fahn, A., E. Werker, and P. Bass. 1986. *Wood Anatomy and Identification of Trees and Shrubs from Israel and Adjacent Regions*. Jerusalem: Israel Academy of Sciences and Humanities.

Fairbairn, A. 2006. Archaeobotany at Kaman-Kalehöyük 2005. *Anatolian Archaeological Studies* 15:133–37.

Fairbairn, A., F.Kulakoğlu, and L. Atici. 2014. Archaeobotanical Evidence for Trade in Hazelnut (*Corylus* sp.) at Middle Bronze Age Kültepe (c. 1950–1830 BC), Kayseri Province, Turkey. *Vegetation History and Archaeobotany* 23(2):167–74.

Fairbairn, A., D. Martinoli, A. Butler, and G. Hillman. 2007. Wild Plant Seed Storage at Neolithic Çatalhöyük East, Turkey. *Vegetation History and Archaeobotany* 16(6):467–79.

Fall, P.L., S.E. Falconer, and J. Klinge. 2015. Bronze Age Fuel Use and Its Implications for Agrarian Landscapes in the Eastern Mediterranean. *Journal of Archaeological Science: Reports* 4:182–91.

Fehr, E., U. Fischbacher, and S. Gachter. 2002. Strong Reciprocity, Human Cooperation, and the Enforcement of Social Norms. *Human Nature* 13(1):1–25.

Figueiral, I., and V. Mosbrugger. 2000. A Review of Charcoal Analysis as a Tool for Assessing Quaternary and Tertiary Environments: Achievements and Limits. *Palaeogeography, Palaeoclimatology, Palaeoecology* 164(1–4):397–407.

Finné, M., K. Holmgren, H.S. Sundqvist, E. Weiberg, and M. Lindblom. 2011. Climate in the Eastern Mediterranean, and Adjacent Regions, During the Past 6000 Years—A Review. *Journal of Archaeological Science* 38(12):3153–73.

Fiorentino, G., V. Caracuta, L. Calcagnile, M. D'Elia, P. Matthiae, F. Mavelli, and G. Quarta. 2008. Third Millennium B.C. Climate Change in Syria Highlighted by Carbon Stable Isotope Analysis of 14C-AMS Dated Plant Remains from Ebla. *Palaeogeography, Palaeoclimatology, Palaeoecology* 266(1–2):51–58.

Fiorentino, G., V. Caracuta, G. Casiello, F. Longobardi, and A. Sacco. 2012. Studying Ancient Crop Provenance: Implications from δ13C and δ15N Values of Charred Barley in a Middle Bronze Age Silo at Ebla (NW Syria). *Rapid Communications in Mass Spectrometry* 26(3):327–35.

Fiorentino, G., J.P. Ferrio, A. Bogaard, J.L. Araus, and S. Riehl. 2015. Stable Isotopes in Archaeobotanical Research. *Vegetation History and Archaeobotany* 24(1):215–27.

Fırıncıoğlu, H.K., B. Şahin, S.S. Seefeldt, F. Mert, B.H. Hakyemez, and M. Vural. 2008. Pilot Study for an Assessment of Vegetation Structure for Steppe Rangelands of Central Anatolia. *Turkish Journal of Agriculture and Forestry* 32:401–14.

Fırıncıoğlu, H.K., S.S. Seefeldt, and B. Şahin. 2007. The Effects of Long-Term Grazing Exclosures on Range Plants in the Central Anatolian Region of Turkey. *Environmental Management* 39:326–37.

Fırıncıoğlu, H.K., S.S. Seefeldt, B. Şahin, and M. Vural. 2009. Assessment of Grazing Effect on Sheep Fescue (*Festuca valesiaca*) Dominated Steppe Rangelands, in the Semi-Arid Central Anatolian Region of Turkey. *Journal of Arid Environments* 73(12):1149–57.

Fisher, C.T. 2005. Demographic and Landscape Change in the Lake Pátzcuaro Basin, Mexico: Abandoning the Garden. *American Anthropologist* 107(1):87–95.

Fisher, W.R. 1908. *Schlich's Manual of Forestry, Volume 5: Forest Utilization*. 2nd ed. London: Bradbury, Agnew, and Company Ltd.

Flannery, K.V. 1969. Origins and Ecological Effects of Early Domestication in Iran and the Near East. In *The Domestication and Exploitation of Plants and Animals*, ed. P.J. Ucko and G.W. Dimbleby, pp. 73–100. Chicago: Aldine.

Fleisher, B. 1990. *Agricultural Risk Management*. Boulder, CO: Lynne Rienner Publishers.

Fleitmann, D., H. Cheng, S. Badertscher, R.L. Edwards, M. Mudelsee, O.M. Göktürk, A. Fankhauser, R. Pickering, C.C. Raible, A. Matter, J. Kramers, and O. Tüysüz. 2009. Timing and Climatic Impact of Greenland Interstadials Recorded in Stalagmites from Northern Turkey. *Geophysical Research Letters* 36(19):L19707.

Flowers, N.M., D.R. Gross, M.L. Ritter, and D.W. Werner. 1982. Variation in Swidden Practices in Four Central Brazilian Indian Societies. *Human Ecology* 10:203–17.

Fogelin, L. 2007. Inference to the Best Explanation: A Common and Effective Form of Archaeological Reasoning. *American Antiquity* 72(4):603–25.

Foley, J.A., R. DeFries, G.P. Asner, C. Barford, G. Bonan, S.R. Carpenter, F.S. Chapin, M.T. Coe, G.C. Daily, H.K. Gibbs, J.H. Helkowski, T. Holloway, E.A. Howard, C.J. Kucharik, C. Monfreda, J.A. Patz, I.C. Prentice, N. Ramankutty, and P.K. Snyder. 2005. Global Consequences of Land Use. *Science* 309(5734):570–74.

Folke, C. 2006. Resilience: The Emergence of a Perspective for Social–Ecological Systems Analyses. *Global Environmental Change* 16(3):253–67.

Folke, C., S. Carpenter, B. Walker, M. Scheffer, T. Elmqvist, L. Gunderson, and C.S. Holling. 2004. Regime Shifts, Resilience, and Biodiversity in Ecosystem Management. *Annual Review of Ecol-*

ogy, Evolution, and Systematics 35:557–81.

Forbes, H. 1998. European Agriculture Viewed Bottom-Side Upwards: Fodder- and Forage-Provision in a Traditional Greek Community. *Environmental Archaeology* 1:19–34.

Ford, R.I. 1979. Paleoethnobotany in American Archaeology. *Advances in Archaeological Method and Theory* 2:285–336.

Foster, D., F. Swanson, J. Aber, I. Burke, N. Brokaw, D. Tilman, and A. Knapp. 2003. The Importance of Land-Use Legacies to Ecology and Conservation. *BioScience* 53(1):77–88.

Fowler, C.S., and D. Rhode. 2011. Plant Foods and Foodways among the Great Basin's Indigenous Peoples. In *Subsistence Economies of Indigenous North American Societies*, ed. B.D. Smith, pp. 233–70. Lanham, MD: Rowman and Littlefield.

Foxhall, L. 2007. *Olive Cultivation in Ancient Greece: Seeking the Ancient Economy.* Oxford: Oxford University Press.

Fraser, R.A., A. Bogaard, T. Heaton, M. Charles, G. Jones, B.T. Christensen, P. Halstead, I. Merbach, P.R. Poulton, D. Sparkes, and A.K. Styring. 2011. Manuring and Stable Nitrogen Isotope Ratios in Cereals and Pulses: Towards a New Archaeobotanical Approach to the Inference of Land Use and Dietary Practices. *Journal of Archaeological Science* 38(10):2790–804.

Fretwell, S.D. 1972. *Populations in a Seasonal Environment.* Princeton: Princeton University Press.

Fritz, G.J. 2005. Paleoethnobotanical Methods and Applications. In *Handbook of Archaeological Methods*, ed. H.D.G. Maschner and C. Chippindale, pp. 773–834. Walnut Creek, CA: Altamira Press.

Fujihara, Y., K. Tanaka, T. Watanabe, T. Nagano, and T. Kojiri. 2008. Assessing the Impacts of Climate Change on the Water Resources of the Seyhan River Basin in Turkey: Use of Dynamically Downscaled Data for Hydrologic Simulations. *Journal of Hydrology* 353(1–2):33–48.

Fuller, D.Q. 2007. Contrasting Patterns in Crop Domestication and Domestication Rates: Recent Archaeobotanical Insights from the Old World. *Annals of Botany* 100(5):903–24.

Fuller, D.Q., R.G. Allaby, and C. Stevens. 2010. Domestication as Innovation: The Entanglement of Techniques, Technology and Chance in the Domestication of Cereal Crops. *World Archaeology* 42(1):13–28.

Fuller, D.Q., and C.J. Stevens. 2009. Agriculture and the Development of Complex Societies: An Archaeobotanical Agenda. In *From Foragers to Farmers: Papers in Honour of Gordon C. Hillman*, ed. A. Fairbarin and E. Weiss, pp. 37–57. Oxford: Oxbow Books.

Füssel, H.-M. 2007. Vulnerability: A Generally Applicable Conceptual Framework for Climate Change Research. *Global Environmental Change* 17:155–67.

Gallant, T.W. 1989. Crisis and Response: Risk-Buffering Behavior in Hellenistic Greek Communities. *Journal of Interdisciplinary History* 19(3):393–413.

———. 1991. *Risk and Survival in Ancient Greece: Reconstructing the Rural Domestic Economy.* Palo Alto: Stanford University Press.

Garnsey, P. 1988. *Famine and Food Supply in the Graeco-Roman World: Responses to Risk and Crisis.* Cambridge: Cambridge University Press.

———. 1999. *Food and Society in Classical Antiquity.* Cambridge: Cambridge University Press.

Garnsey, P., and I. Morris. 1989. Risk and the Polis: The Evolution of Institutionalised Responses to Food Supply Problems in the Ancient Greek State. In *Bad Year Economics: Cultural Responses to Risk and Uncertainty*, ed. P. Halstead and J. O'Shea, pp. 98–105. Cambridge: Cambridge University Press.

Giblin, J.I. 2009. Strontium Isotope Analysis of Neolithic and Copper Age Populations on the Great Hungarian Plain. *Journal of Archaeological Science* 36(2):491–97.

Gintis, H. 2000. Strong Reciprocity and Human Sociality. *Journal of Theoretical Biology* 206(2):169–79.

Gintis, H., S. Bowles, R. Boyd, and E. Fehr. 2003. Explaining Altruistic Behavior in Humans. *Evolution and Human Behavior* 24(3):153–72.

Gintis, H., E.A. Smith, and S.L. Bowles. 2001. Costly Signaling and Cooperation. *Journal of Theoretical Biology* 213(1):103–19.

Glendinning, M. 2005. A Decorated Roof at Gordion: What Tiles Are Revealing About the Phrygian Past. In *The Archaeology of Midas and the Phrygians: Recent Work at Gordion*, ed. L. Kealhofer, pp. 82–100. Philadelphia: University of Pennsylvania Museum of Archaeology and Anthropology.

Göbel, B. 2008. Dangers, Experience, and Luck: Living with Uncertainty in the Andes. In *Culture and the Changing Environment: Uncertainty, Cognition and Risk Management in Cross-Cultural Perspective*, ed. M.J. Casimir, pp. 221–50. New York: Berghahn Books.

Goddard, J., and M. Nesbitt. 1997. Why Draw Seeds? Illustrating Archaeobotany. *Graphic Archaeology* 1997:13–21.

Godwin, H., and A.G. Tansley. 1941. Prehistoric Charcoals as Evidence of Former Vegetation, Soil and Climate. *Journal of Ecology* 29:117–26.

Göktürk, O.M., D. Fleitmann, S. Badertscher, H. Cheng, R.L. Edwards, M. Leuenberger, A. Fankhauser, O. Tüysüz, and J. Kramers. 2011. Climate on the Southern Black Sea Coast During the Holocene: Implications from the Sofular Cave Record. *Quaternary Science Reviews* 30(19–20):2433–45.

Goland, C. 1993. Field Scattering as Agricultural Risk Management: A Case-Study from Cuyo-Cuyo, Department of Puno, Peru. *Mountain Research and Development* 13(4):317–38.

Goldberg, P., and R. Macphail. 2006. *Practical and Theoretical Geoarchaeology*. Oxford: Blackwell Publishing.

Goldman, A.L. 2000. The Roman-Period Settlement at Gordion, Turkey. Ph.D. diss., University of North Carolina, Department of Classics, Chapel Hill, NC.

———. 2005. Reconstructing the Roman-Period Town at Gordion. In *The Archaeology of Midas and the Phrygians: Recent Work at Gordion*, ed. L. Kealhofer, pp. 56–67. Philadelphia: University of Pennsylvania Museum of Archaeology and Anthropology.

———. 2007. From Phrygian Capital to Rural Fort: New Evidence for the Roman Military at Gordion, Turkey. *Expedition* 49(3):6–12.

———. 2010. A Pannonian Auxiliary's Epitaph from Roman Gordion. *Anatolian Studies* 60:129–46.

Goodfriend, G.A. 1991. Holocene Trends in O-18 in Land Snail Shells from the Negev Desert and Their Implications for Changes in Rainfall Source Areas. *Quaternary Research* 35(3):417–26.

———. 1999. Terrestrial Stable Isotope Records of Late Quaternary Paleoclimates in the Eastern Mediterranean Region. *Quaternary Science Reviews* 18(4–5):501–13.

Graf, D.F. 1994. The Persian Royal Road System. In *Achaemenid History VIII: Continuity and Change*, ed. H. Sancisi-Weerdenburg, A. Kuhrt, and M.C. Root, pp. 167–89. Leiden: Nederlands Instituut voor Nabije Oosten.

Grafen, A. 1990. Biological Signals as Handicaps. *Journal of Theoretical Biology* 144(4):517–46.

Grave, P., L. Kealhofer, B. Marsh, G.K. Sams, M.M. Voigt, and K. DeVries. 2009. Ceramic Production and Provenience at Gordion, Central Anatolia. *Journal of Archaeological Science* 36:2162–76.

Grave, P., L. Kealhofer, B. Marsh, T. Sivas, and H. Sivas. 2012. Reconstructing Iron Age Community Dynamics in Eskişehir Province, Central Turkey. *Journal of Archaeological Method and Theory* 19(3):377–406.

Graves, H.S. 1919. *The Use of Wood for Fuel*. Bulletin 753. Washington, DC: U.S. Department of Agriculture.

Gray, R. 1987. Faith and Foraging: A Critique of the "Paradigm Argument from Design." In *Foraging Behavior*, ed. A.C. Kamil, J.R. Krebs, and H.R. Pulliam, pp. 69–140. New York: Plenum Press.

Gremillion, K.J. 1996. Diffusion and Adoption of Crops in Evolutionary Perspective. *Journal of Anthropological Archaeology* 15(2):183–204.

———. 2002. Foraging Theory and Hypothesis Testing in Archaeology: An Exploration of Methodological Problems and Solutions. *Journal of Anthropological Archaeology* 21(2):142–64.

———. 2004. Seed Processing and the Origins of Food Production in Eastern North America. *American Antiquity* 69:215–34.

———. 2006. Central Place Foraging and Food Production on the Cumberland Plateau, Eastern Kentucky. In *Behavioral Ecology and the Transition to Agriculture*, ed. D.J. Kennett and B. Winterhalder, pp. 41–62. Berkeley: University of California Press.

———. 2014. Human Behavioral Ecology and Paleoethnobotany. In *Method and Theory in Paleoethnobotany*, ed. J.M. Marston, J. d'Alpoim Guedes, and C. Warinner, pp. 339–54. Boulder: University Press of Colorado.

———. 2015. Prehistoric Upland Farming, Fuelwood, and Forest Composition on the Cumberland Plateau, Kentucky, USA. *Journal of Ethnobiology* 35(1):60–84.

Gremillion, K.J., L. Barton, and D.R. Piperno. 2014. Particularism and the Retreat from Theory in the Archaeology of Agricultural Origins. *Proceedings of the National Academy of Sciences of the United States of America* 111(17):6171–77.

Gremillion, K.J., and D.R. Piperno. 2009. Human Behavioral Ecology, Phenotypic (Developmental) Plasticity, and Agricultural Origins: Insights from the Emerging Evolutionary Synthesis. *Current Anthropology* 50:615–19.

Gremillion, K.J., J. Windingstad, and S.C. Sherwood. 2008. Forest Opening, Habitat Use, and Food Production on the Cumberland Plateau, Kentucky: Adaptive Flexibility in Marginal Settings. *American Antiquity* 73(3):387–411.

Groesbeck, A.S., K. Rowell, D. Lepofsky, and A.K. Salomon. 2014. Ancient Clam Gardens Increased Shellfish Production: Adaptive Strategies from the Past Can Inform Food Security Today. *PLoS ONE* 9(3):e91235.

Groffman, P.M., J.S. Baron, T. Blett, A.J. Gold, I. Goodman, L.H. Gunderson, B.M. Levinson, M.A. Palmer, H.W. Paerl, G.D. Peterson, N.L. Poff, D.W. Rejeski, J.F. Reynolds, M.G. Turner, K.C. Weathers, and J. Wiens. 2006. Ecological Thresholds: The Key to Successful Environmental Management or an Important Concept with No Practical Application? *Ecosystems* 9(1):1–13.

Gunderson, L.H., C.R. Allen, and C.S. Holling, eds. 2009. *Foundations of Ecological Resilience*. Washington, DC: Island Press.

Gunderson, L.H., and C.S. Holling, eds. 2002. *Panarchy: Understanding Transformations in Human and Natural Systems*. Washington, DC: Island Press.

Gunter, A.C. 1991. *The Gordion Excavations Final Reports Vol. 3: The Bronze Age*. Philadelphia: University of Pennsylvania Museum of Archaeology and Anthropology.

Gürsan-Salzmann, A. 1997. Ethnoarchaeology at Yassıhöyük (Gordion): Space, Activity, Subsistence. In "Fieldwork at Gordion: 1993–1995." *Anatolica* 23:26–31.

———. 2005. Ethnographic Lessons for Past Agro-Pastoral Systems in the Sakarya-Porsuk Valleys.

In *The Archaeology of Midas and the Phrygians: Recent Work at Gordion*, ed. L. Kealhofer, pp. 172–90. Philadelphia: University of Pennsylvania Museum of Archaeology and Anthropology.

Gurven, M. 2004. To Give and to Give Not: The Behavioral Ecology of Human Food Transfers. *Behavioral and Brain Sciences* 27(4):543–83.

———. 2006. The Evolution of Contingent Cooperation. *Current Anthropology* 47(1):185–92.

Haas, J.N., S. Karg, and P. Rasmussen. 1998. Beech Leaves and Twigs Used as Winter Fodder: Examples from Historic and Prehistoric Times. *Environmental Archaeology* 1:81–86.

Haas, J.N., and F.H. Schweingruber. 1994. Wood-Anatomical Evidence of Pollarding in Ash Stems from the Valais, Switzerland. *Dendrochronologia* 11:35–43.

Halbwachs, M. 1992. *On Collective Memory*. Translated by Lewis A. Coser. Chicago: University of Chicago Press.

Haldon, J., N. Roberts, A. Izdebski, D. Fleitmann, M. McCormick, M. Cassis, O. Doonan, W. Eastwood, H. Elton, and S. Ladstätter. 2014. The Climate and Environment of Byzantine Anatolia: Integrating Science, History, and Archaeology. *Journal of Interdisciplinary History* 45(2):113–61.

Hale, J.D. 1933. *Heating Value of Wood Fuels*. Canada: Forest Products Laboratories of Canada, Forest Service, Department of the Interior.

Hall, R.T., and M.B. Dickerman. 1942. *Wood Fuel in Wartime*. Farmer's Bulletin 1912. Washington, DC: U.S. Department of Agriculture.

Halstead, P. 1989. The Economy Has a Normal Surplus: Economic Stability and Social Change among Early Farming Communities of Thessaly, Greece. In *Bad Year Economics: Cultural Responses to Risk and Uncertainty*, ed. P. Halstead and J. O'Shea, pp. 68–80. Cambridge: Cambridge University Press.

———. 1990. Waste Not, Want Not: Traditional Responses to Crop Failure in Greece. *Rural History* 1:147–64.

———. 1998. Ask the Fellows Who Lop the Hay: Leaf-Fodder in the Mountains of Northwest Greece. *Rural History* 9:211–34.

Halstead, P., and G. Jones. 1989. Agrarian Ecology in the Greek Islands: Time Stress, Scale and Risk. *Journal of Hellenic Studies* 109:41–55.

Halstead, P., and J. O'Shea, eds. 1989a. *Bad Year Economics: Cultural Responses to Risk and Uncertainty*. Cambridge: Cambridge University Press.

———. 1989b. Introduction: Cultural Responses to Risk and Uncertainty. In *Bad Year Economics: Cultural Responses to Risk and Uncertainty*, ed. P. Halstead and J. O'Shea, pp. 1–7. Cambridge: Cambridge University Press.

Halstead, P., and J. Tierney. 1998. Leafy Hay: An Ethnoarchaeological Study in NW Greece. *Environmental Archaeology* 1:71–80.

Harissis, H.V., and A.V. Harissis. 2009. *Apiculture in the Prehistoric Aegean: Minoan and Mycenaean Symbols Revisited*. BAR International Series. Oxford: British Archaeological Reports.

Harrower, M.J. 2008. Hydrology, Ideology, and the Origins of Irrigation in Ancient Southwest Arabia. *Current Anthropology* 49:497–510.

Harrower, M.J., E.A. Oches, and J. McCorriston. 2012. Hydro-Geospatial Analysis of Ancient Pastoral/Agro-Pastoral Landscapes Along Wadi Sana (Yemen). *Journal of Arid Environments* 86:131–38.

Hartung, J. 1976. On Natural Selection and the Inheritance of Wealth. *Current Anthropology* 17(4):607–22.

Haught, E. 1958. *Further Study on Compression Wood in Loblolly Pine*. 2nd Report Forest Tree Improvement Program. Raleigh: School of Forestry, North Carolina State College.

Hawkes, K. 1991. Showing Off: Tests of an Explanatory Hypothesis About Men's Foraging Goals. *Ethology and Sociobiology* 12(1):29–54.

Hawkes, K., and R. Bliege Bird. 2002. Showing Off, Handicap Signaling, and the Evolution of Men's Work. *Evolutionary Anthropology* 11(2):58–67.

Hawkes, K., K. Hill. and J.F. O'Connell. 1982. Why Hunters Gather: Optimal Foraging and the Aché of Eastern Paraguay. *American Ethnologist* 9(2):379–98.

Heat Shed, The. n.d. Tree Identification and Heat Value Chart (Advertisement). Revere, PA: The Heat Shed Inc.

Hegmon, M., M.A. Peeples, A.P. Kinzig, S. Kulow, C.M. Meegan, and M.C. Nelson. 2008. Social Transformation and Its Human Costs in the Prehispanic U.S. Southwest. *American Anthropologist* 110(3):313–24.

Henrich, J. 2004. Demography and Cultural Evolution: How Adaptive Cultural Processes Can Produce Maladaptive Losses: The Tasmanian Case. *American Antiquity* 69(2):197–214.

Henrich, J., and R. McElreath. 2002. Are Peasants Risk-Averse Decision Makers? *Current Anthropology* 43(1):172–81.

Henrickson, R.C., and M.J. Blackman. 1996. Large Scale Production of Pottery at Gordion: A Comparison of the Late Bronze and Early Phrygian Industries. *Paléorient* 22(1):67–88.

Henry, A., and I. Théry-Parisot. 2014. From Evenk Campfires to Prehistoric Hearths: Charcoal Analysis as a Tool for Identifying the Use of Rotten Wood as Fuel. *Journal of Archaeological Science* 52:321–36.

Hill, K., H. Kaplan, K. Hawkes, and A.M. Hurtado. 1987. Foraging Decisions among Aché Hunter-Gatherers: New Data and Implications for Optimal Foraging Models. *Ethology and Sociobiology* 8(1):1–36.

Hillman, G. 1984. Interpretation of Archaeological Plant Remains: The Application of Ethnographic Models from Turkey. In *Plants and Ancient Man: Studies in Palaeoethnobotany*, ed. W. van Zeist and W.A. Casparie, pp. 1–43. Rotterdam: A.A. Balkema.

Hillman, G.C., and M.S. Davies. 1990. Measured Domestication Rates in Wild Wheats and Barley under Primitive Cultivation, and Their Archaeological Implications. *Journal of World Prehistory* 4(2):157–222.

Hockett, B., and J. Haws. 2003. Nutritional Ecology and Diachronic Trends in Paleolithic Diet and Health. *Evolutionary Anthropology: Issues, News, and Reviews* 12(5):211–16.

Hoffner, H.A. 1974. *Alimenta Hethaeorum: Food Production in Hittite Asia Minor*. New Haven: American Oriental Society.

Hoffner, H.A. 1995. Oil in Hittite Texts. *Biblical Archaeologist* 58(2):108–14.

Holling, C.S. 1973. Resilience and Stability of Ecological Systems. *Annual Review of Ecology Evolution and Systematics* 4:1–23.

———. 2001. Understanding the Complexity of Economic, Ecological, and Social Systems. *Ecosystems* 4(5):390–405.

Holling, C.S., and L.H. Gunderson. 2002. Resilience

and Adaptive Cycles. In *Panarchy: Understanding Transformations in Human and Natural Systems*, ed. L.H. Gunderson and C.S. Holling, pp. 25–62. Washington, DC: Island Press.

Holling, C.S., L.H. Gunderson, and D. Ludwig. 2002a. In Quest of a Theory of Adaptive Change. In *Panarchy: Understanding Transformations in Human and Natural Systems*, ed. L.H. Gunderson and C.S. Holling, pp. 3–22. Washington, DC: Island Press.

Holling, C.S., L.H. Gunderson, and G. Peterson. 2002b. Sustainability and Panarchies. In *Panarchy: Understanding Transformations in Human and Natural Systems*, ed. L.H. Gunderson and C.S. Holling, pp. 63–102. Washington, DC: Island Press.

Hopkins, L. 2003. *Archaeology at the North-East Anatolian Frontier, VI: An Ethnoarchaeological Study of Sos Höyük and Yiğittaşı Village*. Ancient Near Eastern Studies. Supplement 11. Louvain: Peeters.

Horne, L. 1982. The Demand for Fuel: Ecological Implications of Socio-Economic Change. In *Desertification and Development: Dryland Ecology in Social Perspective*, ed. B. Spooner and H.S. Mann, pp. 201–15. New York: Academic Press.

Hunt, T.L. 2007. Rethinking Easter Island's Ecological Catastrophe. *Journal of Archaeological Science* 34(3):485–502.

Hunt, T.L., and C.P. Lipo. 2010. Ecological Catastrophe, Collapse, and the Myth of "Ecocide" on Rapa Nui (Easter Island). In *Questioning Collapse: Human Resilience, Ecological Vulnerability, and the Aftermath of Empire*, ed. P.A. McAnany and N. Yoffee, pp. 21–44. Cambridge: Cambridge University Press.

Iannone, G., K. Prufer, and D.Z. Chase. 2014. Resilience and Vulnerability in the Maya Hinterlands. *Archeological Papers of the American Anthropological Association* 24(1):155–70.

Izdebski, A., J. Pickett, C.N. Roberts, and T. Waliszewski. 2016. The Environmental, Archaeological and Historical Evidence for Regional Climatic Changes and Their Societal Impacts in the Eastern Mediterranean in Late Antiquity. *Quaternary Science Reviews* 136:189–208.

Jackson, J.B.C., M.X. Kirby, W.H. Berger, K.A. Bjorndal, L.W. Botsford, B.J. Bourque, R.H. Bradbury, R. Cooke, J. Erlandson, J.A. Estes, T.P. Hughes, S. Kidwell, C.B. Lange, H.S. Lenihan, J.M. Pandolfi, C.H. Peterson, R.S. Steneck, M.J. Tegner, and R.R. Warner. 2001. Historical Overfishing and the Recent Collapse of Coastal Ecosystems. *Science* 293(5530):629–38.

Jacomet, S. 2006a. Identification of Cereal Remains from Archaeological Sites, 2nd Edition. Unpublished manuscript, Archaeobotany Lab, IPAS, Basel University.

———. 2006b. Introduction. In *Charcoal Analysis: New Analytical Tools and Methods for Archaeology: Papers from the Table-Ronde Held in Basel 2004*, ed. A. Dufraisse, p. 7. Oxford: Archaeopress.

Johnston, K.J. 2003. The Intensification of Pre-Industrial Cereal Agriculture in the Tropics: Boserup, Cultivation Lengthening, and the Classic Maya. *Journal of Anthropological Archaeology* 22(2):126–61.

Jones, G. 1987. A Statistical Approach to the Archaeological Identification of Crop Processing. *Journal of Archaeological Science* 14(3):311–23.

Jones, G., M. Charles, A. Bogaard, and J.G. Hodgson. 2010. Crops and Weeds: The Role of Weed Functional Ecology in the Identification of Crop Husbandry Methods. *Journal of Archaeological Science* 37(1):70–77.

Jones, G., and P. Halstead. 1995. Maslins, Mixtures and Monocrops: On the Interpretation of Archaeobotanical Crop Samples of Heterogeneous Composition. *Journal of Archaeological Science* 22(1):103–14.

Jones, J.E., A.J. Graham, and L.H. Sackett. 1973. An Attic Country House Below the Cave of Pan at Vari. *The Annual of the British School at Athens* 68:355–452.

Jones, K.T., and D.B. Madsen. 1989. Calculating the Cost of Resource Transportation: A Great Basin Example. *Current Anthropology* 30(4):529–34.

Jones, M. 1988. The Phytosociology of Early Arable Weed Communities with Special Reference to Southern England. In *Der prähistorische Mensch und seine Umwelt: Festschrift für Udelgard Körber-Grohne*, ed. H. Küster, pp. 43–51. Stuttgart: Konrad Theiss Verlag.

Jones, M.D., C.N. Roberts, M.J. Leng, and M. Türkeş. 2006. A High-Resolution Late Holocene

Lake Isotope Record from Turkey and Links to North Atlantic and Monsoon Climate. *Geology* 34(5):361–64.

Joseph, S. 2000. Anthropological Evolutionary Ecology: A Critique. *Journal of Ecological Anthropology* 4(1):6–30.

Kaniewski, D., E. Paulissen, E. Van Campo, M. Al-Maqdissi, J. Bretschneider, and K. Van Lerberghe. 2008. Middle East Coastal Ecosystem Response to Middle-to-Late Holocene Abrupt Climate Changes. *Proceedings of the National Academy of Sciences of the United States of America* 105(37):13941–46.

Kaniewski, D., E. Paulissen, E. Van Campo, H. Weiss, T. Otto, J. Bretschneider, and K. Van Lerberghe. 2010. Late Second–Early First Millennium BC Abrupt Climate Changes in Coastal Syria and Their Possible Significance for the History of the Eastern Mediterranean. *Quaternary Research* 74(2):207–15.

Kaniewski, D., V. De Laet, E. Paulissen, and M. Waelkens. 2007a. Long-Term Effects of Human Impact on Mountainous Ecosystems, Western Taurus Mountains, Turkey. *Journal of Biogeography* 34(11):1975–97.

Kaniewski, D., E. Paulissen, V. De Laet, K. Dossche, and M. Waelkens. 2007b. A High-Resolution Late Holocene Landscape Ecological History Inferred from an Intramontane Basin in the Western Taurus Mountains, Turkey. *Quaternary Science Reviews* 26(17–18):2201–18.

Kaplan, H., and K. Hill. 1985. Food Sharing among Ache Foragers: Tests of Explanatory Hypotheses. *Current Anthropology* 26(2):223–46.

———. 1992. The Evolutionary Ecology of Food Acquisition. In *Evolutionary Ecology and Human Behavior*, ed. E.A. Smith and B. Winterhalder, pp. 167–203. New York: Aldine de Gruyter.

Kaplan, H., K. Hill, and A.M. Hurtado. 1990. Risk, Foraging and Food Sharing among the Ache. In *Risk and Uncertainty in Tribal and Peasant Economies*, ed. E.A. Cashdan, pp. 107–43. Boulder, CO: Westview Press.

Kealhofer, L., ed. 2005a. *The Archaeology of Midas and the Phrygians: Recent Work at Gordion*. Philadelphia: University of Pennsylvania Museum of Archaeology and Anthropology.

———. 2005b. Settlement and Land Use: The Gor-

dion Regional Survey. In *The Archaeology of Midas and the Phrygians: Recent Work at Gordion*, ed. L. Kealhofer, pp. 137–48. Philadelphia: University of Pennsylvania Museum of Archaeology and Anthropology.

Keegan, W.F. 1986. The Optimal Foraging Analysis of Horticultural Production. *American Anthropologist* 88(1):92–107.

Kendall, M. 1938. A New Measure of Rank Correlation. *Biometrika* 30:81–89.

Kennett, D.J., and B. Winterhalder, eds. 2006. *Behavioral Ecology and the Transition to Agriculture*. Berkeley: University of California Press.

Kent, M. 2012. *Vegetation Description and Data Analysis: A Practical Approach*. 2nd ed. Chichester, UK: Wiley-Blackwell.

Khazraee, E., and S. Gasson. 2015. Epistemic Objects and Embeddedness: Knowledge Construction and Narratives in Research Networks of Practice. *The Information Society* 31(2):139–59.

Kimpe, K., P.A. Jacobs, and M. Waelkens. 2001. Analysis of Oil Used in Late Roman Oil Lamps with Different Mass Spectrometric Techniques Revealed the Presence of Predominantly Olive Oil Together with Traces of Animal Fat. *Journal of Chromatography A* 937(1–2):87–95.

———. 2002. Mass Spectrometric Methods Prove the Use of Beeswax and Ruminant Fat in Late Roman Cooking Pots. *Journal of Chromatography A* 968(1–2):151–60.

Kimura, F., A. Kitoh, A. Sumi, J. Asanuma, and A. Yatagai. 2007. Downscaling of the Global Warming Projections to Turkey. The Final Report of the ICCAP Project, Unpublished manuscript, Research Institute for Humanity and Nature, National Institute for the Humanities of Japan.

Kinzig, A.P., P. Ryan, M. Etienne, H. Allison, T. Elmqvist, and B.H. Walker. 2006. Resilience and Regime Shifts: Assessing Cascading Effects. *Ecology and Society* 11(1):20.

Kirch, P.V. 2006. Agricultural Intensification: A Polynesian Perspective. In *Agricultural Strategies*, ed. J. Marcus and C. Stanish, pp. 191–217. Los Angeles: Cotsen Institute of Archaeology at UCLA.

Klinge, J., and P. Fall. 2010. Archaeobotanical Inference of Bronze Age Land Use and Land Cover in the Eastern Mediterranean. *Journal of Archaeological Science* 37(10):2622–29.

Knight, F.H. 1921. *Risk, Uncertainty and Profit.* Boston: Houghton Mifflin Co.

Knipping, M., M. Müllenhoff, and H. Brückner. 2008. Human Induced Landscape Changes around Bafa Gölü (Western Turkey). *Vegetation History and Archaeobotany* 17(4):365–80.

Kohler, E.L. 1995. *The Gordion Excavations (1950–1973) Final Reports, Vol. II: The Lesser Phrygian Tumuli. Part 1: The Inhumations.* Philadelphia: University of Pennsylvania Museum of Archaeology and Anthropology.

Körte, A. 1897. Kleinasiatische Studien II: Gordion und der Zug des Manlius gegen die Galater. *Mitteilungen des Kaiserlich Deutschen Archäologischen Instituts, Athenische Abteilung* 22:1–29.

Körte, A., and G. Körte. 1901. Gordion. *Archäologischer Anzeiger* 1:1–11.

Körte, G., and A. Körte. 1904. *Gordion: Ergebnisse der Ausgrabung im Jahre 1900.* Berlin: G. Reimer.

Kovats, R.S., R. Valentini, L. Bouwer, E. Georgopoulou, D. Jacob, E. Martin, M. Rounsevell, and J.-F. Soussana. 2014. Europe. In *Climate Change 2014: Impacts, Adaptation, and Vulnerability. Part B: Regional Aspects. Contribution of Working Group II to the Fifth Assessment Report of the Intergovernmental Panel on Climate Change,* ed. V.R. Barros, C.B. Field, D.J. Dokken, M.D. Mastrandrea, K.J. Mach, T.E. Bilir, M. Chatterjee, K.L. Ebi, Y.O. Estrada, R.C. Genova, B. Girma, E.S. Kissel, A.N. Levy, S. MacCracken, P.R. Mastrandrea, and L.L. White, pp. 1267–326. Cambridge: Cambridge University Press.

Kraft, J.R., and W.M. Baum. 2001. Group Choice: The Ideal Free Distribution of Human Social Behavior. *Journal of the Experimental Analysis of Behavior* 76(1):21–42.

Kraft, J.R., W.M. Baum, and M.J. Burge. 2002. Group Choice and Individual Choices: Modeling Human Social Behavior with the Ideal Free Distribution. *Behavioural Processes* 57(2):227–40.

Krebs, J.R., and N.B. Davies. 1981. *An Introduction to Behavioural Ecology.* 1st ed. Oxford: Blackwell Scientific Publications.

———. 1993. *An Introduction to Behavioural Ecology.* 3rd ed. Oxford: Blackwell Scientific Publications.

Krebs, J.R., J.T. Erichsen, M.I. Webber, and E.L. Charnov. 1977. Optimal Prey Selection in the Great Tit (*Parus major*). *Animal Behaviour* 25(Feb):30–38.

Krebs, J.R., and R.H. McCleery. 1984. Optimization in Behavioural Ecology. In *Behavioural Ecology: An Evolutionary Approach,* ed. J.R. Krebs and N.B. Davies, pp. 91–121. 2nd ed. Sunderland, MA: Sinauer Associates.

Krishna, S., and S. Ramaswami. 1932. *Calorific Values of Some Indian Woods.* Forest Bulletin 79. Calcutta: Forest Resources Institute.

Kroll, H.J. 1983. *Kastanas: Ausgrabungen in einem Siedlungshügel der Bronze- und Eisenzeit Makedoniens, 1975–1979.* Berlin: V. Spiess.

———. 1991. Südosteuropa (Southeast Europe). In *Progress in Old World Palaeoethnobotany,* ed. W. van Zeist, Kr. Wasylikowa, and K.-E. Behre, pp. 161–77. Rotterdam: A.A. Balkema.

Kron, J.G. 2000. Roman Ley-Farming. *Journal of Roman Archaeology* 13:277–87.

———. 2012a. Agriculture, Roman Empire. In *The Encyclopedia of Ancient History,* ed. Roger S. Bagnall, K. Brodersen, C.B. Champion, A. Erskine, and S.R. Huebner, pp. 217–22. Malden, MA: Blackwell Publishing Ltd.

———. 2012b. Food Production. In *The Cambridge Companion to the Roman Economy,* ed. W. Scheidel, pp. 156–74. Cambridge: Cambridge University Press.

Kuijt, I. 2009. What Do We Really Know About Food Storage, Surplus, and Feasting in Preagricultural Communities? *Current Anthropology* 50(5):641–44.

Kuniholm, P.I. 1977. Dendrochronology at Gordion and on the Anatolian Plateau. Ph.D. diss., University of Pennsylvania, Department of Anthropology, Philadelphia.

Kuniholm, P.I., M. Newton, and R.F. Liebhart. 2011. Dendrochronology at Gordion. In *The New Chronology of Iron Age Gordion,* ed. C.B. Rose and G. Darbyshire, pp. 79–122. Philadelphia: University of Pennsylvania Museum of Archaeology and Anthropology.

Kuzucuoğlu, C., W. Dörfler, S. Kunesch, and F. Goupille. 2011. Mid- to Late-Holocene Climate Change in Central Turkey: The Tecer Lake Record. *The Holocene* 21(1):173–88.

Laland, K.N., and M.J. O'Brien. 2011. Cultural Niche Construction: An Introduction. *Biological Theory* 6(3):191–202.

Langlie, B.S., and E.N. Arkush. 2016. Managing Mayhem: Conflict, Environment, and Subsistence in the Andean Late Intermediate Period, Puno, Peru. In *The Archaeology of Food and Warfare: Food Insecurity in Prehistory*, ed. A.M. VanDerwarker and G.D. Wilson, pp. 259–90. Heidelberg: Springer.

Lawall, M.L. 1997. Greek Transport Amphoras at Gordion. In "Fieldwork at Gordion: 1993–1995." *Anatolica* 23:21–23.

———. 2012. Pontic Inhabitants at Gordion? Pots, People, and Plans of Houses at Middle Phrygian through Early Hellenistic Gordion. In *The Archaeology of Phrygian Gordion: Royal City of Midas*, ed. C.B. Rose, pp. 219–24. Philadelphia: University of Pennsylvania Museum of Archaeology and Anthropology.

Leibold, M.A., M. Holyoak, N. Mouquet, P. Amarasekare, J.M. Chase, M.F. Hoopes, R.D. Holt, J.B. Shurin, R. Law, and D. Tilman. 2004. The Metacommunity Concept: A Framework for Multi-Scale Community Ecology. *Ecology Letters* 7(7):601–13.

Lentz, D.L., and B. Hockaday. 2009. Tikal Timbers and Temples: Ancient Maya Agroforestry and the End of Time. *Journal of Archaeological Science* 36(7):1342–53.

Lepofsky, D., and M. Caldwell. 2013. Indigenous Marine Resource Management on the Northwest Coast of North America. *Ecological Processes* 2(1):1–12.

Lepofsky, D., and J. Kahn. 2011. Cultivating an Ecological and Social Balance: Elite Demands and Commoner Knowledge in Ancient Maʿohi Agriculture, Society Islands. *American Anthropologist* 113(2):319–35.

Lepofsky, D., and K. Lertzman. 2005. More on Sampling for Richness and Diversity in Archaeobiological Assemblages. *Journal of Ethnobiology* 25(2):175–88.

———. 2008. Documenting Ancient Plant Management in the Northwest of North America. *Botany* 86:129–45.

Levi, T., F. Lu, D.W. Yu, and M. Mangel. 2011. The Behaviour and Diet Breadth of Central-Place Foragers: An Application to Human Hunters and Neotropical Game Management. *Evolutionary Ecology Research* 13(2):171–85.

Lewontin, R.C. 1979. Fitness, Survival, and Optimality. In *Analysis of Ecological Systems*, ed. D.J. Horn, G.R. Stairs, and R.D. Mitchell, pp. 3–21. Columbus: Ohio State University Press.

———. 1983. Gene, Organism, and Environment. In *Evolution from Molecules to Men*, ed. D.S. Bendall, pp. 273–85. Cambridge: Cambridge University Press.

Liebhart, R.F., and J.S. Johnson. 2005. Support and Conserve: Conservation and Environmental Monitoring of the Tomb Chamber of Tumulus MM. In *The Archaeology of Midas and the Phrygians: Recent Work at Gordion*, ed. L. Kealhofer, pp. 192–203. Philadelphia: University of Pennsylvania Museum of Archaeology and Anthropology.

Linseele, V., W. Van Neer, S. Thys, R. Phillipps, R. Cappers, W. Wendrich, and S. Holdaway. 2014. New Archaeozoological Data from the Fayum "Neolithic" with a Critical Assessment of the Evidence for Early Stock Keeping in Egypt. *PLoS ONE* 9(10):e108517.

Liu, J., T. Dietz, S.R. Carpenter, M. Alberti, C. Folke, E. Moran, A.N. Pell, P. Deadman, T. Kratz, J. Lubchenco, E. Ostrom, Z. Ouyang, W. Provencher, C.L. Redman, S.H. Schneider, and W.W. Taylor. 2007a. Complexity of Coupled Human and Natural Systems. *Science* 317(5844):1513–16.

Liu, J., T. Dietz, S.R. Carpenter, C. Folke, M. Alberti, C.L. Redman, S.H. Schneider, E. Ostrom, A.N. Pell, J. Lubchenco, W.W. Taylor, Z. Ouyang, P. Deadman, T. Kratz, and W. Provencher. 2007b. Coupled Human and Natural Systems. *AMBIO: A Journal of the Human Environment* 36(8):639–49.

Lortie, C.J., R.W. Brooker, P. Choler, Z. Kikvidze, R. Michalet, F.I. Pugnaire, and R.M. Callaway. 2004. Rethinking Plant Community Theory. *Oikos* 107(2):433–38.

Love, S. 2013a. Architecture as Material Culture: Building Form and Materiality in the Pre-Pottery Neolithic of Anatolia and Levant. *Journal of Anthropological Archaeology* 32(4):746–58.

———. 2013b. The Performance of Building and Technological Choice Made Visible in Mudbrick Architecture. *Cambridge Archaeological Journal* 23(02):263–82.

Ludemann, T. 2002. Anthracology and Forest Sites: The Contribution of Charcoal Analysis to Our

Knowledge of Natural Forest Vegetation in South-West Germany. In *Charcoal Analysis: Methodological Approaches, Palaeoecological Results and Wood Uses: Proceedings of the Second International Meeting of Anthracology, Paris, September 2000*, ed. S. Thiébault, pp. 209–17. Oxford: Archaeopress.

———. 2006. Anthracological Analysis of Recent Charcoal-Burning in the Black Forest, SW Germany. In *Charcoal Analysis: New Analytical Tools and Methods for Archaeology: Papers from the Table-Ronde Held in Basel 2004*, ed. A. Dufraisse, pp. 61–70. Oxford: Archaeopress.

Lupo, K.D. 2007. Evolutionary Foraging Models in Zooarchaeological Analysis: Recent Applications and Future Challenges. *Journal of Archaeological Research* 15(2):143–89.

Lyman, R.L., and K.M. Ames. 2004. Sampling to Redundancy in Zooarchaeology: Lessons from the Portland Basin, Northwestern Oregon and Southwestern Washington. *Journal of Ethnobiology* 24:329–46.

MacArthur, R.H., and E.R. Pianka. 1966. On Optimal Use of a Patchy Environment. *American Naturalist* 100(916):603–9.

Mace, R. 1993. Nomadic Pastoralists Adopt Subsistence Strategies That Maximize Long-Term Household Survival. *Behavioral Ecology and Sociobiology* 33(5):329–34.

MacKinnon, D., and K.D. Derickson. 2013. From Resilience to Resourcefulness: A Critique of Resilience Policy and Activism. *Progress in Human Geography* 37(2):253–70.

Madden, G.J., Blaine F. Peden, and T. Yamaguchi. 2002. Human Group Choice: Discrete-Trial and Free-Operant Tests of the Ideal Free Distribution. *Journal of the Experimental Analysis of Behavior* 78(1):1–15.

Madgwick, R., J. Mulville, and J. Evans. 2012. Investigating Diagenesis and the Suitability of Porcine Enamel for Strontium ($^{87}Sr/^{86}Sr$) Isotope Analysis. *Journal of Analytical Atomic Spectrometry* 27(5):733–42.

Madsen, D.B. 1993. Testing Diet Breadth Models: Examining Adaptive Change in the Late Prehistoric Great Basin. *Journal of Archaeological Science* 20:321–29.

Madsen, E.A., R.J. Tunney, G. Fieldman, H.C. Plotkin, R.I.M. Dunbar, J.-M. Richardson, and D. McFarland. 2007. Kinship and Altruism: A Cross-Cultural Experimental Study. *British Journal of Psychology* 98:339–59.

Makal, M. 1954. *A Village in Anatolia*. Translated by Sir Wyndham Deedes. London: Vallentine, Mitchell, and Co.

Mallory, J.P. 1989. *In Search of the Indo-Europeans: Language, Archaeology and Myth*. London: Thames and Hudson.

Manning, S.W., and B. Kromer. 2011. Radiocarbon Dating Iron Age Gordion and the Early Phrygian Destruction in Particular. In *The New Chronology of Iron Age Gordion*, ed. C.B. Rose and G. Darbyshire, pp. 123–54. Philadelphia: University of Pennsylvania Museum of Archaeology and Anthropology.

Marconetto, M.B. 2002. Analysis of Burnt Building Structures of the Ambato Valley (Catamarca, Argentina). In *Charcoal Analysis: Methodological Approaches, Palaeoecological Results and Wood Uses: Proceedings of the Second International Meeting of Anthracology, Paris, September 2000*, ed. S. Thiébault, pp. 267–71. Oxford: Archaeopress.

Marsh, B. 1999. Alluvial Burial of Gordion, an Iron-Age City in Anatolia. *Journal of Field Archaeology* 26(2):163–75.

———. 2005. Physical Geography, Land Use, and Human Impact at Gordion. In *The Archaeology of Midas and the Phrygians: Recent Work at Gordion*, ed. L. Kealhofer, pp. 161–71. Philadelphia: University of Pennsylvania Museum of Archaeology and Anthropology.

———. 2012. Reading Gordion Settlement History from Stream Sedimentation. In *The Archaeology of Phrygian Gordion: Royal City of Midas*, ed. C.B. Rose, pp. 39–46. Philadelphia: University of Pennsylvania Museum of Archaeology and Anthropology.

Marsh, B., and L. Kealhofer. 2014. Scales of Impact: Settlement History and Landscape Change in the Gordion Region, Central Anatolia. *The Holocene* 24(6):689–701.

Marshall, F., and E. Hildebrand. 2002. Cattle before Crops: The Beginnings of Food Production in Africa. *Journal of World Prehistory* 16(2):99–143.

Marston, J.M. 2003. Plant Remains from Middle and Late Phrygian Domestic Contexts at Gordion, Turkey (1993–1994). MASCA Ethnobotanical

Lab Report 34, on file, Gordion Archive, University of Pennsylvania Museum of Archaeology and Anthropology, Philadelphia.

———. 2009. Modeling Wood Acquisition Strategies from Archaeological Charcoal Remains. *Journal of Archaeological Science* 36(10):2192–200.

———. 2010. Evaluating Risk, Sustainability, and Decision Making in Agricultural and Land-Use Strategies at Ancient Gordion. Ph.D. diss., University of California, Interdepartmental Graduate Program in Archaeology, Los Angeles.

———. 2011. Archaeological Markers of Agricultural Risk Management. *Journal of Anthropological Archaeology* 30:190–205.

———. 2012a. Agricultural Strategies and Political Economy in Ancient Anatolia. *American Journal of Archaeology* 116:377–403.

———. 2012b. Reconstructing the Functional Use of Wood at Phrygian Gordion through Charcoal Analysis. In *The Archaeology of Phrygian Gordion: Royal City of Midas*, ed. C.B. Rose, pp. 47–54. Philadelphia: University of Pennsylvania Museum of Archaeology and Anthropology.

———. 2014. Ratios and Simple Statistics in Paleoethnobotanical Analysis: Data Exploration and Hypothesis Testing. In *Method and Theory in Paleoethnobotany*, ed. J.M. Marston, J. d'Alpoim Guedes, and C. Warinner, pp. 163–79. Boulder: University Press of Colorado.

———. 2015. Modeling Resilience and Sustainability in Ancient Agricultural Systems. *Journal of Ethnobiology* 35:585–605.

Marston, J.M., and S. Branting. 2016. Agricultural Adaptation to Highland Climate in Iron Age Anatolia. *Journal of Archaeological Science: Reports* 9:25–32.

Marston, J.M., and N.F. Miller. 2014. Intensive Agriculture and Land Use at Roman Gordion, Central Turkey. *Vegetation History and Archaeobotany* 23:761–73.

Martin, M. 1980. Making a Living in Turan: Animals, Land, and Wages. *Expedition* 22(4):29–35.

Masi, A., L. Sadori, F.B. Restelli, I. Baneschi, and G. Zanchetta. 2013a. Stable Carbon Isotope Analysis as a Crop Management Indicator at Arslantepe (Malatya, Turkey) During the Late Chalcolithic and Early Bronze Age. *Vegetation History and Archaeobotany* 23(6):1–10.

Masi, A., L. Sadori, I. Baneschi, A.M. Siani, and G. Zanchetta. 2013b. Stable Isotope Analysis of Archaeological Oak Charcoal from Eastern Anatolia as a Marker of Mid-Holocene Climate Change. *Plant Biology* 15:83–92.

Masi, A., L. Sadori, G. Zanchetta, I. Baneschi, and M. Giardini. 2013c. Climatic Interpretation of Carbon Isotope Content of Mid-Holocene Archaeological Charcoals from Eastern Anatolia. *Quaternary International* 303:64–72.

Maynard Smith, J. 1978. Optimization Theory in Evolution. *Annual Review of Ecology and Systematics* 9:31–56.

Maynard Smith, J., and G.R. Price. 1973. Logic of Animal Conflict. *Nature* 246(5427):15–18.

Mayr, E. 1961. Cause and Effect in Biology. *Science* 134(3489):1501–6.

McAnany, P.A., and N. Yoffee, eds. 2010. *Questioning Collapse: Human Resilience, Ecological Vulnerability, and the Aftermath of Empire*. Cambridge: Cambridge University Press.

McCloskey, D.N. 1976. English Open Fields as Behavior toward Risk. *Research in Economic History* 1:144–70.

McCorriston, J. 2006. Breaking the Rain Barrier and the Tropical Spread of Near Eastern Agriculture into Southern Arabia. In *Behavioral Ecology and the Transition to Agriculture*, ed. D.J. Kennett and B. Winterhalder, pp. 217–36. Berkeley: University of California Press.

McGovern, P.E. 2000. The Funerary Banquet of "King Midas." *Expedition* 42(1):21–29.

McGovern, P.E., D.L. Glusker, R.A. Moreau, A. Nuñez, C.W. Beck, E. Simpson, E.D. Butrym, L.J. Exner, and E.C. Stout. 1999. A Funerary Feast Fit for King Midas. *Nature* 402:863–64.

McGovern, T.H., O. Vésteinsson, A. Fridriksson, M. Church, I. Lawson, I.A. Simpson, A. Einarsson, A. Dugmore, G. Cook, S. Perdikaris, K.J. Edwards, A.M. Thomson, W.P. Adderley, A. Newton, G. Lucas, R. Edvardsson, O. Aldred, and E. Dunbar. 2007. Landscapes of Settlement in Northern Iceland: Historical Ecology of Human Impact and Climate Fluctuation on the Millennial Scale. *American Anthropologist* 109(1):27–51.

McNicholl, A. 1983. *Taşkun Kale: Keban Rescue Excavations, Eastern Anatolia*. BAR International Series. Oxford: BAR.

Meiggs, D.C. 2007. Visualizing the Seasonal Round: A Theoretical Experiment with Strontium Isotope Profiles in Ovicaprine Teeth. *Anthropozoologica* 42(2):107–27.

Messner, T.C. and G.E. Stinchcomb. 2014. Peopling the Environment: Interdisciplinary Inquiries into Socioecological Systems Incorporating Paleoclimatology and Geoarchaeology. In *Method and Theory in Paleoethnobotany*, ed. J.M. Marston, J. d'Alpoim Guedes, and C. Warinner, pp. 257–74. Boudler: University Press of Colorado.

Metcalfe, D., and K.R. Barlow. 1992. A Model for Exploring the Optimal Trade-Off between Field Processing and Transport. *American Anthropologist* 94(2):340–56.

METU. 1965. *Yassıhöyük: A Village Study*. Ankara: Middle East Technical University.

Michel, R.H., P.E. McGovern, and V.R. Badler. 1992. Chemical Evidence for Ancient Beer. *Nature* 360(6399):24.

Middleton, G.D. 2012. Nothing Lasts Forever: Environmental Discourses on the Collapse of Past Societies. *Journal of Archaeological Research* 20(3):257–307.

Mieth, A., and H.-R. Bork. 2010. Humans, Climate or Introduced Rats—Which Is to Blame for the Woodland Destruction on Prehistoric Rapa Nui (Easter Island)? *Journal of Archaeological Science* 37(2):417–26.

Migowski, C., M. Stein, S. Prasad, J.F.W. Negendank, and A. Agnon. 2006. Holocene Climate Variability and Cultural Evolution in the Near East from the Dead Sea Sedimentary Record. *Quaternary Research* 66(3):421–31.

Milinski, M., and R. Heller. 1978. Influence of a Predator on Optimal Foraging Behavior of Sticklebacks (*Gasterosteus aculeatus* L.). *Nature* 275(5681):642–44.

Miller, F., H. Osbahr, E. Boyd, F. Thomalla, S. Bharwani, G. Ziervogel, B. Walker, J. Birkmann, S. van der Leeuw, J. Rockström, J. Hinkel, T. Downing, C. Folke, and D. Nelson. 2010. Resilience and Vulnerability: Complementary or Conflicting Concepts? *Ecology and Society* 15(3):11.

Miller, N.F. 1985. Paleoethnobotanical Evidence for Deforestation in Ancient Iran: A Case Study of Urban Malyan. *Journal of Ethnobiology* 5:1–21.

———. 1988. Ratios in Paleoethnobotanical Analysis. In *Current Paleoethnobotany: Analytical Methods and Cultural Interpretations of Archaeological Plant Remains*, ed. C.A. Hastorf and V.S. Popper, pp. 72–96. Chicago: University of Chicago Press.

———. 1990. Clearing Land for Farmland and Fuel: Archaeobotanical Studies of the Ancient Near East. In *Economy and Settlement in the Near East: Analyses of Ancient Sites and Materials*, ed. N.F. Miller, pp. 70–78. Philadelphia: MASCA, University of Pennsylvania Museum.

———. 1991a. Forest and Wood Use at Gordion: Analysis of Wood Charcoal Recovered in 1988 and 1989. MASCA Ethnobotanical Lab Report 10, on file, Gordion Archive, University of Pennsylvania Museum of Archaeology and Anthropology, Philadelphia.

———. 1991b. The Near East. In *Progress in Old World Palaeoethnobotany*, ed. W. van Zeist, K. Wasylikowa, and K.-E. Behre, pp. 133–60. Rotterdam: A.A. Balkema.

———. 1996. Seed Eaters of the Ancient Near East: Human or Herbivore? *Current Anthropology* 37(3):521–28.

———. 1997. Farming and Herding Along the Euphrates: Environmental Constraint and Cultural Choice (Fourth to Second Millennia B.C.). *MASCA Research Papers in Science and Archaeology* 14:123–32.

———. 1998. Patterns of Agriculture and Land Use at Medieval Gritille. In *The Archaeology of the Frontier in the Medieval Near East: Excavations at Gritille, Turkey*, ed. S. Redford, pp. 211–52. Philadelphia: University of Pennsylvania Museum of Archaeology and Anthropology.

———. 1999a. Erozyon, Bioçesitlilik ve Arkeoloji, Gordion'daki Midas Höyügü'nun Korunmasi [Erosion, Biodiversity, and Archaeology: Preserving the Midas Tumulus at Gordion]. *Arkeoloji ve Sanat* 93:12–17.

———. 1999b. Interpreting Ancient Environment and Patterns of Land Use: Seeds, Charcoal and Archaeological Context. *TÜBA-AR* 2:15–27.

———. 2000. Plants in the Service of Archaeological Preservation. *Expedition* 42(1):30–36.

———. 2002. Tracing the Development of the Agropastoral Economy in Southeastern Anatolia and Northern Syria. In *The Dawn of Farming in the*

Near East, ed. R.T.J. Cappers and S. Bottema, pp. 85–94. Berlin: Ex Oriente.

———. 2007a. Roman and Medieval Charcoal from the 2004 Excavation at Gordion, Operations 52, 53, 54, and 55. MASCA Ethnobotanical Lab Report 41, on file, Gordion Archive, University of Pennsylvania Museum of Archaeology and Anthropology, Philadelphia.

———. 2007b. Roman Flotation Samples from the 2004 and 2005 Excavation at Gordion, Operations 44, 52, 53, 54, and 55. MASCA Ethnobotanical Lab Report 45, on file, Gordion Archive, University of Pennsylvania Museum of Archaeology and Anthropology, Philadelphia.

———. 2010. *Botanical Aspects of Environment and Economy at Gordion, Turkey.* Philadelphia: University of Pennsylvania Museum of Archaeology and Anthropology.

———. 2011. Managing Predictable Unpredictability: The Question of Agricultural Sustainability at Gordion. In *Sustainable Lifeways: Cultural Persistence in an Ever-Changing Environment*, ed. N.F. Miller, K.M. Moore, and K. Ryan, pp. 310–24. Philadelphia: University of Pennsylvania Museum of Archaeology and Anthropology.

———. 2012. Working with Nature to Preserve Site and Landscape at Gordion. In *The Archaeology of Phrygian Gordion: Royal City of Midas*, ed. C.B. Rose, pp. 243–58. Philadelphia: University of Pennsylvania Museum of Archaeology and Anthropology.

———. 2013. Agropastoralism and Archaeobiology: Connecting Plants, Animals and People in West and Central Asia. *Environmental Archaeology* 18(3):247–56.

———. 2014. Flotation Charcoal from the Gordion 1988 and 1989 Excavation Seasons. MASCA Ethnobotanical Lab Report 59, on file, Gordion Archive, University of Pennsylvania Museum of Archaeology and Anthropology, Philadelphia.

Miller, N.F., and K. Bluemel. 1999. Plants and Mudbrick: Preserving the Midas Tumulus at Gordion, Turkey. *Conservation and Management of Archaeological Sites* 3:225–37.

Miller, N.F., and D. Enneking. 2014. Bitter Vetch (*Vicia ervilia*): Ancient Medicinal Crop and Farmers' Favorite for Feeding Livestock. In *New Lives for Ancient and Extinct Crops*, ed. P.E. Minnis, pp.

254–68. Tucson: University of Arizona.

Miller, N.F., K.E. Leaman, and J. Unruh. 2006. Serendipity: Secrets of the Mudballs. *Expedition* 48(3):40–41.

Miller, N.F., and J.M. Marston. 2012. Archaeological Fuel Remains as Indicators of Ancient West Asian Agropastoral and Land-Use Systems. *Journal of Arid Environments* 86:97–103.

Miller, N.F., and T.L. Smart. 1984. Intentional Burning of Dung as Fuel: A Mechanism for the Incorporation of Charred Seeds into the Archeological Record. *Journal of Ethnobiology* 4:15–28.

Miller, N.F., M.A. Zeder, and S.R. Arter. 2009. From Food and Fuel to Farms and Flocks: The Integration of Plant and Animal Remains in the Study of the Agropastoral Economy at Gordion, Turkey. *Current Anthropology* 50:915–24.

Mitchell, S. 1993. *Anatolia: Land, Men, and Gods in Asia Minor. Volume I: The Celts in Anatolia and the Impact of Roman Rule.* Oxford: Clarendon Press.

Mohlenhoff, K.A., J.B. Coltrain, and B.F. Codding. 2015. Optimal Foraging Theory and Niche-Construction Theory Do Not Stand in Opposition. *Proceedings of the National Academy of Sciences* 112(24):E3093.

Monson, A. 2012. *From the Ptolemies to the Romans: Political and Economic Change in Egypt.* Cambridge: Cambridge University Press.

Montanari, C., S. Scipioni, G. Calderoni, G. Leonardi, and D. Moreno. 2002. Linking Anthracology and Historical Ecology: Suggestions from a Post-Medieval Site in the Ligurian Apennines (North-West Italy). In *Charcoal Analysis: Methodological Approaches, Palaeoecological Results and Wood Uses: Proceedings of the Second International Meeting of Anthracology, Paris, September 2000*, ed. S. Thiébault, pp. 235–41. Oxford: Archaeopress.

Morehart, C.T. 2012. Mapping Ancient Chinampa Landscapes in the Basin of Mexico: A Remote Sensing and GIS Approach. *Journal of Archaeological Science* 39(7):2541–51.

Morehart, C.T., and S. Morell-Hart. 2015. Beyond the Ecofact: Toward a Social Paleoethnobotany in Mesoamerica. *Journal of Archaeological Method and Theory* 22(2):483–511.

Morell-Hart, S. 2015. Paleoethnobotanical Analysis, Post-Processing. In *Method and Theory in*

Paleoethnobotany, ed. J.M. Marston, J. d'Alpoim Guedes, and C. Warinner, pp. 371–90. Boulder: University Press of Colorado.

Morin, E. 2007. Fat Composition and Nunamiut Decision-Making: A New Look at the Marrow and Bone Grease Indices. *Journal of Archaeological Science* 34(1):69–82.

Morris, I. 2010. *Why the West Rules—for Now: The Patterns of History and What They Reveal About the Future*. New York: Farrar, Strauss, and Giroux.

Morrison, K.D. 1994. The Intensification of Production: Archaeological Approaches. *Journal of Archaeological Method and Theory* 1(2):111–59.

———. 1996. Typological Schemes and Agricultural Change: Beyond Boserup in Precolonial South India. *Current Anthropology* 37:583–608.

Nelle, O. 2002. Charcoal Burning Remains and Forest Stand Structure: Examples from the Black Forest (South-West Germany) and the Bavarian Forest (South-East Germany). In *Charcoal Analysis: Methodological Approaches, Palaeoecological Results and Wood Uses: Proceedings of the Second International Meeting of Anthracology, Paris, September 2000*, ed. S. Thiébault, pp. 201–7. Oxford: Archaeopress.

Nelson, M.C., M. Hegmon, K.W. Kintigh, A.P. Kinzig, B.A. Nelson, J.M. Anderies, D.A. Abbott, K.A. Spielmann, S.E. Ingram, M.A. Peeples, S. Kulow, C.A. Strawhacker, and C. Meegan. 2012. Long-Term Vulnerability and Resilience: Three Examples from Archaeological Study in the Southwestern United States and Northern Mexico. In *Surviving Sudden Environmental Change: Answers from Archaeology*, ed. J. Cooper and P. Sheets, pp. 197–220. Boulder: University Press of Colorado.

Nelson, M.C., K. Kintigh, D.R. Abbott, and J.M. Anderies. 2010. The Cross-Scale Interplay between Social and Biophysical Context and the Vulnerability of Irrigation-Dependent Societies: Archaeology's Long-Term Perspective. *Ecology and Society* 15(3):31.

Nesbitt, M. 1989. Report on Gordion Botanical Samples from Early Phrygian Destruction Layer. Unpublished letter to G. Kenneth Sams, dated January 22, 1989, on file, Gordion Archive, University of Pennsylvania Museum of Archaeology and Anthropology, Philadelphia.

———. 1995. Recovery of Archaeological Plant Remains at Kaman-Kalehöyük. In *Essays on Ancient Anatolia and Its Surrounding Civilizations, vol. 8, Bulletin of the Middle East Culture Centre in Japan*, ed. T. Mikasa, pp. 115–30. Wiesbaden: Harrassowitz.

———. 2006. *Identification Guide for Near Eastern Grass Seeds*. London: Institute of Archaeology, University College London.

Nesbitt, M., and D. Samuel. 1996. Archaeobotany in Turkey: A Review of Current Research. *Orient-Express* 1996:91–96.

Nesbitt, M., and G.D. Summers. 1988. Some Recent Discoveries of Millet (*Panicum miliaceum* L. and *Setaria italica* (L.) P. Beauv.) at Excavations in Turkey and Iran. *Anatolian Studies* 38:85–97.

Netting, R.McC. 1993. *Smallholders, Householders: Farm Families and the Ecology of Intensive, Sustainable Agriculture*. Palo Alto: Stanford University Press.

Neumann, J., and S. Parpola. 1987. Climatic Change and the Eleventh-Tenth-Century Eclipse of Assyria and Babylonia. *Journal of Near Eastern Studies* 46(3):161–82.

Nicoll, K., and C. Küçükuysal. 2013. Emerging Multi-Proxy Records of Late Quaternary Palaeoclimate Dynamics in Turkey and the Surrounding Region. *Turkish Journal of Earth Sciences* 22(1):126–42.

O'Connell, J.F., and K. Hawkes. 1981. Food Choice and Foraging Sites among the Alyawara. In *Hunter-Gatherer Foraging Strategies: Ethnographic and Archeological Analyses*, ed. B. Winterhalder and E.A. Smith, pp. 99–125. Chicago: University of Chicago Press.

Odling-Smee, F.J., K.N. Laland, and M.W. Feldman. 2003. *Niche Construction: The Neglected Process in Evolution*. Princeton: Princeton University Press.

Oral, M.Z. 2002. Selçuk Devri Yemekleri ve Ekmekleri [Seljuk Period Food and Bread]. In *Yemek Kitabı: Tarih-Halkbilimi-Edebiyat [Cookbook: History, Folklore, Literature]*, ed. M.S. Koz, pp. 18–34. Istanbul: Kitabevi.

Orians, G.H., and N.E. Pearson. 1979. On the Theory of Central Place Foraging. In *Analysis of Ecological Systems*, ed. D.J. Horn, G.R. Stairs, and R.D. Mitchell, pp. 155–77. Columbus: Ohio State University Press.

Orland, I.J., M. Bar-Matthews, A. Ayalon, A. Matthews, R. Kozdon, T. Ushikubo, and J.W. Valley. 2012. Seasonal Resolution of Eastern Mediterranean Climate Change since 34 ka from a Soreq Cave Speleothem. *Geochimica et Cosmochimica Acta* 89:240–55.

Orland, I.J., M. Bar-Matthews, N.T. Kita, A. Ayalon, A. Matthews, and J.W. Valley. 2009. Climate Deterioration in the Eastern Mediterranean as Revealed by Ion Microprobe Analysis of a Speleothem That Grew from 2.2 to 0.9 ka in Soreq Cave, Israel. *Quaternary Research* 71(1):27–35.

Ostrom, E. 1990. *Governing the Commons: The Evolution of Institutions for Collective Action.* Cambridge: Cambridge University Press.

Pamir, H.N., and C. Erentöz. 1975. *Explanatory Text of the Geological Map of Turkey, Ankara Volume.* Ankara: Maden Tetkik ve Arama Enstitüsü (Mine Investigation and Research Institute).

Panshin, A.J., and C. de Zeeuw. 1970. *Textbook of Wood Technology.* 3rd ed. New York: McGraw Hill.

Parker, G.A. 1978. Searching for Mates. In *Behavioural Ecology: An Evolutionary Approach,* ed. J.R. Krebs and N.B. Davies, pp. 214–44. 1st ed. Oxford: Blackwell.

Parker, G.A., and R.A. Stuart. 1976. Animal Behavior as a Strategy Optimizer: Evolution of Resource Assessment Strategies and Optimal Emigration Thresholds. *American Naturalist* 110(976):1055–76.

Parr, S.W., and C.N. Davidson. 1922. The Calorific Value of American Woods. *The Journal of Industrial and Engineering Chemistry* 14(10):935–36.

Patterson, W.P., K.A. Dietrich, C. Holmden, and J.T. Andrews. 2010. Two Millennia of North Atlantic Seasonality and Implications for Norse Colonies. *Proceedings of the National Academy of Sciences of the United States of America* 107(12):5306–10.

Payne, S. 1973. Kill-Off Patterns in Sheep and Goats: The Mandibles from Aşvan Kale. *Anatolian Studies* 23:281–303.

Pearsall, D.M. 1983. Evaluating the Stability of Subsistence Strategies by Use of Paleoethnobotanical Data. *Journal of Ethnobiology* 3(2):121–37.

———. 2000. *Paleoethnobotany: A Handbook of Procedures.* 2nd ed. San Diego: Academic Press.

Peeples, M.A., C.M. Barton, and S. Schmich. 2006. Resilience Lost: Intersecting Land Use and Landscape Dynamics in the Prehistoric Southwestern United States. *Ecology and Society* 11(2):22.

Penn Museum. 2015. Digital Gordion. University of Pennsylvania Museum of Archaeology and Anthropology. http://sites.museum.upenn.edu/gordion/. Accessed October 23, 2015.

Peres, T.M. 2010. Methodological Issues in Zooarchaeology. In *Integrating Zooarchaeology and Paleoethnobotany: A Consideration of Issues, Methods, and Cases,* ed. A.M. VanDerwarker and T.M. Peres, pp. 15–36. Berlin: Springer.

Picornell-Gelabert, L., E. Asouti, and E.A. Martí. 2011. The Ethnoarchaeology of Firewood Management in the Fang Villages of Equatorial Guinea, Central Africa: Implications for the Interpretation of Wood Fuel Remains from Archaeological Sites. *Journal of Anthropological Archaeology* 30(3):375–84.

Piperno, D.R. 2011. The Origins of Plant Cultivation and Domestication in the New World Tropics: Patterns, Process, and New Developments. *Current Anthropology* 52(S4):S453–S470.

Piperno, D.R., and D.M. Pearsall. 1998. *The Origins of Agriculture in the Lowland Neotropics.* San Diego: Academic Press.

Pirolli, P., and S. Card. 1999. Information Foraging. *Psychological Review* 106(4):643–75.

Pope, R.D. 2003. Risk and Agriculture: Some Issues and Evidence. In *The Economics of Risk,* ed. D.J. Meyer, pp. 127–67. Kalamazoo, MI: W.E. Upjohn Institute for Employment Research.

Popper, V.S. 1988. Selecting Quantitative Measurements in Paleoethnobotany. In *Current Paleoethnobotany: Analytical Methods and Cultural Interpretations of Archaeological Plant Remains,* ed. C.A. Hastorf and V.S. Popper, pp. 53–71. Chicago: University of Chicago Press.

Pulliam, H.R. 1975. Diet Optimization with Nutrient Constraints. *American Naturalist* 109(970):765–68.

Pyke, G.H. 1984. Optimal Foraging Theory: A Critical Review. *Annual Review of Ecology and Systematics* 15:523–75.

Pyke, G.H., H.R. Pulliam, and E.L. Charnov. 1977. Optimal Foraging: A Selective Review of Theory and Tests. *Quarterly Review of Biology* 52(2):137–54.

Rackham, O. 1980. *Ancient Woodland: Its History,*

Vegetation, and Uses in England. London: E. Arnold.

Ramankutty, N. 2001. Common Property Institutions and Sustainable Governance of Resources. *World Development* 29(10):1649–72.

Ramsay, J., and A.A. Eger. 2015. Analysis of Archaeobotanical Material from the Tüpraş Field Project of the Kinet Höyük Excavations, Turkey. *Journal of Islamic Archaeology* 2(1):35–50.

Redman, C.L. 1999. *Human Impact on Ancient Environments*. Tucson: University of Arizona Press.

———. 2005. Resilience Theory in Archaeology. *American Anthropologist* 107(1):70–77.

Redman, C.L., J.M. Grove, and L.H. Kuby. 2004. Integrating Social Science into the Long-Term Ecological Research (LTER) Network: Social Dimensions of Ecological Change and Ecological Dimensions of Social Change. *Ecosystems* 7(2):161–71.

Redman, C.L., and A.P. Kinzig. 2003. Resilience of Past Landscapes: Resilience Theory, Society, and the Longue Duree. *Conservation Ecology* 7(1):14.

Redman, C.L., M.C. Nelson, and A.P. Kinzig. 2009. The Resilience of Socioecological Landscapes: Lessons from the Hohokam. In *The Archaeology of Environmental Change: Socionatural Legacies of Degradation and Resilience*, ed. C.T. Fisher, J.B. Hill, and G.M. Feinman, pp. 15–39. Tucson: University of Arizona Press.

Reitz, E.J., and E.S. Wing. 1999. *Zooarchaeology*. Cambridge: Cambridge University Press.

———. 2008. *Zooarchaeology*. 2nd ed. Cambridge: Cambridge University Press.

Renfrew, J.M. 1973. *Palaeoethnobotany: The Prehistoric Food Plants of the Near East and Europe*. New York: Columbia University Press.

Reynolds, R.V., and A.H. Pierson. 1942. *Fuel Wood Used in the United States 1630–1930*. Circular 641. Washington, DC: U.S. Department of Agriculture.

Rhode, D. 1990. On Transportation Costs of Great Basin Resources: An Assessment of the Jones-Madsen Model. *Current Anthropology* 31(4):413–19.

Richerson, P.J., and R. Boyd. 2005. *Not by Genes Alone: How Culture Transformed Human Evolution*. Chicago: University of Chicago Press.

Richerson, P.J., R. Boyd, and R.L. Bettinger. 2001. Was Agriculture Impossible During the Pleistocene but Mandatory During the Holocene? A Climate Change Hypothesis. *American Antiquity* 66(3):387–411.

Richerson, P.J., R. Boyd, and J. Henrich. 2003. The Cultural Evolution of Human Cooperation. In *The Genetic and Cultural Evolution of Cooperation*, ed. P. Hammerstein, pp. 357–88. Cambridge, MA: MIT Press.

Riehl, S. 2009. Archaeobotanical Evidence for the Interrelationship of Agricultural Decision-Making and Climate Change in the Ancient Near East. *Quaternary International* 197:93–114.

———. 2014. Significance of Prehistoric Weed Floras for the Reconstruction of Relations between Environment and Crop Husbandry Practices in the Near East. In *Ancient Plants and People: Contemporary Trends in Archaeobotany*, ed. M. Madella, C. Lancelotti, and M. Savard, pp. 135–52. Tucson: University of Arizona Press.

Rindos, D. 1984. *The Origins of Agriculture: An Evolutionary Perspective*. Orlando: Academic Press.

Roberts, C.N., G. Zanchetta, and M.D. Jones. 2010. Oxygen Isotopes as Tracers of Mediterranean Climate Variability: An Introduction. *Global and Planetary Change* 71(3–4):135–40.

Roberts, N. 1990. Human-Induced Landscape Change in South and Southwest Turkey During the Later Holocene. In *Man's Role in the Shaping of the Eastern Mediterranean Landscape*, ed. S. Bottema, G. Entjes-Nieborg, and W. van Zeist, pp. 53–68. Rotterdam: A.A. Balkema.

Roberts, N., S. Black, P. Boyer, W.J. Eastwood, H.I. Griffiths, H.F. Lamb, M.J. Leng, R. Parish, J.M. Reed, D. Twigg, and H. Yığıtbaşıoğlu. 1999. Chronology and Stratigraphy of Late Quaternary Sediments in the Konya Basin, Turkey: Results from the KOPAL Project. *Quaternary Science Reviews* 18(4–5):611–30.

Roberts, N., D. Brayshaw, C. Kuzucuoğlu, R. Perez, and L. Sadori. 2011a. The Mid-Holocene Climatic Transition in the Mediterranean: Causes and Consequences. *The Holocene* 21(1):3–13.

Roberts, N., W.J. Eastwood, C. Kuzucuoğlu, G. Fiorentino, and V. Caracuta. 2011b. Climatic, Vegetation and Cultural Change in the Eastern Mediterranean During the Mid-Holocene Environmental Transition. *The Holocene* 21(1):147–62.

Roberts, N., M.D. Jones, A. Benkaddour, W.J. East-wood, M.L. Filippi, M.R. Frogley, H.F. Lamb, M.J. Leng, J.M. Reed, M. Stein, L. Stevens, B. Valero-Garcés, and G. Zanchetta. 2008. Stable Isotope Records of Late Quaternary Climate and Hydrology from Mediterranean Lakes: The Isomed Synthesis. *Quaternary Science Reviews* 27(25–26):2426–41.

Roberts, N., A. Moreno, B.L. Valero-Garcés, J.P. Corella, M. Jones, S. Allcock, J. Woodbridge, M. Morellón, J. Luterbacher, E. Xoplaki, and M. Türkeş. 2012. Palaeolimnological Evidence for an East–West Climate See-Saw in the Mediterranean since AD 900. *Global and Planetary Change* 84–85:23–34.

Roberts, N., J.M. Reed, M.J. Leng, C. Kuzucuoğlu, M. Fontugne, J. Bertaux, H. Woldring, S. Bottema, S. Black, E. Hunt, and M. Karabıyıkoğlu. 2001. The Tempo of Holocene Climatic Change in the Eastern Mediterranean Region: New High-Resolution Crater-Lake Sediment Data from Central Turkey. *The Holocene* 11(6):721–36.

Robinson, M.E., and H.I. McKillop. 2013. Ancient Maya Wood Selection and Forest Exploitation: A View from the Paynes Creek Salt Works, Belize. *Journal of Archaeological Science* 40(10):3584–95.

Rockström, J., W. Steffen, K. Noone, Å. Persson, F.S. Chapin III, E.F. Lambin, T.M. Lenton, M. Scheffer, C. Folke, H.J. Schellnhuber, B. Nykvist, C.A. de Wit, T. Hughes, S. van der Leeuw, H. Rodhe, S. Sörlin, P.K. Snyder, R. Costanza, U. Svedin, M. Falkenmark, L. Karlberg, R.W. Corell, V.J. Fabry, J. Hansen, B. Walker, D. Liverman, K. Richardson, P. Crutzen, and J.A. Foley. 2009. A Safe Operating Space for Humanity. *Nature* 461(7263):472–75.

Roller, L.E. 1984. Midas and the Gordian Knot. *Classical Antiquity* 3:256–71.

———. 1987. *Gordion Special Studies I: Nonverbal Graffiti, Dipinti, and Stamps.* Philadelphia: University of Pennsylvania Museum of Archaeology and Anthropology.

———. 2009. *The Incised Drawings from Early Phrygian Gordion.* Philadelphia: University of Pennsylvania Museum of Archaeology and Anthropology.

Romano, I.B. 1995. *Gordion Special Studies II: The Terracotta Figurines and Related Vessels.* Philadelphia: University of Pennsylvania Museum of Ar-chaeology and Anthropology.

Roos, C.I., D.M.J.S. Bowman, J.K. Balch, P. Artaxo, W.J. Bond, M. Cochrane, C.M. D'Antonio, R. DeFries, M. Mack, F.H. Johnston, M.A. Krawchuk, C.A. Kull, M.A. Moritz, S. Pyne, A.C. Scott, and T.W. Swetnam. 2014. Pyrogeography, Historical Ecology, and the Human Dimensions of Fire Regimes. *Journal of Biogeography* 41(4):833–36.

Roos, C.I., A.P. Sullivan III, and C. McNamee. 2010. Paleoecological Evidence for Systematic Indigenous Burning in the Upland Southwest. In *The Archaeology of Anthropogenic Environments*, ed. R.M. Dean, pp. 142–71. Carbondale, IL: Center for Archaeological Investigations.

Rose, C.B., and G. Darbyshire, eds. 2011. *The New Chronology of Iron Age Gordion.* Philadelphia: University of Pennsylvania Museum of Archaeology and Anthropology.

Rosen, A.M., and I. Rivera-Collazo. 2012. Climate Change, Adaptive Cycles, and the Persistence of Foraging Economies During the Late Pleistocene/Holocene Transition in the Levant. *Proceedings of the National Academy of Sciences of the United States of America* 109(10):3640–45.

Rosen, A.M. 2007. *Civilizing Climate: Social Responses to Climate Change in the Ancient Near East.* Lanham, MD: Altamira Press.

Rossignol-Strick, M. 1995. Sea-Land Correlation of Pollen Records in the Eastern Mediterranean for the Glacial-Interglacial Transition: Biostratigraphy Versus Radiometric Time-Scale. *Quaternary Science Reviews* 14:893–915.

Rotroff, S.I. 2001. A New Type of Beehive. In "Notes from the Tins: Research in the Stoa of Attalos, Summer 1999." *Hesperia* 70(2):163–82.

———. 2006. *Hellenistic Pottery: The Plain Wares.* The Athenian Agora. Athens: American School of Classical Studies at Athens.

Rubiales, J.M., L. Hernández, F. Romero, and C. Sanz. 2011. The Use of Forest Resources in Central Iberia During the Late Iron Age. Insights from the Wood Charcoal Analysis of Pintia, a Vaccaean Oppidum. *Journal of Archaeological Science* 38:1–10.

Sagona, A., and P. Zimansky. 2009. *Ancient Turkey.* London: Routledge.

Salavert, A., and A. Dufraisse. 2014. Understanding

the Impact of Socio-Economic Activities on Archaeological Charcoal Assemblages in Temperate Areas: A Comparative Analysis of Firewood Management in Two Neolithic Societies in Western Europe (Belgium, France). *Journal of Anthropological Archaeology* 35:153–63.

Salisbury, K.J., and F.W. Jane. 1940. Charcoals from Maiden Castle and Their Significance in Relation to the Vegetation and Climatic Conditions in Prehistoric Times. *Journal of Ecology* 28:310–25.

Salmon, M.H. 1976. "Deductive" Versus "Inductive" Archaeology. *American Antiquity* 41(3):376–81.

Salmon, M.H., and W.C. Salmon. 1979. Alternative Models of Scientific Explanation. *American Anthropologist* 81(1):61–74.

Sams, G.K. 1977. Beer in the City of Midas. *Archaeology* 30(2):108–15.

———. 1994a. Aspects of Early Phrygian Architecture at Gordion. In *Anatolian Iron Ages 3: Proceedings of the Third Anatolian Iron Ages Colloquium Held at Van, 6–12 August 1990*, ed. A. Çilingiroğlu and D. French, pp. 211–14. Ankara: British Institute of Archaeology at Ankara.

———. 1994b. *The Gordion Excavations, 1950–1973: Final Reports Vol. 4: The Early Phrygian Pottery*. Philadelphia: University of Pennsylvania Museum of Archaeology and Anthropology.

———. 2005. Gordion: Explorations over a Century. In *The Archaeology of Midas and the Phrygians: Recent Work at Gordion*, ed. L. Kealhofer, pp. 10–21. Philadelphia: University of Pennsylvania Museum of Archaeology and Anthropology.

Sams, G.K., and R.B. Burke. 2008. Gordion, 2006. *Kazı Sonuçları Toplantısı* 29(2):329–42.

Sams, G K., R.B. Burke, and A.L. Goldman. 2007. Gordion, 2005. *Kazı Sonuçları Toplantısı* 28(2):365–86.

Sams, G.K., and A.L. Goldman. 2006. Gordion, 2004. *Kazı Sonuçları Toplantısı* 27(2):43–56.

Sams, G.K., and M.M. Voigt. 1995. Gordion Archaeological Activities, 1993. *Kazı Sonuçları Toplantısı* 16(1):369–92.

———. 1996. Gordion Archaeological Activities, 1994. *Kazı Sonuçları Toplantısı* 17(1):433–52.

———. 1997. Gordion 1995. *Kazı Sonuçları Toplantısı* 18(1):475–97.

———. 1998. Gordion 1996. *Kazı Sonuçları Toplantısı* 19(1):681–701.

———. 1999. Gordion Archaeological Activities, 1997. *Kazı Sonuçları Toplantısı* 20(1):559–76.

———. 2003. Gordion 2001. *Kazı Sonuçları Toplantısı* 24(2):139–48.

———. 2004. Gordion, 2002. *Kazı Sonuçları Toplantısı* 25(2):195–206.

Samuel, D. 2001. Archaeobotanical Evidence and Analysis. In *Peuplement rural et aménagements hydroagricoles dans la moyenne vallée de l'Euphrate fin VIIe–XIXe siècle*, ed. S. Berthier, pp. 347–481. Damascus: Institut Français d'Études Arabes de Damas.

Sandstrom, P.E. 1994. An Optimal Foraging Approach to Information-Seeking and Use. *Library Quarterly* 64(4):414–49.

———. 2001. Scholarly Communication as a Socio-ecological System. *Scientometrics* 51(3):573–605.

Scarry, C.M. 1993. Agricultural Risk and the Development of the Moundville Chiefdom. In *Foraging and Farming in the Eastern Woodlands*, ed. C.M. Scarry, pp. 157–81. Gainesville: University Press of Florida.

———. 2008. Crop Husbandry Practices in North America's Eastern Woodlands. In *Case Studies in Environmental Archaeology*, ed. E.J. Reitz, C.M. Scarry, and S.J. Scudder, pp. 391–404. 2nd ed. New York: Springer.

Scelza, B.A., D.W. Bird, and R. Bliege Bird. 2014. Bush Tucker, Shop Tucker: Production, Consumption, and Diet at an Aboriginal Outstation. *Ecology of Food and Nutrition* 53(1):98–117.

Scheel-Ybert, R. 2002. Evaluation of Sample Reliability in Extant and Fossil Assemblages. In *Charcoal Analysis: Methodological Approaches, Palaeoecological Results and Wood Uses: Proceedings of the Second International Meeting of Anthracology, Paris, September 2000*, ed. S. Thiébault, pp. 9–16. Oxford: Archaeopress.

Scheel-Ybert, R., and O.F. Dias. 2007. Corondó: Palaeoenvironmental Reconstruction and Palaeoethnobotanical Considerations in a Probable Locus of Early Plant Cultivation (South-Eastern Brazil). *Environmental Archaeology* 12(2):129–38.

Scheffer, M. 2009. *Critical Transitions in Nature and Society*. Princeton: Princeton University Press.

Scheffer, M., and S.R. Carpenter. 2003. Catastrophic Regime Shifts in Ecosystems: Linking Theory to Observation. *Trends in Ecology & Evolution*

18(12):648–56.

Scheidel, W., ed. 2009. *Rome and China: Comparative Perspectives on Ancient World Empires*. Oxford: Oxford University Press.

Schiffer, M.B. 1972. Archaeological Context and Systemic Context. *American Antiquity* 37(2):156–65.

———. 1976. *Behavioral Archeology*. New York: Academic Press.

———. 1987. *Formation Processes of the Archaeological Record*. Albuquerque: University of New Mexico Press.

———. 2011. *Studying Technological Change: A Behavioral Approach*. Salt Lake City: University of Utah Press.

Schilman, B., A. Ayalon, M. Bar-Matthews, E.J. Kagan, and A. Almogi-Labin. 2002. Sea–Land Paleoclimate Correlation in the Eastern Mediterranean Region During the Late Holocene. *Israel Journal of Earth Science* 51:181–90.

Schilman, B., M. Bar-Matthews, A. Almogi-Labin, and B. Luz. 2001. Global Climate Instability Reflected by Eastern Mediterranean Marine Records During the Late Holocene. *Palaeogeography, Palaeoclimatology, Palaeoecology* 176:157–76.

Schoch, W.H., B. Pawlik, and F.H. Schweingruber. 1988. *Botanical Macro-Remains: An Atlas for the Determination of Frequently Encountered and Ecologically Important Plant Seeds*. Berne: Paul Haupt.

Schoch, W., I. Heller, F.H. Schweingruber, and F. Kienast. 2004. Wood Anatomy of Central European Species. http://www.woodanatomy.ch. Accessed January 16, 2016.

Schoon, M., C. Fabricius, J.M. Anderies, and M. Nelson. 2011. Synthesis: Vulnerability, Traps, and Transformations—Long-Term Perspectives from Archaeology. *Ecology and Society* 16(2):24.

Schweingruber, F.H. 1990. *Anatomy of European Woods*. Stuttgart: Haupt.

Schweingruber, F.H., A. Börner, and E.-D. Schulze. 2006. *Atlas of Woody Plant Stems: Evolution, Structure, and Environmental Modifications*. Berlin: Springer.

Scott, A.C., and F. Damblon. 2010. Charcoal: Taphonomy and Significance in Geology, Botany and Archaeology. *Palaeogeography, Palaeoclimatology, Palaeoecology* 291(1):1–10.

Selinsky, P. 2005. A Preliminary Report on the Human Skeletal Material from Gordion's Lower Town Area. In *The Archaeology of Midas and the Phrygians: Recent Work at Gordion*, ed. L. Kealhofer, pp. 117–23. Philadelphia: University of Pennsylvania Museum of Archaeology and Anthropology.

———. 2015. Celtic Ritual Activity at Gordion, Turkey: Evidence from Mortuary Contexts and Skeletal Analysis. *International Journal of Osteoarchaeology* 25(2):213–25.

Shackleton, C.M., and F. Prins. 1992. Charcoal Analysis and the "Principle of Least Effort"—A Conceptual Model. *Journal of Archaeological Science* 19(6):631–37.

Shannon, C.E., and W. Weaver. 1949. *The Mathematical Theory of Communication*. Urbana: University of Illinois Press.

Shannon, C.E. 1948a. A Mathematical Theory of Communication. *The Bell System Technical Journal* 27:379–423.

Shannon, C.E. 1948b. A Mathematical Theory of Communication. *The Bell System Technical Journal* 27:623–66.

Shutes, M.T. 1997. Working Things Out: On Examining the Relationships between Agricultural Practice and Social Rules in Ancient Korinthos. In *Aegean Strategies: Studies of Culture and Environment on the European Fringe*, ed. P.N. Kardulias and M.T. Shutes, pp. 237–57. Lanham, MD: Rowman and Littlefield Publishers, Inc.

Siddique, K.H.M., G.H. Walton, and M. Seymour. 1993. A Comparison of Seed Yields of Winter Grain Legumes in Western Australia. *Australian Journal of Experimental Agriculture* 33(7):915–22.

Simpson, E.H. 1949. Measurement of Diversity. *Nature* 163:688.

Simpson, E. 1990. "Midas' Bed" and a Royal Phrygian Funeral. *Journal of Field Archaeology* 17:69–87.

———. 2007. Wooden Furniture and Small Objects from Gordion. Paper presented at the Archaeology of Phrygian Gordion Conference, University of Pennsylvania Museum, Philadelphia.

———. 2010. *The Furniture from Tumulus MM*. 2 vols. Leiden: Brill.

Simpson, E., and K. Spirydowicz. 1999. *Gordion Wooden Furniture*. Ankara: Museum of Anatolian Civilizations.

Smart, T.L., and E.S. Hoffman. 1988. Environmental Interpretation of Archaeological Charcoal. In *Current Paleoethnobotany: Analytical Methods and Cultural Interpretations of Archaeological Plant Remains*, ed. C.A. Hastorf and V.S. Popper, pp. 167–205. Chicago: University of Chicago Press.

Smith, B.D. 1977. Archaeological Inference and Inductive Confirmation. *American Anthropologist* 79(3):598–617.

———. 2006. Human Behavioral Ecology and the Transition to Food Production. In *Behavioral Ecology and the Transition to Agriculture* ed. D.J. Kennett and B. Winterhalder, pp. 289–303. Berkeley: University of California Press.

———. 2007. Niche Construction and the Behavioral Context of Plant and Animal Domestication. *Evolutionary Anthropology* 16:188–99.

———. 2009a. Core Conceptual Flaws in Human Behavioral Ecology. *Communicative & Integrative Biology* 2(6):533–34.

———. 2009b. Resource Resilience, Human Niche Construction, and the Long-Term Sustainability of Pre-Columbian Subsistence Economies in the Mississippi River Valley Corridor. *Journal of Ethnobiology* 29(2):167–83.

———. 2011. A Cultural Niche Construction Theory of Initial Domestication. *Biological Theory* 6(3):1–12.

———. 2013. Modifying Landscapes and Mass Kills: Human Niche Construction and Communal Ungulate Harvests. *Quaternary International* 297:8–12.

———. 2014. Documenting Human Niche Construction in the Archaeological Record. In *Method and Theory in Paleoethnobotany*, ed. J.M. Marston, J. d'Alpoim Guedes, and C. Warinner, pp. 355–70. Boulder: University Press of Colorado.

———. 2015. A Comparison of Niche Construction Theory and Diet Breadth Models as Explanatory Frameworks for the Initial Domestication of Plants and Animals. *Journal of Archaeological Research* 23:215–62.

Smith, B.D., and M.A. Zeder. 2013. The Onset of the Anthropocene. *Anthropocene* 4:8–13.

Smith, E.A. 1983. Anthropological Applications of Optimal Foraging Theory: A Critical Review. *Current Anthropology* 24(5):625–51.

———. 1991. *Inujjuamiut Foraging Strategies: Evolutionary Ecology of an Arctic Hunting Economy*. New York: Aldine de Gruyter.

Smith, E.A., and R. Bliege Bird. 2000. Turtle Hunting and Tombstone Opening: Public Generosity as Costly Signaling. *Evolution and Human Behavior* 21(4):245–61.

———. 2005. Costly Signaling and Cooperative Behavior. In *Moral Sediments and Material Interests: On the Foundations of Cooperation in Economic Life*, ed. H. Gintis, S. Bowles, R. Boyd, and E. Fehr, pp. 115–48. Cambridge, MA: MIT Press.

Smith, E.A., R. Bliege Bird, and D.W. Bird. 2003. The Benefits of Costly Signaling: Meriam Turtle Hunters. *Behavioral Ecology* 14(1):116–26.

Smith, E.A., and R. Boyd. 1990. Risk and Reciprocity: Hunter-Gatherer Socioecology and the Problem of Collective Action. In *Risk and Uncertainty in Tribal and Peasant Economies*, ed. E.A. Cashdan, pp. 167–91. Boulder, CO: Westview Press.

Smith, E.A., and B. Winterhalder, eds. 1992. *Evolutionary Ecology and Human Behavior*. New Brunswick, NJ: Aldine Transaction.

Smith, M.L. 2006. How Ancient Agriculturalists Managed Yield Fluctuations through Crop Selection and Reliance on Wild Plants: An Example from Central India. *Economic Botany* 60(1):39–48.

Speer, J.H. 2010. *Fundamentals of Tree-Ring Research*. Tucson: University of Arizona Press.

Staubwasser, M., F. Sirocko, P.M. Grootes, and M. Segl. 2003. Climate Change at the 4.2 ka BP Termination of the Indus Valley Civilization and Holocene South Asian Monsoon Variability. *Geophysical Research Letters* 30(8):1425.

Steffen, W., K. Richardson, J. Rockström, S.E. Cornell, I. Fetzer, E.M. Bennett, R. Biggs, S.R. Carpenter, W. de Vries, and C.A. de Wit. 2015. Planetary Boundaries: Guiding Human Development on a Changing Planet. *Science* 347(6223):1259855.

Stephens, D.W. 1981. The Logic of Risk-Sensitive Foraging Preferences. *Animal Behaviour* 29:628–29.

———. 1990. Risk and Incomplete Information in Behavioral Ecology. In *Risk and Uncertainty in Tribal and Peasant Economies*, ed. E.A. Cashdan, pp. 19–46. Boulder, CO: Westview Press.

Stephens, D.W., and E.L. Charnov. 1982. Optimal

Foraging: Some Simple Stochastic Models. *Behavioral Ecology and Sociobiology* 10(4):251–63.

Stephens, D.W., and J.R. Krebs. 1986. *Foraging Theory*. Princeton: Princeton University Press.

Stern, P.C., K.L. Ebi, R. Leichenko, R.S. Olson, J.D. Steinbruner, and R. Lempert. 2013. Managing Risk with Climate Vulnerability Science. *Nature Climate Change* 3(7):607–9.

Steward, J.H. 1955. *Irrigation Civilizations: A Comparative Study*. Washington, DC: Pan American Union.

Stewart, S. 2010. Gordion after the Knot: Hellenistic Pottery and Culture. Ph.D. diss., University of Cincinnati, Department of Classics, Cincinnati.

Stinchcomb, G.E., T.C. Messner, S.G. Driese, L.C. Nordt, and R.M. Stewart. 2011. Pre-Colonial (AD 1100–1600) Sedimentation Related to Prehistoric Maize Agriculture and Climate Change in Eastern North America. *Geology* 39(4):363–66.

Stone, G.D., and C.E. Downum. 1999. Non-Boserupian Ecology and Agricultural Risk: Ethnic Politics and Land Control in the Arid Southwest. *American Anthropologist* 101(1):113–28.

Sullivan, A.P. 1978. Inference and Evidence in Archaeology: A Discussion of the Conceptual Problems. *Advances in Archaeological Method and Theory* 1:183–222.

Sullivan, A.P., III, J.N. Berkebile, K.M. Forste, and R.M. Washam. 2015. Disturbing Developments: An Archaeobotanical Perspective on Pinyon-Juniper Woodland Fire Ecology, Economic Resource Production, and Ecosystem History. *Journal of Ethnobiology* 35(1):37–59.

Sullivan, A.P., III, and K.M. Forste. 2014. Fire-Reliant Subsistence Economies and Anthropogenic Coniferous Ecosystems in the Pre-Columbian Northern American Southwest. *Vegetation History and Archaeobotany* 23(1):135–51.

Sullivan, D.G. 1989. Human-Induced Vegetation Change in Western Turkey: Pollen Evidence from Central Lydia. Ph.D. diss., University of California, Department of Geography, Berkeley.

Tainter, J.A. 1988. *The Collapse of Complex Societies*. Cambridge: Cambridge University Press.

———. 2006a. Archaeology of Overshoot and Collapse. *Annual Review of Anthropology* 35:59–74.

———. 2006b. Social Complexity and Sustainability. *Ecological Complexity* 3(2):91–103.

Théry-Parisot, I., L. Chabal, and J. Chrzavzez. 2010. Anthracology and Taphonomy, from Wood Gathering to Charcoal Analysis: A Review of the Taphonomic Processes Modifying Charcoal Assemblages, in Archaeological Contexts. *Palaeogeography, Palaeoclimatology, Palaeoecology* 291:142–53.

Théry-Parisot, I., and A. Henry. 2012. Seasoned or Green? Radial Cracks Analysis as a Method for Identifying the Use of Green Wood as Fuel in Archaeological Charcoal. *Journal of Archaeological Science* 39(2):381–88.

Thiébault, S. 2005. L'apport du fourrage d'arbre dans l'élevage depuis le Néolithique. *Anthropozoologica* 40(1):95–108.

———. 2006. Wood-Anatomical Evidence of Pollarding in Ring-Porous Species: A Study to Develop? In *Charcoal Analysis: New Analytical Tools and Methods for Archaeology: Papers from the Table-Ronde Held in Basel 2004*, ed. A. Dufraisse, pp. 95–102. Oxford: Archaeopress.

Trigger, B.G. 2006. *A History of Archaeological Thought*. 2nd ed. Cambridge: Cambridge University Press.

Trivers, R.L. 1971. Evolution of Reciprocal Altruism. *Quarterly Review of Biology* 46(1):35–57.

Türk Silahlı Kuvvetleri. 1997. 1:100,000 Maps of Turkey. Türk Silahlı Kuvvetleri (Turkish Armed Forces).

Turner, B.L., II. 2010. Vulnerability and Resilience: Coalescing or Paralleling Approaches for Sustainability Science? *Global Environmental Change* 20(4):570–76.

Turner, B.L., II, R.E. Kasperson, P.A. Matson, J.J. McCarthy, R.W. Corell, L. Christensen, N. Eckley, J.X. Kasperson, A. Luers, and M.L. Martello. 2003. A Framework for Vulnerability Analysis in Sustainability Science. *Proceedings of the National Academy of Sciences of the United States of America* 100(14):8074–79.

Turner, N.J., and F. Berkes. 2006. Developing Resource Management and Conservation. *Human Ecology* 34(4):475–78.

USDA. 2010. Invasive and Noxious Weeds. US Department of Agriculture, Natural Resources Conservation Service. http://plants.usda.gov/java/noxComposite. Accessed March 7 2015.

van der Leeuw, S.E. 2008. Climate and Society: Lessons from the Past 10,000 Years. *AMBIO: A Journal of the Human Environment* 14:476–82.

———. 2009. What Is an "Environmental Crisis" to an Archaeologist? In *The Archaeology of Environmental Change: Socionatural Legacies of Degradation and Resilience*, ed. C.T. Fisher, J.B. Hill, and G.M. Feinman, pp. 40–61. Tucson: University of Arizona Press.

van der Leeuw, S.E., and C.L. Redman. 2002. Placing Archaeology at the Center of Socio-Natural Studies. *American Antiquity* 67(4):597–605.

van der Veen, M. 2007. Formation Processes of Desiccated and Carbonized Plant Remains—the Identification of Routine Practice. *Journal of Archaeological Science* 34:968–90.

———. 2010. Agricultural Innovation: Invention and Adoption or Change and Adaptation? *World Archaeology* 42(1):1–12.

———. 2014. The Materiality of Plants: Plant–People Entanglements. *World Archaeology* 46(5):799–812.

Van Dyke, R.M., and S.E. Alcock, eds. 2003. *Archaeologies of Memory*. Oxford: Blackwell Publishers.

van Zeist, W., and J.A.H. Bakker-Heeres. 1982. Archaeobotanical Studies in the Levant I. Neolithic Sites in the Damascus Basin: Aswad, Ghoraifé, Ramad. *Palaeohistoria* 24:165–256.

———. 1984a. Archaeobotanical Studies in the Levant 2. Neolithic and Halaf Levels at Ras Shamra. *Palaeohistoria* 26:151–70.

———. 1984b. Archaeobotanical Studies in the Levant 3. Late-Palaeolithic Mureybit. *Palaeohistoria* 26:171–99.

———. 1985. Archaeobotanical Studies in the Levant 4. Bronze Age Sites on the North Syrian Euphrates. *Palaeohistoria* 27:247–316.

van Zeist, W., and S. Bottema. 1991. *Late Quaternary Vegetation of the Near East*. Wiesbaden: Dr. Ludwig Reichert Verlag.

van Zeist, W., and G.J. de Roller. 1993. Plant Remains from Maadi, a Predynastic Site in Lower Egypt. *Vegetation History and Archaeobotany* 2:1–14.

———. 2003. The Çayönü Archaeobotanical Record. In *Reports on Archaeobotanical Studies in the Old World*, ed. W. van Zeist, pp. 143–66. Groningen: The Groningen Institute of Archaeology, University of Groningen.

van Zeist, W., P.E.L. Smith, R.M. Palfenier-Vegter, M.

Suwijn, and W.A. Casparie. 1984. An Archaeobotanical Study of Ganj Dareh Tepe, Iran. *Palaeohistoria* 26:201–24.

van Zeist, W., R.W. Timmers, and S. Bottema. 1968. Studies of Modern and Holocene Pollen Precipitation in Southeastern Turkey. *Palaeohistoria* 14:19–39.

van Zeist, W., and H. Woldring. 1978. A Postglacial Pollen Diagram from Lake Van in East Anatolia. *Review of Palaeobotany and Palynology* 26:249–76.

van Zeist, W., H. Woldring, and D. Stapert. 1975. Late Quaternary Vegetation and Climate of Southwestern Turkey. *Palaeohistoria* 17:53–143.

VanDerwarker, A.M., J.B. Marcoux, and K.D. Hollenbach. 2013. Farming and Foraging at the Crossroads: The Consequences of Cherokee and European Interaction through the Late Eighteenth Century. *American Antiquity* 78(1):68–88.

Vassileva, M. 2005. Phrygia, Troy and Thrace. In *Anatolian Iron Ages 5: Proceedings of the Fifth Anatolian Iron Ages Colloquium Held at Van, 6–10 August 2001*, ed. A. Çilingiroğlu and G. Darbyshire, pp. 227–34. London: British Institute at Ankara.

Verheyden, S., F.H. Nader, H.J. Cheng, L.R. Edwards, and R. Swennen. 2008. Paleoclimate Reconstruction in the Levant Region from the Geochemistry of a Holocene Stalagmite from the Jeita Cave, Lebanon. *Quaternary Research* 70(3):368–81.

Vermoere, M. 2004. *Holocene Vegetation History in the Territory of Sagalassos (Southwest Turkey): A Palynological Approach*. Turnhout, Belgium: Brepols Publishers.

Vermoere, M., S. Bottema, L. Vanhecke, M. Waelkens, E. Paulissen, and E. Smets. 2002a. Palynological Evidence for Late-Holocene Human Occupation Recorded in Two Wetlands in SW Turkey. *The Holocene* 12(5):569–84.

Vermoere, M., E. Smets, M. Waelkens, H. Vanhaverbeke, I. Librecht, E. Paulissen, and L. Vanhecke. 2000. Late Holocene Environmental Change and the Record of Human Impact at Gravgaz near Sagalassos, Southwest Turkey. *Journal of Archaeological Science* 27(7):571–95.

Vermoere, M., T. Van Thuyne, S. Six, L. Vanhecke, M. Waelkens, E. Paulissen, and E. Smets. 2002b. Late Holocene Local Vegetation Dynamics in the

Marsh of Gravgaz (Southwest Turkey). *Journal of Paleolimnology* 27(4):429–51.

Vermoere, M., L. Vanhecke, M. Waelkens, and E. Smets. 2001. Modern Pollen Studies in the Territory of Sagalassos (Southwest Turkey) and Their Use in the Interpretation of a Late Holocene Pollen Diagram. *Review of Palaeobotany and Palynology* 114(1–2):29–56.

———. 2003. Modern and Ancient Olive Stands near Sagalassos (South-West Turkey) and Reconstruction of the Ancient Agricultural Landscape in Two Valleys. *Global Ecology and Biogeography* 12(3):217–36.

Voigt, M.M. 1994. Excavations at Gordion 1988–89: The Yassıhöyük Stratigraphic Sequence. In *Anatolian Iron Ages 3: Proceedings of the Third Anatolian Iron Ages Colloquium Held at Van, 6–12 August 1990*, ed. A. Çilingiroğlu and D. French, pp. 265–93. Ankara: British Institute of Archaeology at Ankara.

———. 2002. Gordion: The Rise and Fall of an Iron Age Capital. In *Across the Anatolian Plateau: Readings in the Archaeology of Ancient Turkey*, ed. D.C. Hopkins, pp. 187–96. The Annual of the American Schools of Oriental Research. Boston: American Schools of Oriental Research.

———. 2004. Yassıhöyük Stratigraphic Sequence 2: 1993–1997. Unpublished manuscript, on file, Gordion Archive, University of Pennsylvania Museum of Archaeology and Anthropology, Philadelphia.

———. 2005. Old Problems and New Solutions: Recent Excavations at Gordion. In *The Archaeology of Midas and the Phrygians: Recent Work at Gordion*, ed. L. Kealhofer, pp. 22–35. Philadelphia: University of Pennsylvania Museum of Archaeology and Anthropology.

———. 2007. The Middle Phrygian Occupation at Gordion. In *Anatolian Iron Ages 6: Proceedings of the Sixth Anatolian Iron Ages Colloquium Held at Eskişehir, 16–20 August 2004*, ed. A. Çilingiroğlu and A.G. Sagona, pp. 311–33. Leuven: Peeters.

———. 2009. The Chronology of Phrygian Gordion. In *Tree-Rings, Kings, and Old World Archaeology and Environment: Papers Presented in Honor of Peter Ian Kuniholm*, ed. S.W. Manning and M.J. Bruce, pp. 219–37. Oxford: Oxbow Books.

———. 2011. Gordion: The Changing Political and Economic Roles of a First Millennium City. In *The Oxford Handbook of Ancient Anatolia (10,000–323 BCE)*, ed. S. Steadman and G. McMahon, pp. 1069–94. Oxford: Oxford University Press.

———. 2012. The Violent Ways of Galatian Gordion. In *The Archaeology of Violence: Interdisciplinary Approaches*, ed. S. Ralph, pp. 203–31. Buffalo: State University of New York.

———. 2013. Gordion as Citadel and City. In *Cities and Citadels in Turkey: From the Iron Age to the Seljuks*, ed. S. Redford and N. Ergin, pp. 161–228. Leuven: Peeters.

Voigt, M.M., K. DeVries, R.C. Henrickson, M. Lawall, B. Marsh, A. Gürsan-Salzman, and T.C. Young. 1997. Fieldwork at Gordion: 1993–1995. *Anatolica* 23:1–59.

Voigt, M.M., and R.C. Henrickson. 2000a. The Early Iron Age at Gordion: The Evidence from the Yassıhöyük Stratigraphic Sequence. In *The Sea Peoples and Their World: A Reassessment*, ed. E.D. Oren, pp. 327–60. Philadelphia: University of Pennsylvania Museum of Archaeology and Anthropology.

———. 2000b. Formation of the Phrygian State: The Early Iron Age at Gordion. *Anatolian Studies* 50:37–54.

Voigt, M.M., and T.C. Young. 1999. From Phrygian Capital to Achaemenid Entrepot: Middle and Late Phrygian Gordion. *Iranica Antiqua* 34:191–241.

Walker, B., S.R. Carpenter, J. Rockström, A.-S. Crépin, and G.D. Peterson. 2012. Drivers, "Slow" Variables, "Fast" Variables, Shocks, and Resilience. *Ecology and Society* 17(3):30.

Walker, B., and J.A. Meyers. 2004. Thresholds in Ecological and Social-Ecological Systems: A Developing Database. *Ecology and Society* 9(2):16.

Walker, B., and D. Salt. 2006. *Resilience Thinking: Sustaining Ecosystems and People in a Changing World*. Washington, DC: Island Press.

Walker, T.S., and N.S. Jodha. 1986. How Small Farm Households Adapt to Risk. In *Crop Insurance for Agricultural Development*, ed. P. Hazell, C. Pomareda, and A. Valdés, pp. 17–34. Baltimore: The Johns Hopkins University Press.

Watson, A.M. 1974. Arab Agricultural Revolution and Its Diffusion, 700–1100. *Journal of Economic History* 34(1):8–35.

———. 1983. *Agricultural Innovation in the Early Islamic World: The Diffusion of Crops and Farming Techniques, 700–1100*. Cambridge: Cambridge University Press.

Watson, P.J., S.A. LeBlanc, and C.L. Redman. 1971. *Explanation in Archeology: An Explicitly Scientific Approach*. New York: Columbia University Press.

Waylen, K.A., K.L. Blackstock, and K.L. Holstead. 2015. How Does Legacy Create Sticking Points for Environmental Management? Insights from Challenges to Implementation of the Ecosystem Approach. *Ecology and Society* 20(2):21.

Weiss, H., M.-A. Courty, W. Wetterstrom, F. Guichard, L. Senior, R. Meadow, and A. Curnow. 1993. The Genesis and Collapse of Third Millennium North Mesopotamian Civilization. *Science* 261(5124):995–1004.

Wells, M. 2012. A Cosmopolitan Village: The Hellenistic Settlement at Gordion. Ph.D. diss., University of Minnesota, Department of Classics, Minneapolis.

Werner, E.E., and D.J. Hall. 1974. Optimal Foraging and Size Selection of Prey by Bluegill Sunfish (*Lepomis macrochirus*). *Ecology* 55(5):1042–52.

West, J.B., J.M. Hurley, F.Ö. Dudás, and J.R. Ehleringer. 2009. The Stable Isotope Ratios of Marijuana. II. Strontium Isotopes Relate to Geographic Origin. *Journal of Forensic Sciences* 54(6):1261–69.

White, C.E., M.S. Chesson, and R.T. Schaub. 2014. A Recipe for Disaster: Emerging Urbanism and Unsustainable Plant Economies at Early Bronze Age Ras an-Numayra, Jordan. *Antiquity* 88:363–77.

Whittaker, R.H. 1953. A Consideration of Climax Theory: The Climax as a Population and Pattern. *Ecological Monographs* 23:41–78.

Wick, L., G. Lemcke, and M. Sturm. 2003. Evidence of Lateglacial and Holocene Climatic Change and Human Impact in Eastern Anatolia: High-Resolution Pollen, Charcoal, Isotopic and Geochemical Records from the Laminated Sediments of Lake Van, Turkey. *The Holocene* 13(5):665–75.

Wilkinson, T.J. 1999. Holocene Valley Fills of Southern Turkey and Northwestern Syria: Recent Geoarchaeological Contributions. *Quaternary Science Reviews* 18(4–5):555–71.

———. 2003. *Archaeological Landscapes of the Near East*. Tucson: University of Arizona Press.

———. 2006. From Highland to Desert: The Organization of Landscape and Irrigation in Southern Arabia. In *Agricultural Strategies*, ed. J. Marcus and C. Stanish, pp. 38–68. Los Angeles: Cotsen Institute of Archaeology at UCLA.

Willcox, G. 1974. A History of Deforestation as Indicated by Charcoal Analysis of Four Sites in Eastern Anatolia. *Anatolian Studies* 24:117–33.

———. 2002. Evidence for Ancient Forest Cover and Deforestation from Charcoal Analysis of Ten Archaeological Sites on the Euphrates. In *Charcoal Analysis: Methodological Approaches, Palaeoecological Results and Wood Uses: Proceedings of the Second International Meeting of Anthracology, Paris, September 2000*, ed. S. Thiébault, pp. 141–45. BAR International Series 1063. Oxford: Archaeopress.

Wills, W.H., B.L. Drake, and W.B. Dorshow. 2014. Prehistoric Deforestation at Chaco Canyon? *Proceedings of the National Academy of Sciences* 111(32):11584–91.

Wilmsen, E.N. 1973. Interaction, Spacing Behavior, and Organization of Hunting Bands. *Journal of Anthropological Research* 29(1):1–31.

Wilson, D.G. 1984. The Carbonisation of Weed Seeds and Their Representation in Macrofossil Assemblages. In *Plants and Ancient Man: Studies in Palaeoethnobotany*, ed. W. van Zeist and W.A. Casparie, pp. 201–6. Rotterdam: A.A. Balkema.

Wilson, E.O. 1975. *Sociobiology: The New Synthesis*. Cambridge, MA: Belknap Press of Harvard University Press.

Winterhalder, B. 1986. Diet Choice, Risk, and Food Sharing in a Stochastic Environment. *Journal of Anthropological Archaeology* 5(4):369–92.

———. 1990. Open Field, Common Pot: Harvest Variability and Risk Avoidance in Agricultural and Foraging Societies. In *Risk and Uncertainty in Tribal and Peasant Economies*, ed. E.A. Cashdan, pp. 67–87. Boulder, CO: Westview Press.

———. 1997. Gifts Given, Gifts Taken: The Behavioral Ecology of Nonmarket, Intragroup Exchange. *Journal of Archaeological Research* 5(2):121–68.

———. 2002. Models. In *Darwin and Archaeology: A Handbook of Key Concepts*, ed. J. Hart and J. Terrell, pp. 201–23. Westport, CT: Bergin and Garvey.

Winterhalder, B., and C. Goland. 1997. An Evolutionary Ecology Perspective on Diet Choice, Risk, and Plant Domestication. In *People, Plants, and Landscapes: Studies in Paleoethnobotany*, ed. K.J. Gremillion, pp. 123–60. Tuscaloosa: University of Alabama Press.

Winterhalder, B., and D.J. Kennett. 2006. Behavioral Ecology and the Transition from Hunting and Gathering to Agriculture. In *Behavioral Ecology and the Transition to Agriculture*, ed. D.J. Kennett and B. Winterhalder, pp. 1–21. Berkeley: University of California Press.

Winterhalder, B., and E.A. Smith, eds. 1981. *Hunter-Gatherer Foraging Strategies: Ethnographic and Archeological Analyses*. Chicago: University of Chicago Press.

———. 2000. Analyzing Adaptive Strategies: Human Behavioral Ecology at Twenty-Five. *Evolutionary Anthropology* 9(2):51–72.

Wong, C.I., and D.O. Breecker. 2015. Advancements in the Use of Speleothems as Climate Archives. *Quaternary Science Reviews* 127:1–18.

Woodbridge, J., and N. Roberts. 2011. Late Holocene Climate of the Eastern Mediterranean Inferred from Diatom Analysis of Annually-Laminated Lake Sediments. *Quaternary Science Reviews* 30(23–24):3381–92.

Wright, N.J., A.S. Fairbairn, J.T. Faith, and K. Matsumura. 2015. Woodland Modification in Bronze and Iron Age Central Anatolia: An Anthracological Signature for the Hittite State? *Journal of Archaeological Science* 55:219–30.

Wright, P.J. 2010. Methodological Issues in Paleoethnobotany: A Consideration of Issues, Methods, and Cases. In *Integrating Zooarchaeology and Paleoethnobotany: A Consideration of Issues, Methods, and Cases*, ed. A.M. VanDerwarker and T.M. Peres, pp. 37–64. Berlin: Springer.

Wylie, A. 1985. The Reaction against Analogy. *Advances in Archaeological Method and Theory* 8:63–110.

Yakar, J. 2000. *Ethnoarchaeology of Anatolia: Rural Socio-Economy in the Bronze and Iron Ages*. Tel Aviv: Institute of Archaeology of Tel Aviv University, Publications Section.

Young, R.S. 1950. Excavations at Yassıhöyük-Gordion 1950. *Archaeology* 3:196–201.

———. 1951. Gordion—1950. *University Museum Bulletin* 16(1):3–20.

———. 1953. Progress at Gordion, 1951–1952. *University Museum Bulletin* 17(4):3–39.

———. 1956. The Campaign of 1955 at Gordion: Preliminary Report. *American Journal of Archaeology* 60:249–66.

———. 1957. Gordion 1956: Preliminary Report. *American Journal of Archaeology* 61:319–31.

———. 1958a. The Gordion Campaign of 1957: Preliminary Report. *American Journal of Archaeology* 62:139–54.

———. 1958b. The Gordion Tomb. *Expedition* 1(1):3–12.

———. 1960. Gordion: Phrygian Construction and Architecture. *Expedition* 2(1):2–9.

———. 1962a. The 1961 Campaign at Gordion. *American Journal of Archaeology* 66:153–68.

———. 1962b. Gordion: Phrygian Construction and Architecture II. *Expedition* 4(4):2–12.

———. 1963. Gordion on the Royal Road. *Proceedings of the American Philosophical Society* 107(4):348–64.

———. 1981. *The Gordion Excavations Final Reports, Vol. I: Three Great Early Tumuli*. Philadelphia: University of Pennsylvania Museum of Archaeology and Anthropology.

Zach, R. 1979. Shell Dropping: Decision-Making and Optimal Foraging in Northwestern Crows. *Behaviour* 68:106–17.

Zahavi, A. 1975. Mate Selection—Selection for a Handicap. *Journal of Theoretical Biology* 53(1):205–14.

Zeanah, D.W. 2004. Sexual Division of Labor and Central Place Foraging: A Model for the Carson Desert of Western Nevada. *Journal of Anthropological Archaeology* 23(1):1–32.

Zeder, M.A. 1991. *Feeding Cities: Specialized Animal Economy in the Ancient Near East*. Washington, DC: Smithsonian Institution Press.

———. 2008. Domestication and Early Agriculture in the Mediterranean Basin: Origins, Diffusion, and Impact. *Proceedings of the National Academy of Sciences of the United States of America* 105(33):11597–604.

———. 2011. The Origins of Agriculture in the Near East. *Current Anthropology* 52:S221–S235.

———. 2012. The Broad Spectrum Revolution at 40: Resource Diversity, Intensification, and an Alternative to Optimal Foraging Explanations.

Journal of Anthropological Archaeology 31:241–64.

———. 2014. Alternative to Faith-Based Science. *Proceedings of the National Academy of Sciences of the United States of America* 111(28):E2827.

———. 2015. Core Questions in Domestication Research. *Proceedings of the National Academy of Sciences of the United States of America* 112(11):3191–98.

Zeder, M.A., and S.R. Arter. 1994. Changing Patterns of Animal Utilization at Ancient Gordion. *Paléorient* 22(2):105–18.

Zeder, M.A., D.G. Bradley, E. Emshwiller, and B.D. Smith, eds. 2006. *Documenting Domestication: New Genetic and Archeological Paradigms.* Berkeley: University of California Press.

Zeder, M.A., and H.A. Lapham. 2010. Assessing the Reliability of Criteria Used to Identify Postcranial Bones in Sheep, *Ovis*, and Goats, *Capra. Journal of Archaeological Science* 37(11):2887–905.

Zeder, M.A., and S.E. Pilaar. 2010. Assessing the Reliability of Criteria Used to Identify Mandibles and Mandibular Teeth in Sheep, *Ovis*, and Goats, *Capra. Journal of Archaeological Science* 37(2):225–42.

Zeder, M.A., and B.D. Smith. 2009. A Conversation on Agricultural Origins: Talking Past Each Other in a Crowded Room. *Current Anthropology* 50:681–91.

Zhang, J., S.E. Jørgensen, M. Beklioglu, and O. Ince. 2003. Hysteresis in Vegetation Shift—Lake Mogan Prognoses. *Ecological Modelling* 164(2):227–38.

Zohary, D., and M. Hopf. 2000. *Domestication of Plants in the Old World: The Origin and Spread of Cultivated Plants in West Asia, Europe, and the Nile Valley.* 3rd ed. New York: Oxford University Press.

Zohary, M. 1973. *Geobotanical Foundations of the Middle East.* Stuttgart: G. Fischer.

Zori, C., and E. Brant. 2012. Managing the Risk of Climatic Variability in Late Prehistoric Northern Chile. *Journal of Anthropological Archaeology* 31(3):403–21.

Turkish Summary/Özet

Bu kitap, 3000 yıldan uzun bir süre boyunca Gordion antik kentinde gerçekleşen insan ve doğal çevre etkileşimine değiniyor. Kitapta bu süre boyunca sosyal, siyasi, ekonomik ve çevresel faktörlerin, Gordion sakinleriyle onları kuşatan çevrenin etkileşimine, özellikle tarım ve toprak kullanımı bakımından nasıl etki ettiğini belirlemeyi amaçlıyorum. Çalışmada kullandığım kuramsal çerçeve uyumlayıcı karar verme üzerine davranış ekolojisi perspektifleriyle insan-çevre etkileşiminin mekânsal ve zamansal ölçeklerini dikkate alan sağlamlık yaklaşımını birleştiriyor. Gordion, karmaşık sosyal tarihi, kazısının on yıllardır devam ediyor olması ve bulunan çok sayıda bitki ve hayvan kalıntısını da içeren detaylı arkeolojik analizlerin yapılması bakımından benzersiz şekilde kıymetli bir vaka çalışmasıdır.

Bu çalışma, Gordion'da 1993–2002 yılları arasında kazılan tabakalardan elde edilen iki arkeolojik veri setinin analizini içerir: (1) suda yüzdürme yöntemiyle elde edilen arkeobotanik kalıntılar ile (2) sistematik eleme yöntemiyle ele geçirilen kömürleşmiş ahşap kalıntıları. Bunlardan elde edilen sonuçları 1988–1989 kazı sezonlarında bulunan ve daha önce Miller (2010) tarafından yayınlanan benzer veri setleriyle birleştirdim. Bu sayede Gordion'daki bütün yerleşim dönemlerinde tarım ve toprak kullanımına dair sağlam bir veri seti ortaya çıktı. Gordion'daki bu arkeobotanik buluntu topluluğunu incelerken tarımla ilgili karar almayı saptamak için risk yönetimi modeli uyguladım. Ürün çeşitliliğine, Gordion'da yerleşim dönemlerinin çoğunda, etkili bir risk yönetimi stratejisi olarak başvurulduğunu tespit ettim. Buna karşılık merkezi siyasal ekonomilerle nitelenen Frig ve Roma dönemlerinde, ürün çeşitliliğinde azalma ve sulu tarıma giderek daha çok başvurulmasıyla beraber tarımsal yoğunlukta artış görülmektedir. Aynı dönemlerde, muhtemelen aşırı otlatma ve artan erozyon hızına bağlı olarak otlak arazilerinin sağlığında da bozulmalar meydana gelmiştir. Erozyonun giderek artması Sakarya nehrinin yukarısındaki havzada Frig döneminden beri Gordion taşkın yatağının 5 metrelik alüvyonlu toprakla dolmasına sebep olmuştur.

Sağlamlık fikri uyuşmazlıkların, eşik olaylarının ve miras etkilerinin, daha fazla analiz edilmeyi hak eden en önemli etkileşimler olduğunun altını çizer. Bu fikir ayrıca insan-çevre etkileşiminin temel ilkeleri olan mekânsal, zamansal ve örgütsel ölçeğe dikkat çeker. Frig toprak kullanım sisteminin zamansal ve mekânsal uyuşmazlıklar göstermesi sonucu erozyona yol açtığı; bu erozyonun da toprağın ağaçlardan temizlenmesi, hayvan otlatılmasından sonra toprakta otların yeniden büyümesi, toprak erozyonu ve toprak oluşumu süreçlerinin farklı hızlarda olması sebebiyle meydana geldiği tespit edilmiştir. Roma dönemindeki mekânsal ve örgütsel uyuşmazlık, tarımla ilgili kararların çiftçinin kendisi değil uzaktaki vilayet vergi tahsildarları tarafından verilmesine sebep oluyordu. Anlaşılan o ki vergi tahsildarları da yerel tarım stratejilerini, vergi ödemesi için ekmeklik buğday üretimini zorunlu kılarak kısıtlıyordu. Bu durum tarım sistemini katı, birbirine aşırı derecede bağlı bir duruma soktu ve dışarıdan gelen ekonomik ya da çevresel şoklarla aniden oluşan büyük değişiklikler karşısında Roma dönemi tarım ekonomisinin riskini artırarak onu savunmasız bıraktı.

Geç Tunç Çağı'nda başlayan ve Geç Frig dönemi boyunca devam eden ormanların yok edilmesi, orman süksesyonu ve peyzajın ağaçlardan arındırılması, burada gelecekte yaşayan insanları daha az tercih edilen ahşap kaynaklarıyla kısıtlayarak miras etkisine yol açtı. İkinci miras etkisi ise toprağın ağaçsızlaştırılması, yoğun tarım ve aşırı otlatmadan kaynaklanan toprak erozyonuydu.

Sonuç olarak çevresel değişiklikler zaman içerisindeki toprak kullanımı stratejilerini önemli ölçüde etkiledi; fakat, Gordion bölgesinde tarımsal sürdürülebilirliği ve tarımın ekolojik etkilerini anlayabilmemiz için insan ve çevrenin bunlara tepkisinin hızına ve derecesine dikkat edilmesi büyük önem taşıyor.

Index